World Soils Book Series

Series editor

Prof. Alfred E. Hartemink
Department of Soil Science, FD Hole Soils Laboratory
University of Wisconsin–Madison
Madison
USA

The World Soils Book Series publishes books containing details on soils of a particular country. They include sections on soil research history, climate, geology, geomorphology, major soil types, soil maps, soil properties, soil classification, soil fertility, land use and vegetation, soil management, soils and humans, soils and industry, future soil issues. The books summarize what is known about the soils in a particular country in a concise and highly reader-friendly way. The series contains both single and multi-authored books as well as edited volumes. There is additional scope for regional studies within the series, particularly when covering large land masses (for example, The Soils of Texas, The Soils of California), however, these will be assessed on an individual basis.

More information about this series at http://www.springer.com/series/8915

Hassan El-Ramady · Tarek Alshaal
Noura Bakr · Tamer Elbana
Elsayed Mohamed · Abdel-Aziz Belal
Editors

The Soils of Egypt

Springer

Editors
Hassan El-Ramady
Soil and Water Department, Faculty of Agriculture
Kafrelsheikh University
Kafr El-Sheikh, Egypt

Tarek Alshaal
Soil and Water Department, Faculty of Agriculture
Kafrelsheikh University
Kafr El-Sheikh, Egypt

Noura Bakr
Soils and Water Use Department, Biological and Agricultural Division
National Research Centre
Giza, Egypt

Tamer Elbana
Soils and Water Use Department, Biological and Agricultural Division
National Research Centre
Giza, Egypt

Elsayed Mohamed
Division of Agriculture Applications and Soil Science
National Authority for Remote Sensing and Space Sciences (NARSS)
Cairo, Egypt

Abdel-Aziz Belal
Division of Agriculture Applications and Soil Science
National Authority for Remote Sensing and Space Sciences (NARSS)
Cairo, Egypt

ISSN 2211-1255 ISSN 2211-1263 (electronic)
World Soils Book Series
ISBN 978-3-030-07049-6 ISBN 978-3-319-95516-2 (eBook)
https://doi.org/10.1007/978-3-319-95516-2

This Springer imprint is published by the registered company Springer Nature Switzerland AG
The registered company address is: Gewerbestrasse 11, 6330 Cham, Switzerland

This book is dedicated to our families

All photos by El-Ramady

Foreword I

This book is long overdue. Its most immediate didactic value resides in helping the reader to understand the concept and state of the art of soils from Egypt. The book thus fills an obvious need for a systematic account of concept of soil science in Middle East to facilitate the practice of science.

Specifically, the book covers extremely important chapters detailing the role of soil use and its relation to human food security, soil pollution, land use and vegetation status, soil biology, and land degradation. The concept of right use of land, cultivating the soil using nutrient lacking in animal and humans, would be a very interesting topic to improve human health.

This book is aimed at researchers and professionals, together with postgraduate students. However, I believe that the material will also stimulate advanced undergraduate students and those interested in the application of this knowledge.

I think that the authors can be confident that there will be many grateful readers who will have gained a broader perspective of the disciplines of soil science in Egypt as a result of their efforts.

São Paulo, Brazil
January 2018

Prof. André Rodrigues dos Reis Ph.D.
São Paulo State University (UNESP)

Foreword II

Our Soils: No Longer Crypto-Resources

A careful reflection on the presence and location of the living systems on our planet, including soils, leads us to observe—in compliance with the proportions—that they surround the Earth as a thin coating. A few kilometers below the Earth surface, the temperature is too high, while a few kilometers above the air is too cold and rarefied to allow survival. Between these two limits affirms animal and plant life. The latter, in particular, constitutes the basis for food chains, contributes to the water and biochemical cycles, and protects the soil's life, which is the base of the biosphere.

Soil develops where the combined actions of atmosphere, hydrosphere, and biosphere act on lithosphere and, together with air and water, completes the triad of the natural resources that are essential to the life. In a most general sense, soil is the medium that allows plants and animals to live and develop, and Man to carry out all his activities. From time to time, and in relation to his needs, Man has considered soil as a mean of agricultural and forestry production, a seat of urban and infrastructure settlements, a place of leisure and sport, a source of raw materials, and in summary, a source of goods and services useful to human life.

Soils fulfill basic functions for the human society, not only in practice because they satisfy human's material needs but also in the abstract, stimulate intellectual activity or indulging needs of spiritual well-being. There are cultural evidences regarding the role of soil throughout history and some are still found in the customs, folklore, and traditions of various populations in several countries. In ancient societies, the soil has always had a privileged position in virtue of its fundamental role of providing foodstuffs. Even today, in various parts of the world, social systems reflect the connections between the soil and the environment, and the management of soil fertility is at the heart of this connection.

Regrettably, "*homo technologicus*", descendant of "*homo sapiens*", forced to live in ever-bigger cities, to feed with ready meals, to breathe air conditioning in hyper-technological offices and impersonal lofts, rarely does he stop and reflect on how much his well-being is fundamentally connected to the soil and to the other primary resources. In the next years, maintaining the welfare and the development of the human societies will largely depend on Man's ability to ensure the sustainable use of the natural resources, in particular soils.

We live in an era and in a cultural system that pays particular attention to the human rights, but not equally indicates duties and responsibilities. We are able to fight for defending our privileges but we wimp out from our duties: our relationship with the environment is characterized by a general indifference and a widespread carelessness. Unfortunately, the awareness of the role played by the environmental resources and by the soil in particular is lacking. Soil, being a "**crypto-resource**", a hidden resource, is considered only after catastrophic events and when the failures are evident!

So far, we have shown skill in getting out from situations that the wisdom would have certainly avoided. It is time to move on by the skill to the wisdom, the same wisdom which has driven Aldo Leopold (1886–1948) to argue that "*the oldest task of man is to live on the soil without despoil it.*"

Now it is imperative a cultural leap: we all have to consider the soil as a good for a human society that is continuously changing and in which the boost toward a continuous economic growth and a rapid technological development, coupled with the progressive increase of the information, often causes considerable and unpredictable changes. The achievement of these goals is based on a sound and comprehensive knowledge of the soils of each country.

I wish to welcome the book *The Soils of Egypt* and to congratulate with the authors.

Palermo, Italy
January 2018

Prof. Dr. Carmelo Dazzi
President of the European Society
for Soil Conservation (ESSC)

Preface

It is well known that soils are the main source of our food, feed, fiber, and fuel. So, soils have a holy position in all religions including Islam, Christianity, and Judaism. Furthermore, great civilizations established depending on soils like the Egyptian civilization. Therefore, this book *The Soils of Egypt* has been written. This book includes soil and its potential from many sides in Egypt such as climate changes and water crisis, different expected scenarios of climate change in Egypt, and effects of climate change on crop productivity. Geology and geomorphology also will be presented as well as major soil types and different soil maps. Soil classification also will be among the most important subjects and will be highlighted in this book. On the other hand, different soil properties and how to sustain these land resources will be also highlighted. Due to the role of Egyptian soils in the ancient Egyptian civilization, soil fertility and its security will be presented as one of the most important issues. Pollution also as one of the most important and serious problems in Egypt will be highlighted, due to pollution already has been penetrated different environmental compartments including soils, waters, and air. Soil pollution and its management as well as different soil pollution sources and the degradation of Egyptian soils are needed more explanation. Different land uses including vegetation as well as future soil issues in Egypt will be also reviewed.

This book will present a comprehensive overview and the vital importance of soils to agriculture, ecosystems, and human life in Egypt. The study of soil resources will allow for more researches and management our challenges including improving soil quality and its sustainability, soil carbon sequestration, and wastewater treatment as well as innovative delivery of nutrients and water for crop production. We also do hope that this book will bring enough knowledge for next generations with continuous delivering proper solutions for different environmental challenges, which we are facing now and in the future. Soil resources could be conserved and sustained only by understanding different soil properties and its processes occurring in the soils.

Hassan El-Ramady

Tarek Alshaal

Noura Bakr

Tamer Elbana

Elsayed Mohamed

Abdel-Aziz Belal

Kafr El-Sheikh, Egypt	Hassan El-Ramady
Kafr El-Sheikh, Egypt	Tarek Alshaal
Giza, Egypt	Noura Bakr
Giza, Egypt	Tamer Elbana
Cairo, Egypt	Elsayed Mohamed
Cairo, Egypt	Abdel-Aziz Belal
January 2018	

Acknowledgements

The authors would like to thank **Prof. Dr.** ***Alfred. E. Hartemink***, who proposed the publication of this volume. Special appreciations for the Springer team including **Dr.** ***Robert Doe***, **Ms.** ***Corina van der Giessen***, and **Mr.** ***Suresh Rettagunta*** for their support and help in submission/production process as well as the typesetting and perform copyediting.

Contents

Contributors

Mohamed H. Abbas Soils and Water Department, Faculty of Agriculture, Moshtohor, Benha University, Benha, Egypt

Neama Abdalla Genetic Engineering Division, Plant Biotechnology Department, National Research Center, Giza, Egypt; Plant Biotechnology Department, Genetic Engineering Division, National Research Center, Giza, Egypt

Hamada Abdelrahman Soils Department, Faculty of Agriculture, Cairo University, Giza, Egypt

R. R. Ali Soils and Water Use Dept, National Research Centre, Cairo, Egypt

Tarek Alshaal Faculty of Agriculture, Soil and Water Department, Kafrelsheikh University, Kafr El-Sheikh, Egypt

Megahed Amer Soils, Water and Environment Research Institute (SWERI), Agricultural Research Center (ARC), Giza, Egypt

Mohamed H. Bahnassy Soil and Water Sciences Department, Faculty of Agriculture, Alexandria University, Alexandria, Egypt

Noura Bakr Soils and Water Use Department, National Research Centre (NRC), Cairo, Egypt; Biological and Agricultural Division, Soils and Water Use Department, National Research Centre, Dokki, Giza, Egypt

Abdel-Aziz Belal Division of Agriculture Applications and Soil Science, National Authority for Remote Sensing and Space Sciences (NARSS), Cairo, Egypt

Eric C. Brevik Department of Natural Sciences, Dickinson State University, Dickinson, ND, USA

Éva Domokos-Szabolcsy Agricultural Botany, Plant Physiology and Biotechnology Department, University of Debrecen, Debrecen, Hungary

Tamer Elbana Biological and Agricultural Division, Soils and Water Use Department, National Research Centre, Dokki, Giza, Egypt

Heba Elbasiouny Environmental and Biological Sciences, Al-Azhar University, Tanta, Egypt

Fathy Elbehiry Central Laboratory of Environmental Studies, Kafr El-Sheikh University, Kafr El-Sheikh, Egypt

Ayman M. El-Ghamry Soils Department, Faculty of Agriculture, Mansoura University, Mansoura, Egypt

Omar El-Hady Biological and Agricultural Division, Soils and Water Use Department, National Research Centre, Dokki, Giza, Egypt

Nevien Elhawat Department of Biological and Environmental Sciences, Al-Azhar University, Cairo, Egypt

Ahmed S. El-Henawy Soil and Water Department, Faculty of Agriculture, Kafrelsheikh University, Kafr El-Sheikh, Egypt

Shaimaa Elmahdy Soil and Water Department, Faculty of Agriculture, Kafrelsheikh University, Kafr El Sheikh, Egypt

Samia El-Marsafawy Soils, Water and Environment Research Institute, Agricultural Research Center, Giza, Egypt

Sahar El-Nahrawy Soils, Water and Environment Research Institute (SWERI), ARC, Sakha, Kafr el-Sheikh, Egypt

Hassan El-Ramady Soil and Water Department, Faculty of Agriculture, Kafrelsheikh University, Kafr El-Sheikh, Egypt

Tamer Elsakhawy Soils, Water and Environment Research Institute (SWERI), ARC, Sakha, Kafr el-Sheikh, Egypt

Salah E.-D. Faizy Soil and Water Department, Faculty of Agriculture, Kafrelsheikh University, Kafr El Sheikh, Egypt

Miklós Fári Agricultural Botany, Plant Physiology and Biotechnology Department, University of Debrecen, Debrecen, Hungary

Ihab M. Farid Soils and Water Department, Faculty of Agriculture, Moshtohor, Benha University, Benha, Egypt

Hesham M. Gaber Soil and Water Sciences Department, Faculty of Agriculture, Alexandria University, Alexandria, Egypt

Saber A. Gaheen Soil and Water Department, Kafrelsheikh University, Kafr El-Sheikh, Egypt

Azza Ghazi Soils, Water and Environment Research Institute (SWERI), ARC, Sakha, Kafr el-Sheikh, Egypt

Ehab A. Hendawy Division of Agriculture Applications and Soil Science, National Authority for Remote Sensing and Space Sciences (NARSS), Cairo, Egypt

Mohamed Jalhoum Division of Agriculture Applications and Soil Science, National Authority for Remote Sensing and Space Sciences (NARSS), Cairo, Egypt

Fawzy M. Kishk Soil and Water Sciences Department, Faculty of Agriculture, Alexandria University, Alexandria, Egypt

Elsayed Mohamed Division of Agriculture Applications and Soil Science, National Authority for Remote Sensing and Space Sciences (NARSS), Cairo, Egypt

Ahmed A. Mousa Soils Department, Faculty of Agriculture, Mansoura University, Mansoura, Egypt

Alaa El-Dein Omara Soils, Water and Environment Research Institute (SWERI), ARC, Sakha, Kafr el-Sheikh, Egypt

József Prokisch Nanofood Laboratory, Faculty of Agricultural and Food Sciences and Environmental Management, Debrecen University, Debrecen, Hungary

Ahmed Saleh Division of Agriculture Applications and Soil Science, National Authority for Remote Sensing and Space Sciences (NARSS), Cairo, Egypt

Nicola Senesi Department of Soil, Plant and Food Sciences, University of Bari, Bari, Italy

Hassan Shams El-Din Soils, Water and Environment Research Institute (SWERI), Agricultural Research Center (ARC), Giza, Egypt

Mohamed S. Shams Soil and Water Department, Faculty of Agriculture, Kafrelsheikh University, Kafr El-Sheikh, Egypt

Sarwat Yousef Soil and Water Department, Faculty of Agriculture, Kafrelsheikh University, Kafr El Sheikh, Egypt

Introduction

1

Hassan El-Ramady, Tarek Alshaal, Ahmed S. El-Henawy, and Mohamed S. Shams

Abstract

Egypt is one of the most old countries worldwide. It has a great history and an amazing civilization. This civilization established and flowered on the two sides of the Nile River. Depending on both the freshwater from the Nile and fertile soils, this civilization is extended for thousands of years. The wonderful location of Egypt besides the temperate weather, the metal wealth, and oil, they enforced Egypt to be a prey to the colonists. So, several colonists tried to occupy the Egyptian lands but it has been destroyed and overcame these enemies. Therefore, the Egyptian soils were and still the domain factor in agricultural production. These soils have several functions in our life associated with the goals of UN sustainable development as well as the ecosystem services. These soil-related ecosystem services also could be subdivided into regulating, provisioning, and cultural subgroups. So, this book is an attempt to discuss the Egyptian soils including the following issues: the historical research of Egyptian soils, the climate and its changes, soil pollution, soil maps and major soil types, soil fertility, and its security, as well as future soil issues.

Keywords

Soils • Egypt • Climate • Soil pollution
Soil security • The Nile delta

1.1 Introduction

Egypt is the third-most populous on the African continent after Nigeria and Ethiopia. About 95% of the Egyptian population lives along the banks of the Nile and in the Nile Delta expanding from the north of Cairo and along the Suez Canal (Fig. 1.1). The total Egyptian population reached more than **93 million people in 2017** according to the Central Agency for Public Mobilization and Statistics (**CAPMAS**). The population in Egypt is expected to grow more than **100** million people by **2023**. The distinguished distribution of population in Egypt has a unique phenomenon. This distribution is recognized as a demographic imbalance due to the following reasons:

(1) About 25% of the total population is squeezed into the Greater Cairo (i.e., Cairo, Giza, and Qalyubia),
(2) About 25% lives in the north coastal governorates (i.e., eight governorates starting from North Sinai in the northeastern coast, Port Said, Damietta, Dakahlia, Kafr El Sheikh, Beheira, Alexandria, and ending with Matruh in the northwestern coast),
(3) About 48% is occupied in the Nile Delta (i.e., Port Said, Sharqia, Damietta, Qalyubia, Gharbia, Menufyia, Kafr El Sheikh, Beheira, and Alexandria governorates),
(4) About 20% is existed in the Upper Egypt governorates (Asyut, Sohag, Qena, Aswan, and Luxor), and
(5) Only about **1.1** million is distributed in the three biggest governorates (New valley, Matruh, and Red Sea).

Concerning the soil and agriculture in ancient Egypt, the Egyptian agriculture has a very long story, i.e., more than 5000 years in the arid climate, which mainly has been dependent on the Nile River (El-Ramady et al. 2013). This agriculture was completely controlled by the flooding of the Nile in ancient times. This system is totally different after beginning the nineteenth century (the era of Mohamed Ali), where the construction of different weirs and dams in the river, in order to control the flow of the Nile, starts the modern agriculture in Egypt. This era could be considered the real revolutionary change in Egyptian agriculture and water use as well as the sustainability of Egyptian agriculture (Satoh and Aboulroos 2017).

H. El-Ramady (✉) · T. Alshaal · A. S. El-Henawy · M. S. Shams
Soil and Water Department, Faculty of Agriculture, Kafrelsheikh University, Kafr El-Sheikh, Egypt
e-mail: ramady2000@gmail.com

H. El-Ramady et al. (eds.), *The Soils of Egypt*, World Soils Book Series,
https://doi.org/10.1007/978-3-319-95516-2_1

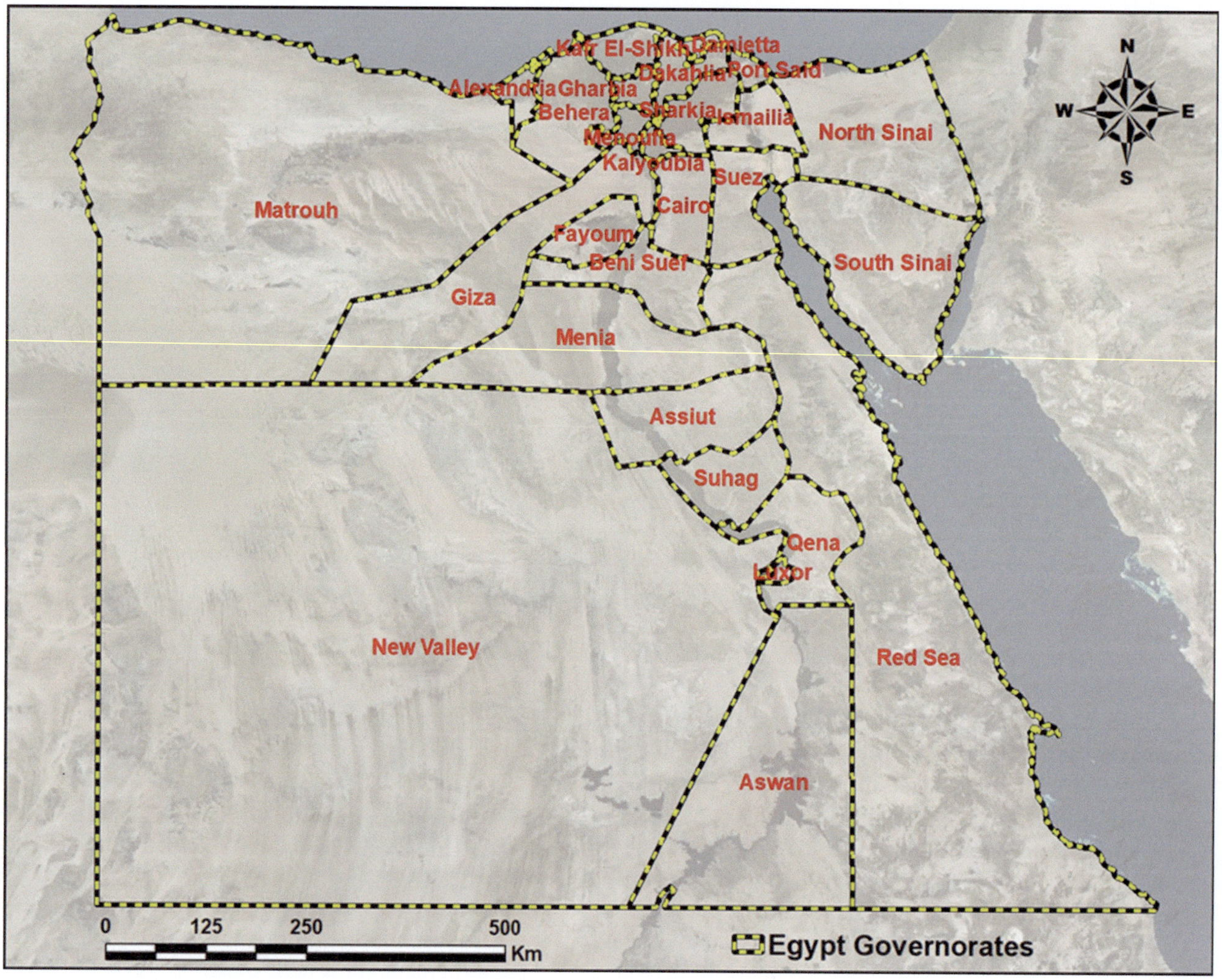

Fig. 1.1 Egypt has 27 governorates distributed in the Nile Delta, in East, and West Deserts from Matruh in the West to Red Sea in the East to Aswan in the South. (This map is created by Dr. Noura Bakr)

Egypt has different names in different languages such as German name (Ägypten), English (Egypt), French (Egypte), Latin (Aegyptus), Italian (Egitto), Spanish (Egipto), Hungarian (Egyiptom), the Ancient Greek (Aígyptos or Αἴγυπτος), and Arabic (Misr or Masr مصر). Concerning the ancient names for Egypt, these names generally recorded the relation between the people, the land, and its agriculture. These include *Ta Akht*, the land of flood and fertile soil; *Ta-meri*, the land cultivated by the hoe; *Kmt*, the black soil; *Tamhi*, the land of the flax plant; *Nht*, the land of the sycomore fig tree; and *Misr*, the safe and civilized country. The name of Egypt was derived from the name of the Earth God, *Ge*, or from *Agpt*, referring to the land covered with floodwaters (Janick 2016).

Egypt is a transcontinental country spanning the southwest corner of Asia Continent and the northeast corner of Africa Continent by a land bridge formed by the Sinai Peninsula. This country could be considered a cradle of civilization, where Ancient Egypt experienced some of the earliest developments in agriculture, writing, organized religion, urbanization, and central government. Therefore, Egypt has one of the longest histories comparing with any modern country and its history could be considered an emerging one of the world's first nation-states in the tenth millennium BC. This history simply could be presented through several monuments including many necropolises, Pyramids, and Great Sphinx in Giza, as well as the ruins of Thebes, Memphis, Karnak, and the Valley of the Kings (Fig. 1.2). This civilization is a great legacy reflecting a significant focus of different archeologists and popular interests worldwide.

Recently, Springer has been published many books about Egypt including the history (past, present, and future) of the irrigated agriculture in Egypt (Satoh and Aboulroos 2017), the Nile River and its great role in the Egyptian life (Negm 2017a), the Nile Delta as one of the Earth's great mega

Fig. 1.2 This place was known with the manufacturing zone of obelisks in Aswan. The ancient Egyptians used to place obelisks in pairs at the entrance of their temples. A number of ancient Egyptian obelisks are known to have survived, plus the unfinished obelisk found partly hewn from its quarry at Aswan. Egyptian obelisks are now dispersed around the world and fewer than half of them remain in Egypt. Photo by El-Ramady (2010)

deltas as well as problems and challenges facing this Nile Delta concerning the agriculture, the shortage of freshwater and other topics (Negm 2017b), the impact of the Grand Ethiopian Renaissance Dam on Egyptian agriculture (Hamada 2017), the future of food gaps in Egypt through the obstacles and opportunities (Ouda et al. 2017), different water resources including the unconventional resources and their impacts on agriculture in Egypt (Negm 2017c), the complicated relationship or nexus among water, energy and food security in the Arab region (Amer et al. 2017) and Middle East (Badran et al. 2017), etc. Therefore, this chapter will highlight on the main geological regions in Egypt, the distribution of different Egyptian soils in these zones, the close relation between soils and climate changes, and soil pollution under different Egyptian conditions.

1.2 Egypt and Its Geological Zones

Egypt has different rocks, mineral, and natural resources. The oldest rocks could be found in the Western Desert, whereas the rocks of the Eastern Desert are largely late Proterozoic in age (Fig. 1.3). It could be divided Egypt into four main areas or regions including (1) the Nile Valley and its Delta, (2) the Eastern Desert, (3) the Western Desert, and (4) the Sinai Peninsula. Each one from the previous areas has particular features. It would be great to highlight some distinguished features for these previous places in the following subsections.

1.2.1 The Nile Valley and Its Delta

The Nile Valley and its Delta (35,000 km^2) were and still the most extensive oasis in the world. This valley and its Delta were created by the longest river in the world (the Nile; 6 853 km). It is well known that the Nile Delta is one of the largest river deltas in the world. It extended from Alexandria in the West to Port Said in the East covering 240 km of Mediterranean coastline and considering the main region for the Egyptian agriculture. It also extends from the North to the South about 160 km in length beginning slightly downriver from Cairo. This Nile Delta is mainly a delta formed in Northern Egypt (Lower Egypt) where the Nile River spreads out and drains into the Mediterranean Sea. The Nile Delta could be characterized with these facts: (1) it covers only about **4%** of Egypt's area but hosts about 48% of the country's population, (2) it consists of about 63% of the Egyptian agricultural land, and (3) it is among the most densely populated agricultural areas in the world recording 1360 inhabitants per km^2 (Negm et al. 2017a, b).

It is well known that the Nile starts its journey and flow into Egypt in the area from the north of Wadi Halfa to Lake Nasser on the Sudanese–Egyptian borders. It is reported that seven branches of the Nile have been ran through the Nile Delta from the first century AD, according to historical accounts. Around the twelfth century, the Nile had only six branches according to the last accounts. Finally, only two main outlets, namely, the east branch (Damietta or Dumyat with 240 km long) and the west branch (Rosetta or Rashid

Fig. 1.3 There are several types of rocks in Egypt, which are formed by different processes as rocks located in the Eastern Desert (Red Sea area in the top four photos) and along the Mediterranean coast as presented in the bottom four photos in Matrouh. Photos by El-Ramady (2010, 2017), respectively

with 239 km long) according to the names of ports located at their respective mouths. The depth of Rosetta branch ranges from 2 to 4 m and its width averages 180 m (Negm et al. 2011). This branch could be used its water in irrigation, drinking, and fishing purposes with a daily flow in average 21,500,000 m^3 per day. The drains receive domestic, industrial, and agricultural wastes (Elhaddad and Al-Zyoud 2017).

Concerning the common soils in both the Nile Valley and its Delta, they are in general alluvial soils. These soils mainly formed from the deposits of the Nile branches, fluviomarine, and lagoon, which located adjacent to the

Fig. 1.4 The area around the monument (the small photo) as a symbol for Egyptian–Soviet friendship or collaboration at the completion of the Aswan High Dam. Photos by El-Ramady (2010)

northern lakes and the coastal plain in the extreme north of the Delta as well as the sandy soils from the coastal plain and beaches (Khalifa and Moussa 2017). The main problem faced by the Nile Delta is represented in no longer receiving an annual supply of nutrients and sediments from upstream because of the construction of the Aswan High Dam (Fig. 1.4); the quality of soils of the floodplains has declined, and large amounts of fertilizers are now used (Negm et al. 2017a, b). For more details about the Nile Delta, see a great book published by Springer carrying this title edited by Negm (2017b), as well as some studies (e.g., Hereher 2010; Mabrouk et al. 2013a, b; Sharaky et al. 2017; Elbeih 2017; Negm and Armanuos 2017; Negm and Eltarabily 2017).

1.2.2 The Western Desert

The Western Desert or the Libyan Desert is a part of the Sahara in Egypt lying between the Nile and Cyrenaica. It is an area of the Sahara, which lies west of the river Nile, up to the Libyan border and south from the Mediterranean Sea to the border with Sudan. Its name derived from the contrast to the Eastern Desert, which extends east from the Nile to Red Sea. The desert covers an area of 680,650 km^2, which represents two-thirds (about **68%**) of the land area of the country. Concerning the geology of the Western Desert, it is mostly rocky desert, though an area of sandy desert, known as the Great Sand Sea, lies to the west against the Libyan border. The highest elevation (1000 m) is recorded for the Gilf Kebir plateau into the far southwest of the country, on the Egypt–Sudan–Libya border. According to the administration position, it is mainly divided among many governorates including the Matrouh Governorate and the New Valley Governorate from there to the Sudan border in the north and west, while in the eastern parts of the Western Desert lie in the Giza, Fayoum, Beni Suef, and Minya Governorates. This Western Desert is barren and uninhabited save for a chain of oases which extend in an arc from Siwa (in the North–West) to Kharga (in the South). On the other hand, part of the Western Desert soils is located by the North Coastal Zone of Egypt. These soils mainly are sandy soils and extend to about 500 km between Alexandria and Salloum city near the Libyan borders covering an area about 10,000 km^2.

It is well known that several natural depressions or oases are scattered in the Western Desert including these famous oases: Siwa, Bahariya, Farafra, Kharga, and Dakhla. The Western Desert also contains promising reclaimed areas including Darb El-Arbain, Toshka, East El-Uwienat, and some wadis of High Dam Lake. Concerning the soils of these natural depressions or oases, they mainly depend on source of parent materials, the erosional patterns, sedimentation environment and eluviation deposit of salts, carbonate, and gypsum. Therefore, the characterization of these soils may show differences with regard to mineral content, their texture, and depth to water table or bedrock, in addition to numerous types of morphopedological features such as accumulation of carbonate and gypsum, salts, shales, and iron oxides (Khalifa and Moussa 2017).

It is worth to mention that there are mainly five famous oases in Egyptian Western Desert: Siwa, Kharga, Dakhla, Baharia, and Farafra. Siwa Oasis is considered an important Egyptian oasis as a deep depression (about 20 m below sea level). It is one of the most isolated Egyptian settlements, with 23,000 people, mostly Berbers who developed a unique culture and a distinct language of the Berber family called Siwi. The Siwa Oasis (derived from the Berber word Isiwan) is an Egyptian oasis located between the Qattara Depression and the Egyptian Sand Sea in the Western Desert, nearly 50 km east of the Libyan border, 300 km from southwest of Mersa Matrouh, and 560 km from Cairo as well as about 80 km in length and 20 km wide (Fig. 1.5). Agriculture through the cultivation of dates and olives is the main activity in Siwi as well as handicrafts like basketry. Tourism has become a vital source of income from recent decades. Much attention has been given to creating hotels that use local materials and display local styles (Aldumairy 2005). Definitely, further studies are needed to investigate several treasures in the Western Desert of Egypt such as Attwa and Nabih (2015), El Ayyat (2015), El Nady and El-Naggar (2016), and Temraz et al. (2017).

1.2.3 The Eastern Desert

The part of Egypt is located east of the Nile River called the Eastern Desert. It is the section of Sahara or Desert between the river Nile and the Red Sea. It extends from Egypt in the north to Eritrea in the south and also comprises parts of Sudan and Ethiopia. The Eastern Desert of Egypt represents about **22%** of the surface area of the country (223,000 km^2) but due to the limited availability of water, this area is undeveloped (Abdel Moneim 2005). The main geographic features of the Eastern Desert are the western Red Sea coastline (with the Red Sea Riviera) and the Eastern Desert mountain range that runs along the coast, the highest peak of which is Shaiyb al-Banat (2187 m). The Eastern Desert is a popular setting for safaris and other excursions. The Eastern Desert includes some governorates, i.e., Suez, Qina, and Red Sea.

There are many mountains and wadis (e.g., W. El Laquita, W. Qena, W. El Assuity, and W. Al-Allaqui) flowing toward the Nile River and the Red Sea. The soils of this desert are mostly deep and very steep, and their soils display young stages of development, whereas soils of the wadis are mainly shallow to deep coarse or moderately fine-textured, with variable content of gravels (Khalifa and Moussa 2017). Recently, several investigations carried out

Fig. 1.5 Siwa is a deep depression, where saline soils could be resulted from high groundwater, and many areas are used in salt production as presented in last or bottom photo (left). The right photo is a view for Siwa Oasis that has been taken over Dakrour Mountain representing a side for the modern city, and other photo represents the old town of Shali in Siwa Oasis. Photos by El-Ramady, April 2017

focusing on different fields of the Eastern Desert of Egypt including remote sensing (e.g., Abou El-Magd et al. 2015; Nigm et al. 2015), geomorphological and geochemical studies (e.g., Abdel Moneim 2005; Mohamed and Abu El-Ela 2011; Emam 2013; Emam and Saad-Eldin 2013), studies of mineralization (e.g., Abd Allah 2012; Mohamed 2013; Omran and Dessouky 2016; Surour 2017), and others.

1.2.4 The Sinai Peninsula

Sinai or the Sinai Peninsula is an Egyptian peninsula situated between the Mediterranean Sea to the north and the Red Sea to the south, serving as a land bridge between Asia and Africa. It is the only part of Egyptian territory located in Asia. Sinai has a land area of about 60,264 km^2 (about **6%** from total area of Egypt) and a population of approximately 639,586 people. Administratively, the peninsula could be divided into two of governorates, e.g., South and North Sinai. The formal name of Sinai is called *Ta Mefkat* or the land of turquoise according to the ancient Egyptians. It is well known that Sinai is triangular in shape. This shape is lying on the southern Mediterranean Sea (in the north) and southwest and southeast shores on Gulf of Suez and Gulf of Aqaba of the Red Sea. A big gap could be realized concerning the investigation of soils of Sinai from different point of views like mineralogy, soil pollution, water harvesting, mining and topography, botanical species, etc. However, an increased interest regarding soil researches in Sinai could be noticed (e.g., Badreldin et al. 2017; El Hady 2017; Sultan et al. 2017; Zaky 2017).

This Peninsula generally is a hot-dry desertic climate. The soils of Sinai could be divided into different zones including (1) soils are mainly moderately deep or deep, gravelly, coarse-textured soil, and located in the south Sinai; (2) soils are deep highly calcareous, gravelly coarse- or moderately- or over fine-textured soils and located in the central part of Sinai; and (3) soils of the alluvial coastal plains, which are located parallel to both the Gulf of Suez and Gulf of Aqaba, are deep, gravelly, coarse-textured soils and deep calcareous, coarse to moderately textured soils (Khalifa and Moussa 2017).

1.3 Egypt and Its Soils

Soil is the main supporter, as well known, for all terrestrial life forms as well as the crucial maintainer for this life. Soils also perform several critical functions for the global population including (1) dividing of precipitation into surface and ground waters, (2) supply and storage of both nutrients and waters for plant growth, (3) disposal and renovation of anthropogenic wastes, (4) habitat for different soil organisms, and (5) support for buildings, roads, and other infrastructures. Soils are also a major reservoir of the global carbon and have the ability to serve as a sink for atmospheric carbon in order to reduce greenhouse gasses with proper management. Soils have enough resilience but also definitely are subject to degradation if managed improperly (West et al. 2017). Day by day, several books, reviews, and articles are published to focus on soils and their roles in our terrestrial life forms such as Piccoli et al. (2017), Kaczynski et al. (2017), Vimal et al. (2017), etc.

The management of soil and water resources in Egypt faces several problems such as the scarcity of water and the majority of the Egyptian lands are desert. So, Egypt could be considered as one of the poorest countries in the world from the cropland point of view (Khalifa and Moussa 2017). Generally, most of the cultivated lands in Egypt are located close to different banks of the Nile River, in the Nile Delta, and the main canals of the Nile Valley. Furthermore, the average per capita share in agricultural land has steadily decreased as follows: 0.12, 0.10, 0.06, and 0.04 ha in 1950, 1960, 1990, and 2013, respectively. Moreover, many reasons enforced the quality of agricultural soils to degradation or deterioration including (1) low investment and management of the agricultural drainage since the 1950s, (2) decrease the soil fertility due to crop intensification, (3) the absence of silt from the Nile after the construction of High Dam, and (4) the rising level of groundwater (Khalifa and Moussa 2017).

1.4 Soils and Climate Changes

Climate change was and still one of the promising issues in soil sciences, where the forming and genesis processes of soils highly depend on many elements of weather. In other words, all processes in soil system mainly depend on humidity or precipitation or the water availability and temperature as well as on biological activities, which control the soil characteristics and their functions for both the environment and human societies (Blum 2005). Therefore, any change in climate will affect significantly the soil. Several publications nowadays have been focused on the effects of climate changes on soils such as carbon sequestration for climate changes (Ussiri and Lal 2017), water and land security in dry regions under climate changes (Ouessar et al. 2017), the adaptation of climate changes in Africa (Filho et al. 2017) or in Pacific Countries (Filho 2017), global soil security, energy and climate changes (Field et al. 2017; McCarl 2017), and others (e.g., Barmentlo et al. 2017; Bosch et al. 2017; Coyle et al. 2017; Lin et al. 2017; Rao et al. 2017).

Climate changes can influence different soil eco-services at both local and regional scales. Therefore, soil and other environmental sciences have the ability to determine future climate change impacts on soils and terrestrial ecosystems. Furthermore, it could be monitored the following research issues regarding the cause–response relationships between specific soil changes and climatic changes as follows:

(1) The impacts of climate change including changes in temperature, humidity, precipitation, and wind velocity in soil processes;
(2) Different indicators for the soil changes and their development with time;
(3) The impacts of changes in the soil on different eco-services including biomass production, air–water interactions, biodiversity, and human health;
(4) The impacts of the changes in the soil on both social and economic systems as well as their feedback; and
(5) Different strategies and operational procedures for the mitigation of impacts on ecological, social, and economic systems (Blum 2005).

1.5 Soil Pollution in Egypt

Pollution of agricultural soils in Egypt has become one of the serious challenges facing the Egyptian nation. Both rapid urbanization and industrialization in Egypt have been led to a very high accumulation of pollutants including trace elements (Cd, Cr, Pb, Zn, Fe, Cu, and others) and organic pollutants in water, soils, sediments, air, etc. Concerning pollution of soils, it was and still a very hot spot attracting several researchers worldwide to find the proper and sustainable method in remediating this problem. Therefore, the quality of soil is very important and dangerous for human health and environmental aspects.

Soil is a complex, dynamic, and open heterogeneous system. This system needs in general hundreds or thousands of years for genesis and mature. This heterogeneity of soils results from chemical, physical, and biological characteristics as well as soil constituents under macro-, micro-, and nanoscales. Furthermore, due to variability in seasonal rainfall, temperature, parent materials, and vegetation, different soil types could be found and each of these soils has distinguished physical, mineralogical, biological, and chemical properties (Saha et al. 2017; Shankar and Shikha 2017). This heterogeneity of soil also armed it with a high resilience power helped soils to overcome many agro-ecological problems like soil pollution. Therefore, it should sustain and maintain land resources, because these resources are common habitat of several macro- and microorganisms. These soil faunae have variable degree of sensitivity toward pollutants (Saha et al. 2017; Bashkin 2017).

There are many global environmental problems in general including global warming, acid rain, ozone depletion, pollution, over-population, depletion of natural resources, waste disposal, deforestation, and loss of global biodiversity (Singh and Singh 2017). So, soil pollution is considered one of the dangerous global problems. Several hazards could be listed resulted from soil pollution including (1) loss or decline in soil productivity; (2) reduce in soil biodiversity; (3) killing some plants, fishes, and other aquatic organisms in rivers and lakes (Fig. 1.6); (4) loss in crop diversity; and (5) pollutants may impair soil stability that will be harmful to human health. It is estimated that pollution of water and soil could reduce the yield of crops by about 15–25% over the total cropped area and the years (Saha et al. 2017; Singh and

Fig. 1.6 The El-Gharbia drain (Kitchener) is one of the largest open drainage systems in the Nile Delta (about 59 km) and collects about 12% of the disposal water including agricultural and industrial effluents (from factories in Tanta, El-Mahalla El-Kubra, and Kafr El-Zayat to deliver into the Mediterranean Sea in Baltim and causes a pollution of soils around it). Photos by El-Ramady (2015)

Singh 2017). Therefore, these pollutants should be remediated using bioremediation (Arora et al. 2017; Kalia and Kumar 2017; Prashanthi et al. 2017), phytoremediation (Ansari et al. 2017; Bauddh and Singh 2017; Dhillon and Bañuelos 2017), and nano-remediation (El-Ramady et al. 2017a, b).

Concerning the self-regulation of soils, soils have a great ability to regulate itself by regulating the bio-physicochemical properties as well as processing their own materials, solar energy, and water (Hillel 2004). Soils can also regulate as a buffer of these phenomena: (1) help the rain- and snowfall to reach the groundwater instead of runoff or flowing over the land, which in turn maintains the steady flow of springs and streams; (2) soils have the capacity to absorb and then store water (moisture); (3) the ability to release water gradually; and (4) rain-falling over the global lands could runoff completely, in the absence of soils, producing violent floods rather than sustained river flow. Therefore, these soils refer to the weathered rocks and minerals as well as fragmented outer layer of our planet's land surfaces. These soils also initially formed through different physical disintegrations and chemical alterations of minerals and rocks via different physical and biogeochemical processes (Hillel 2004; Lal 2005). Furthermore, soils govern the most processes in terrestrial life because it is the most important component for all natural resources. These processes include production of biomass, water purification, detoxification of pollutants, recycling of elements, and resilience and restoration of agroecosystems (Hillel 2004; Lal 2005; Chesworth 2008; Lin et al. 2017).

Therefore, comprehensive studies are needed regarding the interrelationship between the Nile water and different pollutants and their effects on the life of inhabitants in the Nile valley. These studies also should mainly focus on the control programs for the actual hazardous pollutants. In other words, the environmental quality control toward reducing the activities of pollution should be included in all resources utilization programs and economic development.

1.6 Conclusion

Therefore, the main target of this book is to present some information about the soils of Egypt. These information include the research history and education of soils in Egypt as well as the great role of Egyptian soils in the Egyptian ancient civilization. As well reported by the Greek historian Herodotus, Egypt was and still the gift of the Nile, and this Nile had the crucial role in the development of this Egyptian civilization. Climate changes and their effects were confirmed all over the world and Egypt as well. In general, climate change may cause increase in sea level and then destroy many cities like Alexandria. Geology and geomorphology of Egyptian soils, soil maps, and major soil types as well as land use and vegetation will be overviewed also in this book. It could be concluded that soils are the most important part in agroecosystem. Soil is an open, dynamic, and complex system. It has the ability to renew our land resources through removing and detoxification of toxic compounds as well as the great resilience in tolerance of many environmental risks. Therefore, soils should be regarded as a common concern of humankind because soil is the substantive constitutive element of land as well as international regulations are required to protect these soils.

Acknowledgements Authors thank the outstanding contribution of STDF research teams (Science and Technology Development Fund, Egypt) and MBMF/DLR (the Federal Ministry of Education and Research of the Federal Republic of Germany) (Project ID 5310) for their help. Great support from this German-Egyptian Research Fund (GERF) is gratefully acknowledged.

References

Abd Allah AG (2012) Two genetic types of volcanic-hosted massive sulfide mineralizations from the Eastern Desert of Egypt. Arab J Geosci 5:217–231. https://doi.org/10.1007/s12517-010-0157-7

Abdel Moneim AA (2005) Overview of the geomorphological and hydrogeological characteristics of the Eastern Desert of Egypt. Hydrogeol J 13:416–425. https://doi.org/10.1007/s10040-004-0364-y

Abou El-Magd I, Mohy H, Basta F (2015) Application of remote sensing for gold exploration in the Fawakhir area, Central Eastern Desert of Egypt. Arab J Geosci 8:3523–3536. https://doi.org/10.1007/s12517-014-1429-4

Aldumairy AA (2005) Siwa: past and present (Dar El-Koub No. 18208-2005, ISBN 977-17-2554-1)

Amer K, Adeel Z, Böer B, Saleh W (2017) The water, energy, and food security Nexus in the Arab Region. Water security in a new world series. Springer International Publishing AG. https://doi.org/10.1007/978-3-319-48408-2

Ansari AA, Gill SS, Gill R, Lanza GR, Newman L (2017) Phytoremediation: management of environmental contaminants, vol 5. Springer International Publishing, AG. https://doi.org/10.1007/978-3-319-52381-1

Arora S, Singh AK, Singh YP (2017) Bioremediation of salt affected soils: an Indian perspective. Springer International Publishing AG. https://doi.org/10.1007/978-3-319-48257-6

Attwa M, Nabih M (2015) Prediction of permeability anisotropy using Dar-Zarrouk parameters of deep resistivity logs: a case study in the northern Western Desert of Egypt. Arab J Geosci 8:10497–10508. https://doi.org/10.1007/s12517-015-1987-0

Badreldin N, Xing Z, Goossens R (2017) The application of satellite-based model and bi-stable ecosystem balance concept to monitor desertification in arid lands, a case study of Sinai Peninsula. Model Earth Syst Environ 3:21. https://doi.org/10.1007/s40808-017-0300-5

Badran A, Murad S, Baydoun E, Daghir N (2017) Water, energy and food sustainability in the middle east: the sustainability triangle. Springer International Publishing AG. https://doi.org/10.1007/978-3-319-48920-9

Barmentlo SH, van Gestel CAM, Álvarez-Rogel J, González-Alcaraz MN (2017) Influence of climate change on the multi-generation toxicity to *Enchytraeus crypticus* of soils polluted by metal/metalloid mining wastes. Environ Pollut 222:101–108

Bashkin VN (2017) Biogeochemical technologies for managing pollution in polar ecosystems environmental pollution series, vol 26. Springer International Publishing Switzerland. https://doi.org/10.1007/978-3-319-41805-6,

Bauddh K, Singh B (2017) Phytoremediation potential of bioenergy plants. Springer Nature Singapore Pvt Ltd. https://doi.org/10.1007/978-981-10-3084-0

Blum WEH (2005) Soils and climate change. J Soils Sediments 5 (2):67–68. https://doi.org/10.0165/jss2005.02.006

Bosch A, Schmidt K, He J-S, Doerfer C, Scholten T (2017) Potential CO_2 emissions from defrosting permafrost soils of the Qinghai-Tibet Plateau under different scenarios of climate change in 2050 and 2070. CATENA 149 (Part 1):221–231

Chesworth W (2008) Encyclopedia of soil science. Encyclopedia of earth sciences series. Springer Science + Business Media B.V. https://doi.org/10.1007/978-1-4020-3995-9,

Coyle DR, Nagendra UJ, Taylor MK, Campbell JH, Cunard CE, Joslin AH, Mundepi A, Phillips CA, Callaham MA Jr (2017) Soil fauna responses to natural disturbances, invasive species, and global climate change: current state of the science and a call to action. Soil Biol Biochem 110:116–133

Dhillon KS, Bañuelos GS (2017) Overview and prospects of selenium phytoremediation approaches. In: Pilon-Smits EAH et al (eds) Selenium in plants, plant ecophysiology, vol 11. Springer International Publishing AG. https://doi.org/10.1007/978-3-319-56249-0_16

El Ayyat AM (2015) Lithostratigraphy, sedimentology, and cyclicity of the Duwi Formation (late Cretaceous) at Abu Tartur plateau, Western Desert of Egypt: evidences for reworking and redeposition. Arab J Geosci 8:99–124. https://doi.org/10.1007/s12517-013-1169-x

El Hady SM (2017) A novel procedure for processing of the xenotime mineral concentrate of southwestern Sinai. Korean J Chem Eng 34 (3):1–7. https://doi.org/10.1007/s11814-017-0095-0

El Nady MM, El-Naggar AY (2016) Petroleum geochemistry of crude oils and oil: source rock correlation of some wells in the North Western Desert, Egypt. Arab J Geosci 9:627. https://doi.org/10.1007/s12517-016-2617-1

Elbeih SF (2017) Mapping of groundwater in Egypt using RS/GIS: case studies, areas surrounding Nile Valley and its Delta. In: Negm AM (ed) The Nile Delta. The handbook of environmental chemistry series, vol 55. Springer International Publishing AG, pp: 63–88. https://doi.org/10.1007/698_2016_52

Elhaddad E, Al-Zyoud S (2017) The quality assessment of pollution of Rosetta branch, Nile River, Egypt. Arab J Geosci 10:97. https://doi.org/10.1007/s12517-017-2870-y

El-Ramady H, Alshaal T, Abowaly M, Abdalla N, Taha HS, Al-Saeedi AH, Shalaby T, Amer M, Fári M, Domokos-Szabolcsy É, Sztrik A, Prokisch J, Selmar D, Pilon Smits EAH, Pilon M (2017a) Nanoremediation: towards sustainable crop production. In: Ranjan S et al (eds) Nanoscience in food and agriculture. Sustainable agriculture reviews, vol 5, 26. Springer International Publishing AG, pp 335–363. https://doi.org/10.1007/978-3-319-58496-6_12

El-Ramady H, Alshaal T, El-Henawy A, Abdalla N, Taha HS, Elmahrouk M, Shalaby T, Elsakhawy T, Omara AE-D, El-Marsafawy S, Elhawat N, Shehata S, Domokos-Szabolcsy É (2017b) Environmental nanoremediation under changing climate. Env Biodiv Soil Secur 1 (in press) https://doi.org/10.21608/jenvbs.2017.1550.1009

El-Ramady HR, El-Marsafawy SM, Lewis LN (2013) Sustainable agriculture and climate changes in Egypt. In: Lichtfouse E (ed) Sustainable agriculture reviews series, vol 12. Springer Science + Business Media Dordrecht, pp 41–95. https://doi.org/10.1007/978-94-007-5961-9_2,

Emam A (2013) Mineralogy and microchemistry of the Egat gold deposit, South Eastern Desert of Egypt. Arab J Geosci 6:4619–4634. https://doi.org/10.1007/s12517-012-0711-6

Emam A, Saad-Eldin M (2013) Distribution and environmental geochemistry of some heavy metals in stream sediments of Wadi Allaqi. South Eastern Desert Egypt Arab J Geosci 6:1325–1332. https://doi.org/10.1007/s12517-011-0437-x

Field DJ, Morgan CLS, McBratney AB (2017) Global soil security. Progress in soil science series. Springer International Publishing Switzerland. https://doi.org/10.1007/978-3-319-43394-3

Filho WL (2017) Climate change adaptation in pacific countries: fostering resilience and improving the quality of life. Climate change management series. Springer International Publishing AG. https://doi.org/10.1007/978-3-319-50094-2

Filho WL, Belay S, Kalangu J, Menas W, Munishi P (2017) Climate change adaptation in africa: fostering resilience and capacity to adapt. Climate change management series. Springer International Publishing AG. https://doi.org/10.1007/978-3-319-49520-0

Hamada YM (2017) The grand ethiopian renaissance dam, its impact on egyptian agriculture and the potential for alleviating water scarcity. Environment and policy series, vol 55. Springer International Publishing AG. https://doi.org/10.1007/978-3-319-54439-7

Hereher ME (2010) Vulnerability of the nile delta to sea level rise: an assessment using remote sensing. Geomat Nat Haz Risk 1(4):315–321. https://doi.org/10.1080/19475705.2010.516912

Hillel D (2004) Encyclopedia of soils in the environment, 1st edn, vol 4. Academic Press

Janick J (2016) Origins of agriculture in Egypt. In: Selin H (ed) Encyclopaedia of the history of science, technology, and medicine in non-western cultures. Springer Science + Business Media Dordrecht, pp 3460–3478. https://doi.org/10.1007/978-94-007-7747-7

Kaczynski R, Siebielec G, Hanegraaf MC, Korevaar H (2017) Modelling soil carbon trends for agriculture development scenarios at regional level. Geoderma 286:104–115

Kalia VC, Kumar P (2017) Microbial applications: bioremediation and bioenergy, vol 1. Springer International Publishing AG. https://doi.org/10.1007/978-3-319-52666-9

Khalifa HE, Moussa HA (2017) Soil and agriculture after the Aswan High Dam. In: Satoh M, Aboulroos S (eds) Irrigated agriculture in Egypt. Springer International Publishing Switzerland, pp 81–124. https://doi.org/10.1007/978-3-319-30216-4_5

Lal R (2005) Encyclopedia of soil science, 2nd edn. Taylor & Francis, pp 838–841https://doi.org/10.1081/e-ess-120042700

Lin Y, Deng H, Du K, Rafay L, Zhang G, Li J, Chen C, Wu C, Lin H, Yu W, Fan H, Ge Y (2017) Combined effects of climate, restoration measures and slope position in change in soil chemical properties and nutrient loss across lands affected by the Wenchuan Earthquake in China. Sci Total Environ 596–597:274–283

Mabrouk B, Arafa S, Farahat H, Badr M, Gampe D, Ludwig R (2013a) Impact of climate change on water and agriculture: challenges and possible solutions for the Nile Delta. In: EGU general assembly conference abstracts, vol 15, p 6293

Mabrouk MB, Jonoski A, Solomatine D, Uhlenbrook S (2013b) A review of seawater intrusion in the Nile Delta groundwater system—the basis for assessing impacts due to climate changes and water resources development. Hydrol Earth Syst Sci Discuss 10 (8):10873–10911. https://doi.org/10.5194/hessd-10-10873-2013

McCarl BA (2017) Economics, energy, climate change, and soil security. In: Field DJ et al (eds) Global soil security, progress in soil science. Springer International Publishing Switzerland. https://doi.org/10.1007/978-3-319-43394-3_17

Mohamed MA (2013) Evolution of mineralizing fluids of cassiterite–wolframite and fluorite deposits from Mueilha tin mine area, Eastern desert of Egypt, evidence from fluid inclusion. Arab J Geosci 6:775–782. https://doi.org/10.1007/s12517-011-0402-8

Mohamed MA, Abu El-Ela FF (2011) Geochemistry and fluid inclusions study of highly fractionated garnet-bearing granite of Gabal Abu Diab, central Eastern desert of Egypt. Arab J Geosci 4:763–773. https://doi.org/10.1007/s12517-009-0088-3

Negm A, Tarek M, Mohamed N, Wessam Y (2011) Impact of future discharges on Damietta branch morphology. In: Fifteenth international water technology conference. IWTC Alexandria, Egypt

Negm A, Abdel-Aziz TM, Salem MN, Yousef W (2017) Morphology of the Nile river due to a flow rate over the maximum current: case study damietta branch. In: Negm AM (ed) The Nile river, Hdb Env Chem. Springer International Publishing AG, pp 239–257. https://doi.org/10.1007/698_2017_3

Negm AM (2017b) The Nile delta. The handbook of environmental chemistry series, vol 55. Springer International Publishing AG. https://doi.org/10.1007/978-3-319-56124-0

Negm AM (2017a) The Nile river. The handbook of environmental chemistry series, vol 56. Springer International Publishing AG. https://doi.org/10.1007/978-3-319-59088-2

Negm AM (2017c) Unconventional water resources and agriculture in Egypt. The handbook of environmental chemistry series, vol 57. Springer International Publishing AG

Negm AM, Armanuos AM (2017) GIS-based spatial distribution of groundwater quality in the western Nile Delta, Egypt. In: Negm AM (ed) The Nile Delta. The handbook of environmental chemistry series, vol 55. Springer International Publishing AG, pp 55, 89–120. https://doi.org/10.1007/698_2016_66

Negm AM, Eltarabily MGA (2017) Modeling of fertilizer transport through soil, case study: Nile Delta. In: Negm AM (ed) The Nile Delta. The handbook of environmental chemistry series, vol 55. Springer International Publishing AG, pp 121–158. https://doi.org/10.1007/698_2016_88

Negm AM, Saavedra O, El-Adawy A (2017) Nile Delta biography: challenges and opportunities. In: Negm AM (ed) The Nile Delta. The handbook of environmental chemistry series, vol 55. Springer International Publishing AG, pp 3–18. https://doi.org/10.1007/698_2016_62

Nigm AM, Elkhodary ST, Elgendi NH, Abdelwahab FM (2015) Mapping the subsurface structural features of Al-Laqitah area, Central Eastern desert of Egypt using aeromagnetic data. Arab J Geosci 8:3537–3549. https://doi.org/10.1007/s12517-014-1444-5

Omran AA, Dessouky OK (2016) Ra's Abdah of the north Eastern desert of Egypt: the role of granitic dykes in the formation of radioactive mineralization, evidenced by zircon morphology and chemistry. Acta Geochim 35(4):368–380. https://doi.org/10.1007/s11631-016-0103-9

Ouda SAH, Zohry A, Alkitkat H, Morsy M, Sayad T, Kamel A (2017) Future of food gaps in Egypt: obstacles and opportunities. Springer briefs in agriculture series. Springer International Publishing AG. https://doi.org/10.1007/978-3-319-46942-3

Ouessar M, Gabriels D, Tsunekawa A, Evett S (2017) Water and land security in Drylands: response to climate change. Springer International Publishing AG. https://doi.org/10.1007/978-3-319-54021-4

Piccoli I, Schjønning P, Lamandé M, Furlan L, Morari F (2017) Challenges of conservation agriculture practices on silty soils. Effects on soil pore and gas transport characteristics in North–Eastern Italy. Soil Tillage Res 172:12–21

Prashanthi M, Sundaram R, Jeyaseelan A, Kaliannan T (2017) Bioremediation and sustainable technologies for cleaner environment. Environmental science and engineering series. Springer International Publishing AG. https://doi.org/10.1007/978-3-319-48439-6

Rao Z, Guo W, Cao J, Shi F, Jiang H, Li C (2017) Relationship between the stable carbon isotopic composition of modern plants and surface soils and climate: a global review. Earth Sci Rev 165:110–119

Saha JK, Selladurai R, Coumar MV, Dotaniya ML, Kundu S, Patra AK (2017) Soil pollution: an emerging threat to agriculture. Environmental chemistry for a sustainable world series, vol 10. Springer Nature Singapore pvt Ltd. https://doi.org/10.1007/978-981-10-4274-4

Satoh M, Aboulroos S (2017) Irrigated Agriculture in Egypt: Past, Present and Future. Springer International Publishing, Switzerland. https://doi.org/10.1007/978-3-319-30216-4

Shankar S, Shikha (2017) Management and remediation of problem soils, solid waste and soil pollution. In: Singh RL (ed) Principles and applications of environmental biotechnology for a sustainable future, applied environmental science and engineering for a sustainable future. Springer Science + Business Media Singapore. https://doi.org/10.1007/978-981-10-1866-4_5

Sharaky A, Salem T, Aal AA (2017) Assessment of water quality and bed sediments of the Nile river from Aswan to Assiut, Egypt. In: Negm AM (ed) The Nile River. Springer International Publishing. GHdb Env Chem 56:207–238. https://doi.org/10.1007/698_2016_118

Singh RL, Singh PK (2017) Global environmental problems. In: Singh RL (ed) Principles and applications of environmental biotechnology for a sustainable future. Applied environmental science and engineering for a sustainable future. Springer Science + Business Media Singapore. https://doi.org/10.1007/978-981-10-1866-4_2

Sultan Y, El–Shafei MK, Arnous MO (2017) Tectonic evolution of kid metamorphic complex and the recognition of Najd fault system in South East Sinai, Egypt. Int J Earth Sci (Geol Rundsch). https://doi.org/10.1007/s00531-017-1461-1

Surour AA (2017) Chemistry of serpentine Bpolymorphs in the Pan-African serpentinites from the Eastern desert of Egypt, with an emphasis on the effect of superimposed thermal metamorphism. Miner Petrol 111:99–119. https://doi.org/10.1007/s00710-016-0460-4

Temraz MG, Mousa DA, Lotfy MA (2017) Evaluation of the shale beds within Alam El Bueib Formation as an unconventional reservoir. J Petrol Explor Prod Technol, Western Desert, Egypt. https://doi.org/10.1007/s13202-017-0353-z

Ussiri DAN, Lal R (2017) Carbon sequestration for climate change mitigation and adaptation. Springer International Publishing AG. https://doi.org/10.1007/978-3-319-53845-7

Vimal SR, Singh JS, Arora NK, Singh S (2017) Soil-plant-microbe interactions in stressed agriculture management: a review. Pedosphere 27(2):177–192

West LT, Singer MJ, Hartemink AE (2017) The soils of the USA. Springer International Publishing AG. https://doi.org/10.1007/978-3-319-41870-4

Zaky KS (2017) Deformation bands in normal fault damage Zones, Southwestern Sinai, Suez Rift. Egypt Geotectonics 51 (2):176–191

Soil Research History

2

Hassan El-Ramady, Tarek Alshaal, and Saber A. Gaheen

Abstract

No doubt that soil is not only a vital component in the agroecosystem but also was the backbone of the Egyptian civilization. This civilization was supported by the Nile and siltation of Egyptian soils. This link between soils and this Egyptian civilization was also built on the history of both humans and soils. Therefore, soils have gained great interest since the ancient Egyptians and could be noticed that from their paintings on their temples. Concerning the education of soils in Egypt, it started since a long time from the Pharaohs and continues till now in different institutions including universities and scientific research centers such as agricultural research center, desert research center, Egyptian atomic energy authority, national research center, national authority for remote sensing and space sciences, desert development center, and Egyptian soil science society. The teaching of soil sciences also could be found in the technical education (the agricultural sector) in Egyptian secondary schools. Therefore, the aim of this chapter is an attempt for understanding the ancient history of research and education of soils in Egypt.

Keywords

Soil history • Soil and civilization • Soil research
Soil education • Technical education

2.1 Introduction

There is an intimate linkage between the history of both humans and soils. This is back to the great respect, sacred, and worshipped by many ancient cultures for soil as a Mother of Earth. According to the Koran, it states that "with water we have made all living things". Water and soil are important constituents of the land and very essential for our life (Lal 2005). It is well known that we all come from the soil and will return to it, as stated in the Koran and other religions. Well known also that our deceased bodies will be decomposed by certain soil organisms, nutrients will be released for the vegetation growth, and thereby all our life will be continued. Many cultures have learnt their people's value of the fertile soil and all nations have developed some sense concerning the importance of soils (Lehmann et al. 2003; Winiwarter and Blum 2009; Winiwarter 2014; Boer et al. 2017). So, soils in both ancient and modern societies have been established and stressed on the value of soils. This is back to the global soil resource has certain characteristics including nonrenewable over the human time frame, finite, unevenly distributed among biomes and eco-regions, liable to misuse and mismanagement and highly variable over time and space (Lal 2005).

Concerning the soils of Egypt, cultivable soils depend on the availability of the water supply and the type of rocks outside the areas of Nile silt deposits. Almost one-third of the total land surface of Egypt consists of Nubian sandstone. This sandstone extends over the southern sections of both the Western and Eastern deserts. Whereas, limestone deposits of Eocene age, which range from 35 to 55 million years old, cover a further one-fifth of the land surface (Mansour 2015). These deposits include central Sinai and the central portions of both the Western and Eastern deserts. The northern part of the Western Desert consists of limestone dating from the Miocene Epoch (25–5 million years ago). The mountains of the Red Sea, the Sinai, and the southwest part of the Western Desert, which represents one-eighth of the total area of Egypt, consist of ancient igneous and metamorphic rocks (Sultan et al. 2017). The current cultivated lands in the delta and the Nile valley are mainly rich in silt, which has been carried down from the Ethiopian Highlands by the Nile's upper tributary system, consisting of the Blue Nile and the Atbara rivers. These silt deposits ranges from 7 to 10 m in Aswan and the northern

H. El-Ramady (✉) · T. Alshaal · S. A. Gaheen
Soil and Water Department, Kafrelsheikh University,
Kafr El-Sheikh, Egypt
e-mail: ramady2000@gmail.com

H. El-Ramady et al. (eds.), *The Soils of Egypt*, World Soils Book Series,
https://doi.org/10.1007/978-3-319-95516-2_2

delta, respectively. In general, the soils are sandy texture outside the cultivated areas in the Delta, whereas very high clay content (from 40 to 70% in the south to the north, respectively) exists in this Delta producing some infertile black-alkali soils as well as saline soils (Negm 2017). Regarding the soils of the Nile Delta, they are recent alluvial soils having in general a light to heavy clay texture. On the other hand, the marine and alluvial deposits form soils located near the north coast and lakes. The desert sandy plains, which are flat to undulating topography, could occur near to the desert on both sides of the Delta. The soils of the beaches and coastal plains are mainly sandy texture with low to medium longitudinal sand dunes (Omran 2017). Therefore, this chapter will highlight on the historical research of soils in Egypt, the role of these soils in the establishment and flowering the Egyptian civilization in past and the education of soils under different educational levels.

2.2 Soil-Forming Factors

The forming of soils is an important issue in soil sciences. This affects not only on the course of the genesis processes but also the present and future performance of the soil systems (Fig. 2.1). It is well known that the main factors forming soils operate collectively including parent material, topography (relief), climate, biota (organisms or vegetation), and time (Jenny 1941). These factors have been included in the known equation as fundamental equation of soil formation which may be written as:

$$S = f(\mathrm{Cl}, O, R, P, T),$$

where S = any soil property, Cl = climate, O = organism or biota, R = relief or topography, P = parent material, T = time, and f = function.

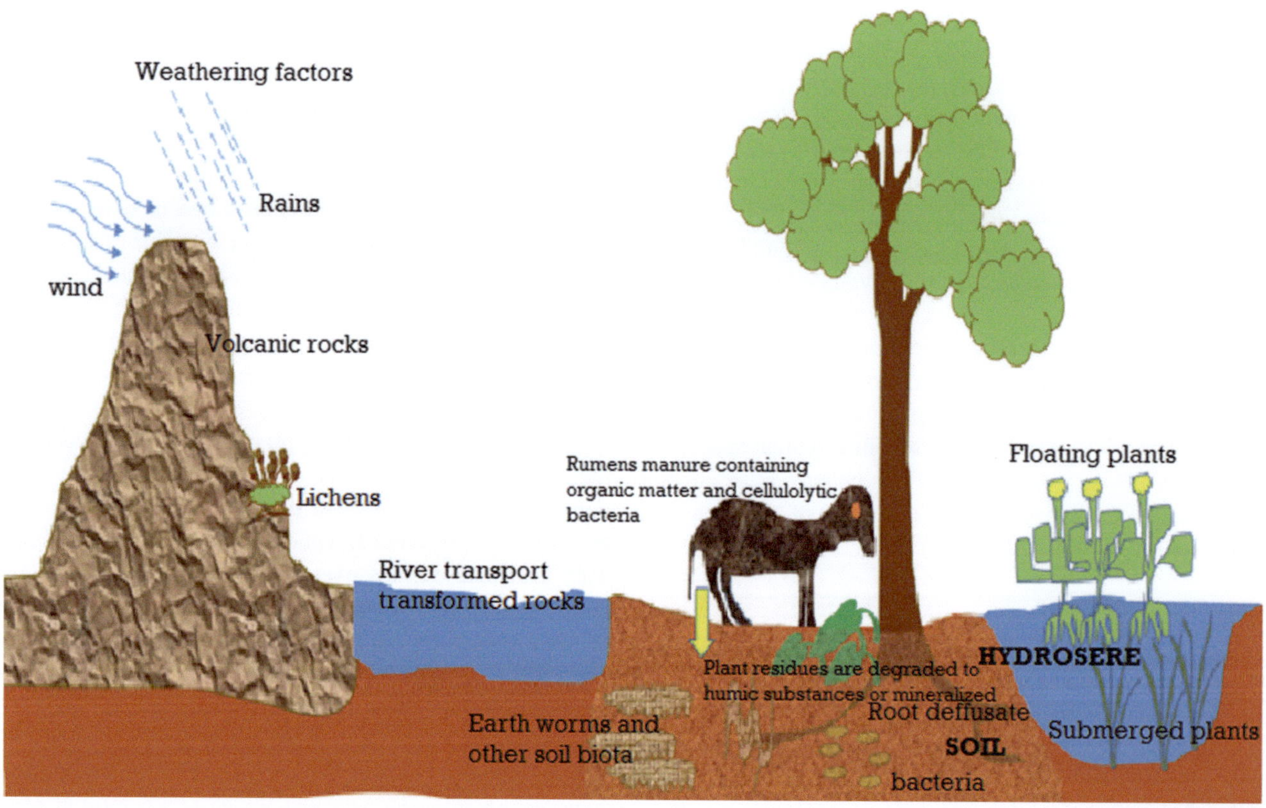

Fig. 2.1 The forming of soils simply starts from the growth of lichens or algae under rainfall conditions on rocks transforming these rocks into soils. The forming of soils also affects not only on the course of the genesis processes but also the present and future performance of the soil systems. This figure has been drawn by Dr. Tamer Elsakhawy, SWERI, Sakha, Kafr El-Sheikh

These factors are interacting over time causing a range of different soil processes (e.g., agri-illuviation, gleization, podzolization, calcification, silicification, salinization, deposition, oxidation, etc.) that result in a diversity of soil properties (e.g., high clay content in the subsoil). Human or anthropogenic activities, which cause several changes in soils, are often considered the sixth factor (Bockheim and Hartemink 2017a; Funakawa and Kilasara 2017; Yassoglou et al. 2017a). Due to the arid environments in Egypt, some distinguished pedogenetic and climatic processes could happen. The role of all soil-forming factors varies with the kind of these factors, where climate as an arid condition and parent material have a distinguished function.

The forming process of soils definitely impacts the characterization of soils. It is thought that organisms and climate are more active factors and more remarkable due to their direct effects. Regarding parent material, relief or topography and time, they called the passive factors because of their effects being less apparent. Any soil property could be estimated from the variables of soil-forming such as parent material, climate, organisms, relief, and time (Osman 2013). Therefore, several soils in Egypt could be characterized including (1) the old Nile Valley (fertile deep alluvial soils), (2) soils of the river terraces at different reliefs (deep soils with gravelly and reddish subsoils), and (3) soils of the desert (sandy, calcareous, and noncalcareous soils), which characterized with low soil fertility (Omran 2017). The common kinds of transportation of parent materials depend on the agents of transportation process and places of deposition. In general, the transportation agents include wind (dune sand and loess), water (i.e., alluvial, lacustrine, marine or deposited in riverbeds, lakes, and sea or oceans, respectively), glacier (till and moraine), or gravity (or colluvial). Wind and water transportation for parents are common in Egypt (Osman 2013). More details concerning geology and geomorphology will be presented in the Chap. 4.

2.3 Soil and Egyptian Civilization

It is well known that Egyptian civilization has a very long history. This civilization has been recorded and it dates back to around 12,000 years along the Nile as proved from Paleolithic–Neolithic artifacts (flint hand axes, arrowheads, and hammers). This period could be divided into two parts; a continuous 6000-year record with a unique and productive agriculture at its base (Janick 2002) and from 4000 to 3000 BCE the people of the Nile valley formed a government, constructed the first pyramids, and established a highly advanced agricultural technology (Janick 2016). It would be nice to cite these wonderful words from Janick (2016): "*The artistic genius engendered by the Egyptian craftsmen, the superb condition of many burial chambers and the dry climate have made it possible to reconstruct a history of agricultural technology*".

According to the common proverb, all roads lead back to the soil regarding the long-term survival of human society (Lal 2005). Concerning the Egyptian ancient civilization, the civilization of Egypt had thrived for several millennia in the same location due to the Nile and the Egyptian soils. What is the difference between this civilization and other and why this Egyptian civilization did not suffer from the demise? The answer lies in the persistence of the Nile, where the irrigated farming as well as fertile soils resulted from silt. This silt is brought by the Blue Nile from the volcanic highlands of Ethiopia mixing with the organic matter brought by the White Nile from its swampy sources (Hillel 2004). So, the flooding of the Nile comes in Egypt after killing the heat of summer, the weeds as well as aerating the soil in the pre-winter planting of grain. It is worth to mention that black soil "*Khami*" comes from ancient Greek word "*Khemia*". This name has been transmuted into our term "chemistry", as reported by Hillel (2004).

Regarding the relationship between soils and civilization, some publications reviewed about this relation such as Hyams (1952), Hillel (1991), Hughes (1992), FAO (1999), Hillel (2004), Lal (2005), El-Ramady et al. (2013), Winiwarter (2014), Soliman (2015), etc. It is reported that Egyptian civilization flowered more than 3000 BC and no other ancient civilization lasted so long (Hughes 1992). The optimal combination of soil, water, nutrients, and organic matter as well as the regular annual regime of the Nile River were the main reason for flowering of this civilization (Hillel 2004). Therefore, it is well reported by the Greek historian Herodotus that "*Egypt is the gift of the Nile.*" Several reasons confirmed this previous proverb including (1) the Nile supplied all people with nourishment and food, (2) the ancient Egyptian have had a stable climate without freezing or storms, and (3) the Nile flooded and its banks brought silt to the fields at the same time every year (Hughes 1992).

It is worth to mention that the history and knowledge of Egyptian agriculture and horticulture has gleaned from the archeological record supported by temple inscriptions, surviving written Egyptian documents and commentaries from antiquity. The supporting technology has been vividly reconstructed from the artistic record, painted, and carved in tombs and temples dating from before 3000 BCE. Agricultural activities were favorite themes of artists who drew or sculpted lively scenes of daily life that adorn the tombs of the pharaohs (Janick 2016). Definitely, the development of Egyptian agriculture was not created in a vacuum, because Egypt has an aggressive culture and Egypt has ruled several parts worldwide starting from present-day Libya in the west, Syria in the northeast, Ethiopia and Somalia and perhaps portions of sub-Saharan Africa in the south. Egypt has continuously incorporated technology as well as new crops from the Fertile Crescent as well as Africa (Janick 2016).

Therefore, it could be concluded that the Egyptian civilization has been thrived over 5000 years ago depending totally on the Nile River. This civilization includes all Egyptian life specially agriculture, astronomy, mummy (embalming), medicine, etc. The Egyptians developed their irrigation system, cultivated several crops including wheat, barley, flax, and papyrus more than 3100 BC ago. Egyptian soils were also the safeguard for this civilization besides the Nile River. Once the Nile was flooded, the water containing silt would be covered soils of basin by basin.

2.4 Soil and Its Research History

There is no doubt that, soil research history has a long story starting from the emergence of the universe. Soils are the main source for human sustenance. The study of soil history needs to incorporate several fields including historical, pedological, social and archeological data (Winiwarter 2014). Due to the complexity of soils and their interactions with humans, there are many forms that represent these interactions between soils and human (Brevik and Burgess 2012, 2014; Winiwarter 2014; Brevik and Sauer 2015). Regarding soils and humans, about 20 books already have been published by Springer such as movement of trace elements from soils to humans (Kabata-Pendias and Mukherjee 2007), soil quality and human health (Lichtfouse 2012) and the international yearbook of soil law and policy (Ginzky et al. 2017). It is reported that soils contribute with directly about 78% of average per capita calorie from crops worldwide and nearly 20% comes indirectly from terrestrial food sources (Brevik and Burgess 2014).

Concerning the research history of soils, there is not a global comprehensive history of soils has been yet written till now due to the enormous difference of soils from country to country and from society to other. So, several societies have developed testing methods for soil quality, soil classification systems, landscaping techniques including terracing process to enhance the utility of their soils and a multitude of measures for soil fertility maintenance. Furthermore, all research histories of soils are generally local due to that soils are so varied as well as the environmental history of soils is still in its infancy. Therefore, a very daunting task was and still will face the environmental historians in providing long-term data on sustainable and unsustainable use of soils in the past and in the future for the last and next years and decades (Winiwarter 2014).

On the international level, several countries have published a book about their soils through the famous book series "*World Soils Book Series*". These books contain many chapters about new challenges of soils and future subjects in soil sciences (e.g., Terribile et al. 2013; Carating et al. 2014a; Merino et al. 2016; Drohan 2017; Bockheim and Hartemink 2017b, c; Yassoglou et al. 2017a). In general, there are many scientific issues that could be considered important for soils in Egypt such as the following:

- Soil salinization under arid and semiarid conditions
- Soil-water efficiency under intensive agriculture systems
- Biogeochemical cycles in soils (paddy and salt-affected soils)
- Nutrient imbalances in agro-soils (paddy and salt-affected soils)
- Carbon sequestration in Egyptian soils
- Pollution and imbalances of nutrients in soils under changing climates
- Soil-water efficiency in marginal lands
- Desertification, degradation, and conservation of Egyptian soils
- Soil pollution under intensive agriculture in greenhouses
- Soil reclamation and decontamination of contaminated areas
- Nanoparticles or nanomaterials in soil and water remediation
- Precision farming and GIS and/or remote sensing techniques
- Sustainable bioenergy production from marginal lands

It could be concluded that the soil research should be followed the selecting and addressing future soil issues for any specific country and this must be depended on adapting soil issues to the current and the future needs of this country.

2.5 Soil and Education in Egypt

Soil and its education were and still one of the most important issues for the global sustainable development. Definitely, there is no sustainable development without sharing soils and this development should start from soils and get back soils. So, many agricultural philosophers and writers have edited or wrote several books or reviews or even articles regarding soils as well as teaching enormous disciplines about soils and related subjects. Several courses also have been taught in many schools, institutes, and universities in Egypt (Fig. 2.2). From the walled paintings in many Egyptian ancient temples, it could be extracted that ancient Egyptians have recorded and considered the potential of soils as well as agriculture (Fig. 2.3). They have taught their sons and daughters how to respect the soils as well as how to maintain these soils.

There are several writers who have documented all events of ancient Egyptian life including the agriculture and soils

Fig. 2.2 Soil study and research did not stop and will be continued in Egypt, where the higher photos (from right to left, respectively) represent a field trip of students at Kafr El-Sheikh University (Soil and Water Department) to Rigwa Farm (May 2017) and Sakha station (SWERI during 2016); the lower photos (from right to left, respectively) represent a field trip of FAO to Nubaria (October 2012) and last one was a photo for the ESSS trip during November 2012. Photos by El-Ramady and Alshaal

Fig. 2.3 The Hatshepsut's mortuary temple at El-Deir El-Bahri as an example of the building of the perfect symmetry that predates the Parthenon by thousand years. Photo by El-Ramady 2010

throughout the ages. The agriculturist Spanish **Ibn-Al-Awam** was and still one of the most important writers, who wrote several volumes on agricultural issues during the twelfth century. He is also called **Abu Zakariya Ibn al-Awwam**, an Arab agriculturist who flourished at Seville in southern Spain. In his wonderful encyclopedia about soil science, Lal (2005) mentioned about **Ibn-Al-Awam** the following statement.

He wrote the wonderful encyclopedia "***Kitab al-Felhah***" or "***Book of Agriculture***" was translated into Spanish in

Table 2.1 Details of wonderful encyclopedia the ***Kitab al-Felhah***" or "***Book of Agriculture***" for **Ibn-Al-Awam**

Chapter (s)	Detail of subjects
1–4	Different types of soils, irrigation, and fertilizers as well as planning a garden layout
5–9	Growing fruit trees, including grafting, pruning, growing from cuttings, etc. and dozens of different fruit trees are treated individually
10–14	Ploughing, the choice of seeds, grain farming, the seasons and their tasks, leguminous plants, small allotments, aromatic plants, and industrial plants
15–20	Many plants are treated individually on how to cultivate them
21	Methods of preserving and storing foods after harvest are devoted to a topic which comes up intermittently elsewhere
22–30	The symptoms of many diseases of trees and vines are indicated, as are methods of cure
31–34	Livestock include discussion of the diseases and injuries to horses and cattle

1802, and was brought to public attention in the Encyclopedia of Islam (1760–1777). In this book, the author writes "*The first step in the science of the agriculture is the recognition of soils and of how to distinguish that which is of good quality and that which is of inferior quality*" (vol. 1, p. 23). This book is divided into 34 chapters (30 chapters about crops and 4 chapters about livestock). The complete chapters were distributed as follows in Table (2.1).

Concerning the modern education, it was introduced under the great auspices of Muhammad Ali Pasha (Father of Modern Egypt) during the early 1800s. He started a dual system of education at the same time including one serving the masses attending traditional Islamic schools (called Kuttab) and another called school (Madrasa in Arabic) for the elite civil servants. Regarding the Kuttab, it taught the students, in general, the basics related to writing through memorizing and reciting Qur'anic verses, while the school system offered a more modern educational pedagogy. This era had a real development in the agricultural sector including soils and building modern irrigation system. Muhammad Ali changed the agricultural system from one harvest a year to three harvests annually using the modern irrigation.

It is worth to mention thatmore details about the role of ancient Egyptian and Islamic civilization in the history of medicine, science and technology could be found in the *Encyclopedia of the History of Science, Technology, and Medicine in Non-Western Cultures*, which was published by Springer (Selin 2016). This encyclopedia includes more than 200 titles or subchapters regarding Islamic civilization and how this civilization helped the humanity in all fields to develop and progress its life through several Islamic astronomers, thinkers, writers, engineers, historians, geographers, philosophers, etc. These subjects include agriculture, medicine, fishing, surveying, gardens, basketry, maps and mapmaking, irrigation, ship- and boat-building, meteorology, algebra, mathematics, astronomy, number theory, etc. it could list some of the famous Islamic scientists from this encyclopedia as follows:

- **Abū al-Barakāt** al-Baghdādī (d. 1164 or 1165): thinkers of the medieval period.
- **Abū Kāmil**, Shujā ‹ ibn Aslam (ca. 850–ca. 930) or "*the Egyptian Reckoner*" (al-ḥāsib al-miṣrī) was, according to the encyclopedist Ibn Khaldūn's report on algebra in his Muqaddima, chronologically the second greatest algebraist after al-Khwārizmī.
- **Abū Mashar al-Balkī's** (787–886 AD): his work is the Kitāb al-Fihrist (the Muslim calendar) and known in the West as Albumasar.
- **Abū'l-Fidā**, Shahanshāh ibn Ayyūb, Imād al-Dīn al-Ayyūbī (born in Damascus, Syria in AD 1273): historian, and geographer belonging to the Ayyūbid family.
- **Abū'l-ṣalt** or Abū'l-ṣalt al-Dānī (an Andalusian polymath born in Denia in 1067) in 1096 went to Egypt where he lived for 16 years. He wrote about pharmacology (a treatise on simple drugs, Kitāb al-adwiya al-mufrada, translated into Latin by Arnold of Vilanova), music, geometry, Aristotelian physics, and astronomy.
- **Abū'l-Wafā** īl ibn al-'Abbās (born in Būzjān, now in Iran, on 10 June 940. In 959, he moved to Baghdad and wrote important works on arithmetic, trigonometry, and astronomy.
- Muḥyī al-Dīn **ibn al- Arabī** (Born in Murcia, Spain in 1165 and died in 1240): one of the most influential Muslim thinkers of the past 700 years. His famous book is al-Futūḥāt al-makkiyya (The Meccan Openings).
- **Ibn al-Bannā**, al-Marrākushī, al-Azdī (born in Marrakesh, Morocco in 1256 and died in 1321): he is credited with having written more than 80 works. Among them Kitāb al anwā › (about asterisms and stars used in meteorology and navigation).
- **Ibn al-Bayṭār** al-Mālaqī, Ḍiyā, ibn Aḥmad (AD 1190–1248): a pharmacologist born in Málaga, Spain, in 1224,

arrived in Cairo, his wrote a book *The Complete Book on Simple Drugs* (Al-Mughnī fī'l-adwiya al-mufrada).

- Al-ḥasan ibn al-ḥasan **ibn al-Haytham** (born in 965 and died in Cairo in 1040): he proposed a hydraulic project to control the waters of the Nile, but it was rejected by the Caliph. He has 96 titles half of them about the field of mathematics, 14 on optics like Book of Optics (Kitāb al-Manāẓir) and 23 on astronomy.
- **Ibn al-Nafīs**, Alā al-Dīn Abu'l-ḥasan Alī Ibn Abī al-ḥazm al-Qurashī (born in 1213 near Damascus, Syria and died in 1288 in Cairo): He chose to live, practice and teach medicine in Egypt. He is an Arab physician mostly famous for being the first to describe the pulmonary circulation of the blood. Among his medical works is the *Comprehensive Book on the Art of Medicine*.
- **Ibn Baṭṭūṭa** (born in 1304 in Tangier, Morocco and died in 1369): was the greatest Muslim traveler of his time. At age 21, he headed toward Mecca both to make a pilgrimage and to study under some notable Muslim scholars in Egypt, Syria, and Hejaz.
- Abd al-Raḥmān **ibn Khaldūn** (born in Tunisia in 1332 and died in Cairo in 1406): he is historiographer, historian, and forerunner of the modern disciplines of sociology and demography. He has the well-known book "Introduction" (Muqaddima) in massive philosophical history of civilization.
- Abu'l-ḥasan, Alī **Ibn Riḍwān** (born in AD 998 at Giza and died in 1061 Baghdad, Iraq): he is physician-philosopher and wrote several books about the medicine in Egypt like his treatise "*On the Prevention of Bodily Ills in Egypt*" or Fī Daf Maḍārr al-Abdān fī Arḍ Miṣr.
- Abu'l-Walīd Muḥammad ibn Aḥmad ibn Muḥammad **ibn Rushd** (or Averroes, native of Cordoba, Spain AD 1126–1198): his famous book of colleges (Kitāb al-Kulliyyāt), which consists of seven books: *The Anatomy of Organs*, *Health*, *Disease*, *Symptoms*, *Drugs and Foods*, *Hygiene* and *Recovery from Disease*.
- Abū Alī al-ḥusain ibn Abdallāh **ibn Sīnā** (980–1037): also known as Avicenna. He demonstrated an incredible genius for learning and having mastered the Quran and the sciences of his time became a physician at age 16. He mastered other sciences such as psychology, astronomy, chemistry, and pharmacology by age 18 and toward the end of his life said he had learned everything he knew by then.
- Abū Mūsā/Abū Abd Allāh **Jābir ibn Hayyān** (born at Tous, Iran and died in 815 AD at Kufa, Iraq): the most important and most enigmatic figures of the history of Islamic science. He was a great chemist, alchemist, astronomer, astrologer, engineer, geographer, philosopher, physicist, pharmacist, and physician.

Therefore, it could be concluded that the education of soils in different schools and institutions started several thousand years ago starting from the Pharaohs. There is no complete record of the history of soil educations in Egypt, although education and research on soils has been initiated by the Pharaohs. Almost all Egyptian universities (governmental) include faculty of agriculture, in which contains soils department as well as some scientific centers or institutes (Table 2.2).

These departments have main mission in teaching soil sciences and other related subjects. Besides the Egyptian universities, there is the agricultural or technical education (the secondary schools for agriculture). These schools also have a great concern about soil sciences and other related courses. A large number of secondary agricultural schools have been established (around 133 agricultural high schools in 2010/2011) scattering in different cities and districts. This agricultural education provides theoretical aspects related to soils and other agricultural sciences taught in the classroom in addition to the practical aspects taught in laboratories and workshops as well as farms. On the other hand, there are some institutions related to soil sciences and is supervised by the Ministry of Agriculture (e.g., ARC, DRC, etc.) or the Ministry of Higher Education (e.g., NRC, NARSS, etc.) or the international center like Desert Development Center (DDC) supervised by the American University in Cairo or the Egyptian society for soil sciences (ESSS) supervised by the Ministry of Social Affairs.

2.5.1 Agricultural Research Center (ARC)

The agricultural research center (ARC) is considered one of the oldest agricultural centers in Middle East. It is the major engine of agricultural development in Egypt since its inception in 1971. It has been mandated to design and implement agricultural research programs within the framework of Egypt's strategy for economic and social development. The ARC is a semiautonomous, state-owned research institution governed by a Board of Directors and chaired by the Ministry of Agriculture and Land Reclamation (MALR). It embodies 16 services of the Central Administration, 16 research institutes, 8 central laboratories, 10 regional stations, 46 specific research stations and 23 research administrations throughout Egypt's governorates and employs more than 9000 researchers and scientists. The type of research conducted in ARC is mainly applied research to serve the Egyptian Economy, society welfare and prosperity of the Egyptian people. Within the national agricultural strategies ARC assumes the following major functions:

Table 2.2 List of governmental Egyptian universities, some scientific research centers in ascending order including established year and official website and have a soil and water program

University	Starting year	Official website
The governmental universities		
Cairo University (CU)	1908	https://cu.edu.eg
Alexandria University (AlexU)	1938	http://alexu.edu.eg
Ain Shams University (ASU)	1950	www.asu.edu.eg
Assiut University (AUN)	1957	http://www.aun.edu.eg
Suez Canal University (SCU)	1964	http://scuegypt.edu.eg
Mansoura University (MANS)	1972	http://www.mans.edu.eg
Tanta University	1972	http://www.tanta.edu.eg
Zagazig University	1974	http://www.zu.edu.eg
Helwan University	1975	http://www.helwan.edu.eg
Al-Azhar University	1975	www.azhar.eg
Minia University	1976	http://www.minia.edu.eg
Minufiya University	1976	http://mu.menofia.edu.eg
South Valley University (SVU)	1995	http://www.svu.edu.eg
Fayoum University	2005	http://www.fayoum.edu.eg
Benha University (BU)	2005	http://www.bu.edu.eg
Beni-Suef University (BSU)	2005	http://www.bsu.edu.eg
Sohag University	2006	http://www.sohag-univ.edu.eg
Kafrelsheikh University (KFS)	2006	http://www.kfs.edu.eg
Port Said University (PSU)	2010	http://www.psu.edu.eg
Damanhour University	2010	http://www.damanhour.edu.eg
Damietta University (DU)	2012	http://www.du.edu.eg
Aswan University (ASWU)	2012	http://www.aswu.edu.eg
Suez University	2012	http://suezuniv.edu.eg
University of Sadat City (USC)	2013	http://www.usc.edu.eg
Arish University (ARU)	2016	http://www.aru.edu.eg
Some important scientific centers are interesting in soil sciences		
Agricultural research center (ARC)	1909	http://www.arc.sci.eg
Desert research center (DRC)	1950	http://www.gndri.net/institution_drc_cairo_egypt_gn.php
Egyptian atomic energy authority	1955	http://www.eaea.org.eg
National research center (NRC)	1956	http://www.nrc.sci.eg
National authority for remote sensing and space sciences (NARSS)	1971	http://www.narss.sci.eg
Desert Development center (DDC)	1979	http://schools.aucegypt.edu/research/ddc

(1) Developing appropriate agriculture technology needed to ensure the continuous development of agricultural production,
(2) Producing new varieties of various field and horticultural crops besides new breeds of livestock, affects agricultural production positively,
(3) Planning and implementing research, extension and training programs based on the principle of integrated research, extension teams in all aspects of agricultural activities to effect sustainable upgrading of Egyptian agriculture,
(4) Optimizing use of agricultural natural resources, reducing production cost, increasing competitiveness of Egyptian agricultural productions, finding practical solutions to agricultural production constraints, pressuring environment and achieving sustainability of Egyptian agricultural development, and

(5) Transfer of new technologies to the farming community through extension service, and monitoring their adoption by the end users.

ARC has recently oriented its research programs more toward achieving productivity gains in view of the already limited land and water resource base. New varieties, including shorter-duration ones, have been developed and released; new animal breeds were produced to maximize productivity. Soil amelioration and rationalized use of agro-chemicals is a major focus area of the ARC. Scientists in ARC also participate in product standard-setting to enhance Egypt's competitiveness in the world markets. The main regional stations include Giza, Sakha, Sids, Al-Serro, Ismaelia, Kassasine, Al-Arish, Mallawi, Shandaweel, and Al-Mattana. Concerning different steps for establishment of ARC, it could be presented in Table (2.3). These institutes include:

1. Agricultural Economics Research Institute (AERI)
2. Agricultural Engineering Research Institute (AENRI)
3. Agricultural Extension and Rural Development Research Institute (AERDRI)
4. Agricultural Genetic Engineering Research Institute (AGERI)
5. Animal Health Research Institute (AHRI)
6. Animal Production Research Institute (APRI)
7. Animal Reproduction Research Institute (ARRI)
8. Cotton Research Institute (CRI)
9. Field Crops Research Institute (FCRI)
10. Food Technology Research Institute (FTRI)
11. Horticulture Research Institute (HRI)
12. Plant Pathology Research Institute (PPATHRI)
13. Plant Protection Research Institute (PPRI)
14. **Soil, Water and Environment Research Institute (SWERI)**
15. Sugar Crops Research Institute (SCRI)
16. Veterinary Serum and Vaccine Research Institute (VSVRI)

Over the past two decades, numerous achievements have been realized by ARC including the improvement of agronomic practices, development of new varieties, maintenance of the national herds, livestock development and better food processing techniques. New crops and animal breeds have also been introduced and researches have been dedicated to problem-solving, side by side with basic sciences. The main target of ARC was and still is to maximize the economic return from both land and water unit. Concerning the Soil, Water and Environment Research Institute (SWERI), it is the main institute in ARC for soils investigation as well as providing farmers with services related to analysis of soils, water, plants, fertilizers, food, and feed. The Soil and Water Research Institute (SWRI) was established and organized as an independent institute in 1971. The Environment Researches are founded beside soil and water in 1994 to be Soil, Water, and Environment Research Institute (SWERI). This institute includes 11 departments with stress on different soil sciences such as soil physics and chemistry, agricultural microbiology, soil fertility and plant nutrition, soil survey and classification, water requirements and field irrigation, etc. These departments include:

1. Agricultural Microbiology Research
2. Environment Research (established since 1994)
3. Field Drainage Research (established since 1975)
4. General Department for Fertilizers Control
5. Improvement and Conservation of Cultivated Soils Research (established since 1971)
6. Sandy and Calcareous Soils Research (established since 1972)
7. Soil Fertility and Plant Nutrition Research (established since 1939)
8. Soil Physics and Chemistry

Table 2.3 Some important steps for establishment the agricultural research center

Event	Year
Establish department of soil chemistry by the ministry of public works	1903
The Egyptian agricultural authority has established an agricultural research station in Bahteem	1909
A special laboratory to provide farmers and institutions with services related to analysis of water, soils, fertilizers, plants, food, and feed was established	1911
Along with five other major agricultural centers in the world, the department was famous for its long-term fertilizer trials in Bahteem	1912
Department of agriculture was established with soil chemistry as one of its major branches of the ARC	1913
Department of crop physiology and plant nutrition was established with two major branches of plant nutrition and water requirements	1957
General administration of soil was founded including 4 divisions for soil survey and improvement, soil research, agricultural microbiology and soil analysis	1960
Soil and water research institute (SWRI) was organized as an independent institute of ARC in the ministry of agriculture, animal wealth, fisheries and land reclamation	1971

9. Soil Salinity and Alkalinity Res. Lab (established since 1951)
10. Soil Survey and Classification Research (established since 1955)
11. Water Requirements and Field Irrigation Research (established since 1956)

It is worth mentioning that besides SWERI there are some institutes related to soil sciences such as Field Crops Research Institute (FCRI), Horticulture Research Institute (HRI), Cotton Research Institute (CRI), etc. Enormous studies about soils from different issues have been published in the SWERI and other institutes related to soils in the ARC in both local and international scientific journals.

2.5.2 Desert Research Center (DRC)

Desert research center (DRC) is one of the oldest scientific research centers in Egypt. In 1927, king Fouad decided to establish the DRC for exploration of the Egyptian desert in order to evaluate its rich natural resources. It was officially opened under the name of the Institute of Fouad I for Desert Research (IFDR) and inaugurated by him on December 30, 1950. This institute has been acquired its name "*Desert Research Center*" with new laws, structures, and bylaws in 1990. This center belongs to the Minister of Agriculture and Land Reclamation. The main mission of this center is to sustain the natural resources in the Egyptian deserts and to improve the livelihood of local communities. The headquarters of this center is located in Cairo and some stations are geographically distributed in Egypt to meet the challenges of each different ecosystem. These stations can be found in Sinai, Marouh, Toshka, Siwa, New valley and Halayeb/Shalateen (Fig. 2.4). Concerning Sinai Station, it includes four locations (1) Sheikh Zuweid station (located in North Sinai for conservation and utilization of Plant Genetic resources), (2) El-Maghara station (located in south of El-Arish in the middle of the Sinai for water harvesting and rangeland management), (3) Ras sudr station (located in south Sinai for Bio-saline agriculture and (4) Baloza station at El-Qantra for rangeland management and sand dune fixation. Concerning the main objectives of the DRC, it could be included the following issues:

- Investigating desert potential for agriculture development.
- Providing proper help and advice to target groups, whether local Bedouins or investors, to best utilize the available natural resources.
- Managing desert and newly reclaimed lands for agriculture use and developments.
- Studying means to ameliorate and combat drought, desertification and sand dunes movements.
- Monitoring and assessing desertification causes using field operations from Sinai to the new valley to the fringes of the high dam lakes.
- Excavation and developing desert natural resources including water, soil, plant and nonconventional energies.
- Monitoring land resources (water and soils) in the desert and newly reclaimed areas.
- Introducing nonconventional crops to be cultivated under drought and salinity stresses.
- Increasing productivity of livestock and poultry under desert conditions.
- Concluding socioeconomic studies in desert regions.
- Publishing and exchanging relevant scientific researches with different local, regional and international scientific institutions.
- Providing extension and training programs for personnel and investors involved in desert development activities.

This center constitutes staff of researchers (professor, assistant research professors, researchers, research assistant, and assistant researchers) as well as administrators, technicians, and laborers like all scientific centers. This center also constitutes of number of administrative offices, in addition to eight Stations and five Associated Units (Tissue Culture Lab, Geographical Information Systems (GIS), Satellite Receiving Station, Private Service Unit (PSU), and Library). This center also includes 32 laboratories and 4 major divisions, 14 departments. These departments include:

I. **Water Resources and Desert Soils Division**
   - Geology Research Dept.
   - Geophysical Exploration Dept.
   - Renewable Energy Research Dept.
   - Hydrology Research Dept.
   - Hydro-geo-chemistry Research Dept.
   - Pedology Research Dept.
   - Soil Physics and Chemistry Research Dept.
   - Soil Fertility and Microbiology Research Dept.
   - Soil Water Conservation Dept.

II. **Ecology and Dry Land Agriculture Division**
   - Plant Genetic Resources Dept.
   - Plant Production Research Dept.
   - Plant Ecology and Range Management Dept.
   - Sand Dune Research Dept.
   - Medicinal and Aromatic Plants Dept.
   - Plant Protection Dept.

Fig. 2.4 The desert research center station in Matruh (the top photo), where the main building of desert research center in Cairo in the bottom. Photo by El-Ramady, April 19, 2017 and December 24, 2013, respectively

III **Animal and Poultry Production Division**
- Animal and Poultry Breeding Dept.
- Animal and Poultry Nutrition Research Dept.
- Animal and Poultry Physiology Research Dept.
- Wool Production and Technology Research Dept.
- Animal Health Research Dept.

IV. **Socioeconomic Studies Division**

Thousands of studies definitely have been published about the soils under Egyptian desert conditions including different issues (Abdel Kawy and Abou El-Magd 2013; Shendi et al. 2013; Darwish and Pöllmann 2015; Abuzaid and Fadl 2016; Rashed 2016; Mousa 2017). It is worth to mention that Prof. Mahmoud Zahran wrote several books about the vegetation of Egypt in 1992 till 2009 focusing on the deserts in Egypt and their plant biodiversity (e.g., Zahran and Willis 2009). Some Egyptian leaders in this field also should be mentioned such as Prof. Kamal El-Din Hassan Batanouny, Prof. Mohamed Kassas (from Cairo University, Faculty of Science), Prof. Loutfy Boulos, etc.

2.5.3 Egyptian Atomic Energy Authority (EAEA)

It is well known that the Egyptian Atomic Energy Authority (EAEA) has been established in 1955. This authority aimed to establish a national research and development in the basic and applied peaceful nuclear researches. This EAEA can play a vital role in protecting the Egyptian environment including soil and water from radiological contamination. This contamination may result from accidents or normal nuclear activities leading to contamination of soil, air, or water. Therefore, these contaminating activities also may be resulted from local wide use of radioisotopes in Egypt or crossing the boundaries from the neighbor countries. The EAEA deals with the handling of radioactive wastes, collecting the radio-wastes from different radioisotopes, treatment, transportation, conditioning and ultimate storage of these wastes. This EAEA can develop and implement different radiological analyses techniques to detect any environmental pollution. These analyses include measuring isotopes in different plant nutrition research. Providing consultation and services in the field of radiation decontamination could be achieved as well as detection of radiation of imported foodstuff and other items.

The EAEA now consists of four major scientific centers, namely, the Nuclear Research Center (NRC), The National Center for Radiation Research and Technology (NCRRT), the Hot Laboratories and Waste Management Center (HLWMC) and National Centre for Nuclear Safety and Radiation Control (NCNSRC).

(a) **Nuclear Research Center**

This center is the oldest and the biggest research institute of the EAEA. It is located in Inshas city and its activities are directed toward the basic nuclear sciences, reactors, nuclear reactor materials, electronic instrumentation, and applications of radioisotopes in medicine, industry, agriculture, etc. This institute includes different studies regarding the use of isotopes in plant nutrition, foodstuffs as well as monitor and detecting of radiation of imported foodstuff and other items.

(b) **Hot Laboratory and Waste Management Center**

This center was established in 1972 aiming to promote research and development using ionizing radiation in different fields of ionizing radiation in medical, industrial, agricultural, environmental, and other applications. The NCRRT contains three major gamma irradiator facilities, namely Egypt's Mega Gamma-1 Irradiator, Alexandria Gamma Irradiator in addition to Electron Beam Accelerator. These facilities are used for sterilization of healthcare products, preservation of agricultural products, industrial applications, and scientific research. It also contains four research gamma irradiators.

(c) **National Centre for Radiation Research and Technology**

The center was established in 1980. Its purpose is to develop of expertise in fields of radioactive waste management, production of radioisotopes for various medical and industrial applications, as well as production of labeled pharmaceutical compounds that are widely used for medical diagnosis.

(d) **National Centre for Nuclear Safety and Radiation Control**

This center is mainly dealing with both the nuclear safety and the control of radiation. Definitely, this will be included the medicinal, industrial, environmental and agricultural sectors.

Therefore, the EAEA is divided into four research centers, where these four centers are subdivided into major research divisions. The activities of this EAEA can run into four major fields through the integrated and concerted planning between these centers to include (1) research and technological projects, (2) radiation protection and safety, (3) society services activities, and (4) regional and international cooperation. More than 1400 qualified academic scientists in various fields including nuclear sciences, agriculture, and engineering supported by about 2300 technical staff, 1300 administrative staff, represent the driving force for research and development in the nuclear sciences and technology. Several studies have been published regarding the use of isotopes in different agricultural sciences including soil sciences (El-Sorogy et al. 2013; Gomaah et al. 2016; Isawi et al. 2016; Mohamed et al. 2016; Ibrahim and Lyons 2017).

2.5.4 National Research Center (NRC)

The National Research Centre (NRC) is considered one of the most important scientific centers in Egypt and the Middle East. It is established in 1956 as an Egyptian research and development center for multiple disciplines including biology, chemistry, agriculture, medicine, engineering and genetics (Fig. 2.5). The main target of NRC is to foster different basic and applied scientific researches, particularly in agriculture, industry, public health, and other sectors of Egyptian national economy. The NRC is the largest institution affiliated with the ministry of Higher Education and Scientific Research. Considering the number of scientists,

Fig. 2.5 National Research Centre is considered one of the most important scientific centers in Egypt and Middle East, established in 1956. Photo by El-Ramady, September 2017

NRC is the largest national multidisciplinary R and D institute among all institutions affiliated to the ministry of Scientific Research, Egypt. It has a research staff of about 5000 scientists and is headed by a president with two vice presidents for (1) research affairs and international relations and (2) technical affairs. When it established, NRC started with four divisions, the agricultural division was one of them. During that time, the division involved three departments, soil and fertilizers, plant protection and agricultural technology departments. The NRC consists of 14 divisions and 111 departments covering the major areas of environment, industry, agriculture, health, basic sciences, and engineering. Concerning different divisions of the NRC, it could be listed as follows:

- **Agriculture and Biology Research Division**
- Chemical Industries Research Division
- Engineering Research Division
- Environmental Sciences Research Division
- Food Industry and Nutrition Division
- Genetic Engineering and Biotechnology Division
- Human Genetics and Genome Researches
- Inorganic Chemical Industries and Mineral Resources Division
- Medical Sciences Division
- Oral and Dental Research Division
- Pharmaceutical Industries Division
- Physics Division
- Textile Industries Division
- Veterinary Research Division

The most important division related to the soil sciences is agriculture and biology research division involves 16 department and the soils and water use department is one of them. The soils and water use department is full of great scientists who effectively contribute in many soils- and water-related research. Among other; Prof. Gamal Abdel Samie, Prof. Abbas Rasheed, and Prof. Mohamed Yousry are highly appreciated as great scientists who are well known at both national and international levels. This division also was established in 1968 with a certain mission including the achievement of sustainable development and promoting the agricultural sector through increasing agricultural production by using modern technologies. This division also includes some departments such as:

- Animal Production Dept.
- Botany Dept.
- Fertilization Technology Dept.
- Field Crops Dept.
- Pests and Plant Protection Dept.
- Plant Nutrition Dept.
- Plant Pathology Dept.
- Pomology Dept.

- Soils and Water Use Dept.
- Vegetable Researches Dept.
- Water Relations and Field Irrigation Dept.

From these previous departments, it could be concluded that general tasks of this division in the frame of these previous departments including the following issues:

- Recycling agricultural residues and promotion of new and renewable energy.
- Reduction the pollution rates in soils and waters by improving the efficiency of soil and water use in quantity and quality.
- Integrated pest management of insects and pathogens.
- Encouraging communication among the exporters, businessmen, and investors in the field of agriculture.
- Promoting collaborative research programs with international agricultural bodies and foreign agriculture universities.
- Supporting transfer of technology to the agriculture sector across different geographical areas in Egypt to provide sustainable agriculture growth.
- Identifying future research needs, priorities and assessment of problems to achieve sustainable productivity.

Concerning Prof. Dr. Gamal Abdel Samie, he was born in 1921, graduated from College of Agriculture, Cairo University with Bachelor degree, and then received his MSc and Ph.D. degrees in soil science from Oregon State University, USA. In 1954, he was selected to establish the Desert Research Institute. In 1959, he awarded a permanent position at NRC as a first head of the soil unit, then he continued as the head of soils department then a chairman of the Agriculture and Biological Research Division before his appointment as Vice President of the Academy of Scientific Research and Technology in 1974 until retirement in 1981. During his career, he worked in different aspect related to soils research, supervised many M.Sc. and Ph.D. theses in soil science and related disciplines, published many scientific papers, and wrote different reports. At the international level, he worked with different international organizations, such as; Food and Agricultural organization (FAO) and United Nations Environment Program (UNEP) and travels to many countries such as USA, Nigeria, Australia, European Countries, and Middle East. After retirement he continued his career as scientific counselor to the President of the Academy as well as part-time Research Professor at NRC.

Prof. Dr. Abbas Rasheed was born in 1936, graduated from College of Agriculture, Ain Shams University with Bachelor degree in soils, and then he awarded the diploma in Radioisotopes in Agriculture in 1959 and he received his Ph. D. degree in soil science from Newcastle University, UK in 1965. He served as the head of soils and water use department (1983–1988), then he was a chairman of the Agriculture and Biological Research Division for many years. At the international level, he was the director of Inter-African Bureau of soils of Scientific, Technical and Research Commission (STRC) of the Organization of African Unity (OAU), Addis Ababa, Ethiopia (1971–1983). He supervised many M.Sc. and Ph.D. theses in soil science and related disciplines and published many scientific papers and technical reports.

Prof. Dr. Mohamed Yousry was born in 1942, graduated from College of Agriculture, Ain Shams University with Bachelor degree in soils, and then he awarded his M.Sc. from Cairo University and he received his Ph.D. degree in soil science from Ain Shams University in 1972. He Granted several scholarships and post-doctoral fellowships and visits; Technical University in Berlin and Munich, FRG and Ghent University, Belgium. He was appointed as the President of the Academy of Scientific Research and Technology for many years.

2.5.5 National Authority for Remote Sensing and Space Sciences

National Authority for Remote Sensing and Space Sciences (NARSS) is the pioneering Egyptian institution for remote sensing and space sciences. It was established in 1971 as an American–Egyptian joint project with affiliation to the Egyptian Academy of Scientific Research and Technology. This authority was moved in its affiliation to the State Ministry of Scientific Research in 1994. The main target of this authority was and still is to promote using of the space technology in developing Egypt as well as to pursue, transfer and provide the most advanced technology in the fields of remote sensing and peaceful application of space sciences and build the self capability to utilize these technologies to support the national activities for sustainable development. The NARSS includes eight main divisions as follows:

- Agriculture Applications, Soils, and Marine
- Aviation and Aerial Photography
- Data Reception, Analysis and Receiving Station Affairs
- Engineering Applications and Water Resources
- Environmental Studies and Land use
- Geological applications and Mineral Resources
- Scientific Training and Continuous Studies
- Space Sciences and Strategic Studies

The most important division related to our book is Agriculture Applications, Soils, and Marine (AASMD). This division has the mandate to carry out consultancy and research for different fields of agriculture, soil, and marine using remote sensing and geographical information system (GIS) techniques. It consists of three departments including agricultural applications, soils, and marine with the following objectives:

- Strengthen the utility of aerospace technologies (RS, GIS, GPS, etc.) and their application in the agricultural sector development plans of Egypt.
- Provide technical and expert assistance in related areas to both governmental and private sectors.
- Establish an Integrated National Agricultural Resources Information System (INARIS).
- Promoting the regional and international cooperation in the field of agriculture and marine fields to build the national capacity.

Concerning different investigations about the use of GIS and remote sensing in soil sciences, several studies have been published including different areas in Egypt such as the Nile Delta (Abdel Kawy and Ali 2012; El-Asmar et al. 2013; Elbasiouny et al. 2014; Moghanm 2015a, b; Shokr et al. 2016; AbdelRahman et al. 2017; Farg et al. 2017), Sinai (Mohamed et al. 2014; Omran 2016a; Yossif and Ebied 2016), or different topics related to agriculture (El-Baroudy 2013; Belal et al. 2014; Mohamed et al. 2015; Omran 2016b; El-Zeiny and El-Kafrawy 2017).

2.5.6 Desert Development Center (DDC)

The Desert Development Center (DDC) was established in 1979 by the American University in Cairo as a center of excellence in applied research and training to promote sustainable development in Egypt's reclaimed desert areas. The DDC is a home to academics, engineers, administrators, technicians, and farmers who are devoted to working toward the social, economical, and environmental sustainable desert development. The main target of research activities of DDC is to understand current desert settlement conditions and their technologies and practices as well as to improve them in a sustainable fashion. To match the complexity of sustainable desert development, research of DDC is both cross-sectoral and integrated including social, technical, and economic research activities.

The DDC has an outstanding residential training facility with a capacity for 150 men and women at its South Tahrir Research Station. There are analytical laboratories and nurseries at the Sadat City Research Station. DDC research programs include a wide range of academic disciplines, from the natural and social sciences through applied fields such as construction and environmental engineering, agriculture, management and planning, and architecture. The DDC trains about 1000 desert settlers in its courses on sustainable agricultural practices each year, with another 500 or so university students receiving specialized training as part of their academic work. The DDC has two research stations in South Tahrir and Sadat City. They carry out applied research on topics ranging from livestock development under desert conditions to testing the energy efficiency of different domestic and farm architectural forms, increased crop productivity, profitability, and the conservation of natural resources. The DDC horticulture, animal sciences, and soil quality researchers work to identify the ideal breeds and varieties for Egypt's desert environment. Concerning the agricultural researches in DDC, this center continuously carries out applied research in the following fields:

I. **Land improvement and soil fertility**
   - Crop rotations
   - Fertilizer requirements and methods of application
   - Soil testing and plant analysis
   - Clonal propagation based on plant tissue culture techniques
   - Application of bioanalytical methods and genetic mapping of nematodes as modern tool for organic vegetable farms
   - Zero, minimum, and conventional tillage treatments

II. **Irrigation and water use efficiency**
   - Desert-appropriate irrigation techniques (sprinkler, drip, Bi-wall, and bubbler)
   - Water and soil management practice under desert conditions
   - Drainage solutions

III. **Farming systems**
   - Citrus improvement for desert conditions
   - Crop/livestock integration for desert settlements
   - Desert-appropriate farm management systems
   - New intensive agriculture techniques such as NFT
   - Appropriate seeding and other agronomic practices
   - Controlled environment agriculture for propagation of forest and horticulture species
   - Maintenance of equipment for agriculture, irrigation, and renewable energy
   - Monitoring of activities using simple or computer/data logger equipment

IV. **Forestry and desertification control**
   - Windbreak trees
   - Weed control and other plant protection measures
   - Desert architecture and renewable energy
   - Application of solar energy for power production
   - Application of solar energy for water heating

- Application of wind energy or wind/PV system for water pumping and power
- Application of biogas technology for cooking, heating and compost
- Passive solar architecture designs for desert housing
- Use of adobe bricks and other conventional or new materials for building

V. **Resource management and local institutions for new desert settlements**
- Community-based water demand management
- Sustainable integrated farming systems in new communities
- Farmer-initiatives for environmentally sustainable resource management

Therefore, it could be concluded that more soil laboratories and institutes have been established in Egypt during the last decades. Several researches have been carried out in various topics related to soils under many conditions. These topics include soil degradation and land protection, soil genesis and classification, soil mapping and land evaluation, soil use and land management, soil fertility and fertilization, soil quality and groundwater pollution and application of wastes in soils, etc.

2.5.7 Egyptian Soil Science Society (ESSS)

The Egyptian society of soil sciences (ESSS) is one of the oldest scientific societies in Egypt. It was established in 1950 for promoting the wise application of water, soil, and environmental management practices. These practices could improve and safeguard the quality of both land and water resources, which continue to meet the needs of society and agricultural sector. It seeks also for a world in which all water, soil, and environment resources will be used in a sustainable, productive, and ecologically sound manner. This society includes researchers from the Egyptian universities, scientific institutes, and research centers in Egypt as well as foreigners.

The ESSS has many tasks in serving soil sciences in Egypt including (1) the providing their members with professional recognition and activities, (2) the publication of scientific articles through the Egyptian Journal of Soil Science (EJSS; since 1961) with what might be available from periodicals and audio visual aids, (3) organizing conferences and scientific symposia as well as international and local exhibitions in the fields of scientific activities of the society, (4) organizing training or workshops in the areas of soil, water, environment and related sciences for the preparation of cadres capable to absorb and implement the application of modern technologies, and (5) supporting the communication with other international societies of soil sciences such as the international union of soil sciences (IUSS) and other societies. The ESSS is an official member in the IUSS and only **Prof. Hassan Hamdi** (The Ex-President of ESSS) was listed to the IUSS Honorary Members in 2002.

It is worth to mention that International Society of Soil Science (IUSS) is the main and official society of soil sciences, which established on 19th May 1924. This union is the principal and main reference for everything in soil sciences and these soil sciences could be investigated through the scientific activities of IUSS, which are undertaken through divisions, commissions, and working groups as follows:

Division 1	Soils in Space and Time
Division 2	Soil properties and processes
Division 3	Soil Use and Management
Division 4	The Role of Soils in Sustaining Society and the Environment

2.6 Conclusion

It could be concluded that soils are the main source for our life starting from our creation, our food, feed, fiber, and fuels till our death and decomposition of our bodies. So, all religions have documented that soils should be respected from all people, pupils, students, scientists, farmers, workers, etc. There are several institutions in Egypt having a main task in dealing and maintaining soils such as universities and scientific research centers. On the national level, many laws have been enacted to protect and respect soils, but still more enforcements are needed. Like the advanced nations, soils have been protected by strict laws as well as the continuous performance.

Acknowledgements Authors thank the outstanding contribution of STDF research teams (Science and Technology Development Fund, Egypt) and MBMF/DLR (the Federal Ministry of Education and Research of the Federal Republic of Germany), (Project ID 5310) for their help. Great support from this German-Egyptian Research Fund (GERF) is gratefully acknowledged.

References

Abdel Kawy WAM, Abou El-Magd IH (2013) Use of satellite data and GIS for assessing the agricultural potentiality of the soils South Farafra Oasis, Western Desert. Egypt Arab J Geosci 6:2299–2311. https://doi.org/10.1007/s12517-012-0518-5

Abdel Kawy WM, Ali RR (2012) Assessment of soil degradation and resilience at northeast Nile Delta, Egypt: The impact on soil productivity. Egypt J Remote Sens Space Sci 15(1):19–30

AbdelRahman MA, Shalaby A, Belal AB (2017) A GIS based model for land evaluation mapping: a case study north delta Egypt. Egypt J Soil Sci (in press). https://doi.org/10.21608/ejss.2017.255.1043

Abuzaid AS, Fadl ME (2016) Land evaluation of eastern suez canal, egypt using remote sensing and GIS. Egypt J Soil Sci 56(3):537–547

Arnalds O (2015) The soils of Iceland. World soils book series. Springer Science + Business Media Dordrecht. https://doi.org/10.1007/978-94-017-9621-7

Bašić F (2013) The soils of Croatia. World soils book series. Springer Science + Business Media Dordrecht. https://doi.org/10.1007/978-94-007-5815-5

Belal AB, El-Ramady H, Mohamed ES, Saleh A (2014) Drought risk assessment using remote sensing and GIS techniques. Arab J Geosci 7:35. https://doi.org/10.1007/s12517-012-0707-2

Bockheim JG (2015) The soils of Antarctica. World soils book series. Springer International Publishing Switzerland. https://doi.org/10.1007/978-3-319-05497-1

Bockheim JG, Hartemink AE (2017a) The soils of Wisconsin. World soils book series. Springer International Publishing AG. https://doi.org/10.1007/978-3-319-52144-2

Bockheim JG, Hartemink AE (2017b) History of soil studies. In: The soils of wisconsin. World soils book series. Springer International Publishing AG, pp 7–21. https://doi.org/10.1007/978-3-319-52144-2_2

Bockheim JG, Hartemink AE (2017c) Current and future soil research. In: The soils of wisconsin. World soils book series. Springer International Publishing AG, pp 223–228. https://doi.org/10.1007/978-3-319-52144-2_18

Boer BW, Ginzky H, Heuser IL (2017) International soil protection law: history, concepts and latest developments. In: Ginzky H et al (eds) International yearbook of soil law and policy 2016, international yearbook of soil law and policy. Springer International Publishing AG. https://doi.org/10.1007/978-3-319-42508-5_7

Brevik EC, Burgess LC (2012) Soils and human health. CRC Press

Brevik EC, Burgess LC (2014) The influence of soils on human health. Nat Educ Knowl 5(12):1

Brevik EC, Sauer TJ (2015) The past, present, and future of soils and human health studies. Soil 1:35–46. https://doi.org/10.5194/soil-1-35-2015

Carating RB, Galanta RG, Bacatio CD (2014a) Soil issues and challenges. In: Carating RB et al. (ed) The soils of the Philippines. World soils book series. Springer Science + Business Media Dordrecht, pp 273–304. https://doi.org/10.1007/978-94-017-8682-9_6

Carating RB, Galanta RG, Bacatio CD (2014b) The soils of the Philippines. World soils book series. Springer Science + Business Media Dordrecht. https://doi.org/10.1007/978-94-017-8682-9

Casanova M, Salazar O, Seguel O, Luzio W (2013) The soils of Chile. World soils book series. Springer Science + Business Media Dordrecht. https://doi.org/10.1007/978-94-007-5949-7

Chen Z-S, Hseu Z-Y, Tsai C-C (2015) The soils of Taiwan. World soils book series. Springer Science + Business Media Dordrecht. https://doi.org/10.1007/978-94-017-9726-9

Costantini EAC, Dazzi C (2013) The soils of Italy. World soils book series. Springer Science + Business Media Dordrecht. https://doi.org/10.1007/978-94-007-5642-7

Darwish MAG, Pöllmann H (2015) Trace elements assessment in agricultural and desert soils of Aswan area, south Egypt: geochemical characteristics and environmental impacts. J Afr Earth Sci **112** (Part A):358–373

Drohan PJ (2017) Future challenges for soil science research, education, and soil survey in the USA. In: West LT et al (eds) The soils of the USA. World soils book series. Springer International Publishing AG, pp 373–383. https://doi.org/10.1007/978-3-319-41870-4_19

El-Baroudy AA (2013) Evaluating environmental sensitivity to desertification in El-fayoum depression. Egypt Egypt J Soil Sci 53(3): 445–460

El-Asmar HM, Hereher ME, El Kafrawy SB (2013) Surface area change detection of the Burullus Lagoon, North of the Nile Delta, Egypt, using water indices: a remote sensing approach. Egypt J Remote Sens Space Sci 16(1):119–123

Elbasiouny H, Abowaly M, Abu Alkheir A, Gad A (2014) Spatial variation of soil carbon and nitrogen pools by using ordinary Kriging method in an area of north Nile Delta. Egypt CATENA 113:70–78

El-Ramady HR, El-Marsafawy SM, Lewis LN (2013) Sustainable agriculture and climate changes in Egypt. In: Lichtfouse E (ed) Sustainable agriculture reviews series, vol 12. Springer Science + Business Media Dordrecht, pp 41–95. https://doi.org/10.1007/978-94-007-5961-9_2

El-Sorogy AS, Nour H, Essa E, Tawfik M (2013) Quaternary coral reefs of the Red Sea coast, Egypt: diagenetic sequence, isotopes and trace metals contamination. Arab J Geosci 6:4981–4991. https://doi.org/10.1007/s12517-012-0806-0

El-ZeinyA S El-Kafrawy (2017) Assessment of water pollution induced by human activities in Burullus Lake using landsat 8 operational land imager and GIS. Egypt J Remote Sens Space Sci 20 (Supplement 1):s49–s56

Espinosa J, Moreno J, Bernal G (2018) The soils of Ecuador. World soils book series. Springer International Publishing AG. https://doi.org/10.1007/978-3-319-25319-0

FAO (1999) Assessment of the world food security situation. Report #CFS: 99 = 2 of the 25th Session of the Committee on World Food Security, Rome, 31 May–2 Italy

Farg E, Arafat S, Abd El-Wahed MS, El-Gindy A (2017) Evaluation of water distribution under pivot irrigation systems using remote sensing imagery in eastern Nile delta. Egypt J Remote Sens Space Sci 20(Supplement 1):s13–s19

Funakawa S, Kilasara M (2017) Soil-forming factors determining the distribution patterns of different soils in Tanzania with special reference to clay mineralogy. In: Funakawa S (ed) Soils, ecosystem processes, and agricultural development, Springer Japan KK, pp 65–83. https://doi.org/10.1007/978-4-431-56484-3_4

Gallardo JF (2016) The soils of Spain. World soils book series. Springer International Publishing Switzerland. https://doi.org/10.1007/978-3-319-20541-0

Ginzky H, Heuser IL, Qin T, Ruppel OC, Wegerdt P (2017) International yearbook of soil law and policy 2016. International yearbook of soil law and policy series, vol 2016. Springer International Publishing AG. https://doi.org/10.1007/978-3-319-42508-5

Gomaah M, Meixner T, Korany EA, Garamoon H, Gomaa MA (2016) Identifying the sources and geochemical evolution of groundwater using stable isotopes and hydrogeochemistry in the Quaternary aquifer in the area between Ismailia and El Kassara canals. Northeast Egypt Arab J Geosci 9:437. https://doi.org/10.1007/s12517-016-2444-4

Hillel DJ (1991) Out of the earth: civilization and the life of the soil. The Free Press, New York, p 321

Hillel DJ (2004) Civilization: role of soils. In: Encyclopedia of soils in the environment, 1st edn. Academic Press, pp 199–204

Hughes JD (1992) Sustainable agriculture in ancient Egypt. Agricultural history, vol 66 (2), History of Agriculture and the Environment, Spring, Agricultural History Society, pp 12–22. http://www.jstor.org/stable/3743841. Accessed 16 June 2012

Hyams E (1952) Soil and Civilization. Thames and Hudson, London, p 312

Ibrahim RGM, Lyons WB (2017) Assessment of the hydrogeochemical processes affecting groundwater quality in the eocene limestone aquifer at the desert fringes of El Minia governorate. Egypt Aquat Geochem 23:33–52. https://doi.org/10.1007/s10498-016-9298-y

Imamul Huq SM, Shoaib JM (2013) The soils of Bangladesh. World soils book series. Springer Science + Business Media Dordrecht. https://doi.org/10.1007/978-94-007-1128-0

Isawi H, El-Sayed MH, Eissa M, Shouakar-Stash O, Shawky H, Abdel Mottaleb MS (2016) Integrated geochemistry, isotopes, and geostatistical techniques to investigate groundwater sources and salinization origin in the sharm EL-Shiekh area, South Sinia. Egypt Water Air Soil Pollut 227:151. https://doi.org/10.1007/s11270-016-2848-5

Janick J (2002) Ancient Egyptian agriculture and the origins of horticulture. Acta Hort 582:23–39

Janick J (2016) Origins of agriculture in Egypt. In: Selin H (ed) Encyclopaedia of the history of science, technology, and medicine in Non-western cultures. Springer Science + Business Media Dordrecht, pp 3460–3478. https://doi.org/10.1007/978-94-007-7747-7

Jenny H (1941) Factors of soil formation a system of quantitative pedology. McGraw-Hill, New York

Kabata-Pendias A, Mukherjee AB (2007) Trace elements from soil to human. Springer Berlin Heidelberg. https://doi.org/10.1007/978-3-540-32714-1

Kapur S, Akça E, Günal H (2018) The soils of Turkey. World soils book series. Springer International Publishing AG. https://doi.org/10.1007/978-3-319-64392-2

Krasilnikov P, Gutiérrez-Castorena MC, Ahrens RJ, Cruz-Gaistardo CO, Sedov S, Solleiro-Rebolledo E (2013) The soils of Mexico. World soils book series. Springer Science + Business Media Dordrecht. https://doi.org/10.1007/978-94-007-5660-1

Lal R (2005) Encyclopedia of soil Science, 2nd edn, Taylor & Francis, pp 838–841. https://doi.org/10.1081/e-ess-120042700

Lehmann J, Dirse CK, Bruno G, William WI (2003) Amazonian dark earths: origin, properties, management. Kluwer Academic Publishers, Dordrecht

Lichtfouse E (2012) Organic fertilisation, soil quality and human health. Sustainable agriculture reviews, vol 9. Springer Science + Business, Media B.V., Springer Dordrecht Heidelberg. https://doi.org/10.1007/978-94-007-4113-3

Mansour AS (2015) Sedimentology and environmental interpretation of carbonate deposits associating the evaporites-bearing Miocene succession, northern Western Desert. Egypt Carbonates Evaporites 30:167–179. https://doi.org/10.1007/s13146-014-0200-y

Merino A, Moreno G, Navarro FB, Gallardo JF (2016) Future issues. In: Gallardo JF (ed) The soils of Spain. World soils book series. Springer International Publishing Switzerland. https://doi.org/10.1007/978-3-319-20541-0_6

Moghanm FS (2015a) Assessment of sustainable agricultural land management by using GIS techniques in North Delta, Egypt. Egypt J Soil Sci 55(4):409–424. https://doi.org/10.21608/ejss.2015.1559

Moghanm FS (2015b) Land capability assessment by using geographic information system in North Delta, Egypt. Egypt J Soil Sci 55 (2):115–113. https://doi.org/10.21608/ejss.2015.310

Mohamed ES, Belal A, Shalaby A (2015) Impacts of soil sealing on potential agriculture in egypt using remote sensing and GIS techniques. Eurasian Soil Sci 48(10):1159–1169

Mohamed ES, Saleh AM, Belal AA (2014) Sustainability indicators for agricultural land use based on GIS spatial modeling in north of Sinai-Egypt. Egypt J Remote Sens Space Sci 17(1):1–15

Mohammed AM, Krishnamurthy RV, Kehew AE, Crossey LJ, Karlstrom KK (2016) Factors affecting the stable isotopes ratios in groundwater impacted by intense agricultural practices: a case study from the Nile Valley of Egypt. Sci Total Environ 573:707–715

Mousa AA (2017) Effect of using some soil conditioners on salt affected soil properties and its productivity at El-Tina Plain area—North Sinai—Egypt. Egypt J Soil Sci (in Press). https://doi.org/10.21608/ejss.2017.1526

Negm AM (2017) The Nile Delta. The handbook of environmental chemistry series, vol 55. Springer International Publishing AG. https://doi.org/10.1007/978-3-319-56124-0

Omran EE (2016a) A stochastic simulation model to early predict susceptible areas to water table level fluctuations in North Sinai, Egypt. Egypt J Remote Sens Space Sci 19(2):235–257

Omran EE (2016b) Environmental modelling of heavy metals using pollution indices and multivariate techniques in the soils of Bahr El Baqar, Egypt. Model Earth Syst Environ 2:119. https://doi.org/10.1007/s40808-016-0178-7

Omran EE (2017) Land and groundwater resources in the Egypt's Nile Valley, Delta, and its Fringes. In Negm AM (ed) Groundwater in the Nile Delta, The handbook of environmental chemistry. Springer International Publishing AG. https://doi.org/10.1007/698_2017_64

Osman KT (2013) Factors and processes of soil formation. In Osman KT (ed) Soils: principles, properties and management. Springer Science + Business Media Dordrecht, pp 17–30. https://doi.org/10.1007/978-94-007-5663-2_3

Pavlović P, Kostić N, Karadžić B, Mitrović M (2017) The soils of Serbia. World soils book series. Springer Science + Business Media Dordrecht. https://doi.org/10.1007/978-94-017-8660-7

Rashed HSA (2016) Change detection in land degradation and environmental hazards sensitivity in some soils of Siwa Oasis. Egypt J Soil Sci 56(3):433–451

Selin H (2016) Encyclopaedia of the history of science, technology, and medicine in non-western cultures. 3rd edn. Springer Science + Business Media Dordrecht. https://doi.org/10.1007/978-94-007-7747-7

Shendi MM, Abdelfattah MA, Harbi A (2013) Spatial monitoring of soil salinity and prospective conservation study for Sinnuris District Soils, Fayoum, Egypt. In Shahid SA et al (eds) Developments in soil salinity assessment and reclamation: innovative thinking and use of marginal soil and water resources in irrigated agriculture. Springer Science + Business Media Dordrecht. https://doi.org/10.1007/978-94-007-5684-7_13

Shishkov T, Kolev N (2014) The soils of Bulgaria. World soils book series. Springer Science + Business Media Dordrecht. https://doi.org/10.1007/978-94-007-7784-2

Shokr MS, El Baroudy AA, Fullen MA, El-beshbeshy TR, Ramadan AR, Abd El Halim A, Guerra AJT, Jorge MCO (2016) Spatial distribution of heavy metals in the middle Nile delta of Egypt. Int Soil Water Conserv Res 4(4):293–303

Soliman I (2015) Diagnosis and challenges of sustainable agricultural development in Egypt. In Petit M et al (eds) Sustainable agricultural development, cooperative management. Springer International Publishing Switzerland, pp 19–64. https://doi.org/10.1007/978-3-319-17813-4_2

Sultan YM, El–Shafei MK, Arnous MO (2017) Tectonic evolution of kid metamorphic complex and the recognition of Najd fault system in South East Sinai, Egypt. Int J Earth Sci (Geol Rundsch) 106:2817–2836. https://doi.org/10.1007/s00531-017-1461-1

Terribile F, Basile A, Bonfante A, Carbone A, Colombo C, Langella G, Iamarino M, Manna P, Minieri L, Vingiani S (2013) Future soil issues. In Costantini EAC Dazzi C (eds) The soils of Italy. World soils book series. Springer Science + Business Media Dordrecht. https://doi.org/10.1007/978-94-007-5642-7_11

Vrščaj B, Repe B, Simončič P (2017) The soils of Slovenia. World soils book series. Springer Science + Business Media Dordrecht. https://doi.org/10.1007/978-94-017-8585-3

West LT, Singer MJ, Hartemink AE (2017) The soils of the USA. World soils book series. Springer International Publishing AG. https://doi.org/10.1007/978-3-319-41870-4

Winiwarter V (2014) Environmental history of soils. In Agnoletti M, Neri Serneri S (eds) The basic environmental history, environmental history, vol 4, Springer International Publishing Switzerland. https://doi.org/10.1007/978-3-319-09180-8_3

Winiwarter V, Blum WEH (2009) Religious aspects of soil-human relationships. In Rattan L et al (ed) Encyclopedia of soil science. http://www.tandfonline.com/doi/abs/10.1081/E-ESS-120044721. UtgObPvZfAk. Accessed 16 Jan 2014

Yassoglou N, Tsadilas C, Kosmas C (2017a) Soil research. In: Yassoglou N et al (eds) The soils of Greece. World soils book series. Springer International Publishing AG, pp 97–100. https://doi.org/10.1007/978-3-319-53334-6_11

Yassoglou N, Tsadilas C, Kosmas C (2017b) The soils of Greece. World soils book series. Springer International Publishing AG. https://doi.org/10.1007/978-3-319-53334-6

Yossif TMH, Ebied GM (2016) Monitor of changes of land uses and soil characteristics at North West area of sinai using satellite imagery. Egypt Egypt J Soil Sci 56(3):471–787. https://doi.org/10.21608/ejss.2016.3245

Zahran MA, Willis AJ (2009) The vegetation of Egypt. Plant and vegetation series, 2nd edn, vol 2. Springer Netherlands. https://doi.org/10.1007/978-1-4020-8756-1

Egyptian Natural Resources

3

Noura Bakr and Mohamed H. Bahnassy

Abstract

Abstract: The Arab Republic of Egypt is the main northeastern entrance of Africa continent. The total area of Egypt territory is around one million square kilometer and occupied by approximately 95 million inhabitants who live on about 4% of this area. This chapter presents an overview of the land, water, and human resources in Egypt. Egypt land is geographically divided into five sections: Western Desert, Eastern Desert, Sinai Peninsula, Nile Valley, and Nile Delta. These sections involve 33 soil units in which sandy areas and loamy sand/sandy loam soils cover over 57% of their surface. Despite that, agriculture is the fundamental economic activity in Egypt. The Egyptian agricultural lands could be classified as Oldlands and Newlands. The Oldlands are assigned for the fertile and intensively cultivated lands in Nile Valley and Delta that have been farmed since ancient time till now, whereas the newly reclaimed areas that have been cultivated relatively recently or in the process of reclamation now are known as Newlands. The natural water resources in Egypt are inadequate since the primary water source is the annually fixed share of Nile River of 55.5 BCM that is utilized for almost all human activities despite the high growth rate of population. The Egyptian government adopts strategies to utilize the non-conventional water resources (mainly recycled agricultural drainage, shallow groundwater, and treated wastewater) in irrigation as the agricultural sector consumes 80–85% of available freshwater in Egypt. At the administrative level, Egypt involves 27 governorates in which New Valley governorate represented 42% of total Egypt's area and occupied with only 0.25% of the Egypt total population. Conversely, the Great Cairo (involves Cairo, Giza, and Kalyoubia governorates) occupied by around 25% of the total population. Agricultural sector supports the livelihood to approximately two-thirds of the Egyptians and considers the primary source of income to about 60% of Egyptians.

Keywords

Land resources • Water resources • Human resources Soil map

3.1 Introduction

Egypt, a common name of Arab Republic of Egypt, is a part of Sahara (largest hot desert in the world) in the Northern Africa region (Fig. 3.1). The Nile River and its deposits make Egypt rich with its black soils with the fertile deposits that cover the area around the Nile Valley and Delta and make obvious green area over the extended desert (Fig. 3.1). This green area supported the settlement of ancient Egyptians who built one of the oldest and greatest civilizations in the world history (Zahran and Willis 2009; Negm et al. 2017).

The ancient Egyptian civilization is obviously depended on agriculture as the ancient Egyptians settled around the Nile over 3000 B.C. (Zahran and Willis 2009; Negm et al. 2017; Embabi 2018). Since then, the agricultural sector in Egypt has been continuously evolved and developed, and the investments in agriculture have been exponentially increased with time. With the vast population growing (Egypt is the largest Arab world country in the population), more pressure is added to the Egypt's natural resources especially land and water resources.

Egypt covers an area of little over one million square kilometers (1,004,458 km^2) with around 95 million inhabitants (CAPMAS 2017a) who are settled on about 4% of the Egypt territory (in Nile Valley and its Delta) and the rest of Egypt's land (around 96%) is preserved as desert land (Bakr

N. Bakr (✉)
Soils and Water Use Department, National Research Centre (NRC), Cairo, Egypt
e-mail: nourabakr@yahoo.com

M. H. Bahnassy
Soil and Water Sciences Department, Faculty of Agriculture, Alexandria University, Alexandria, Egypt

H. El-Ramady et al. (eds.), *The Soils of Egypt*, World Soils Book Series,
https://doi.org/10.1007/978-3-319-95516-2_3

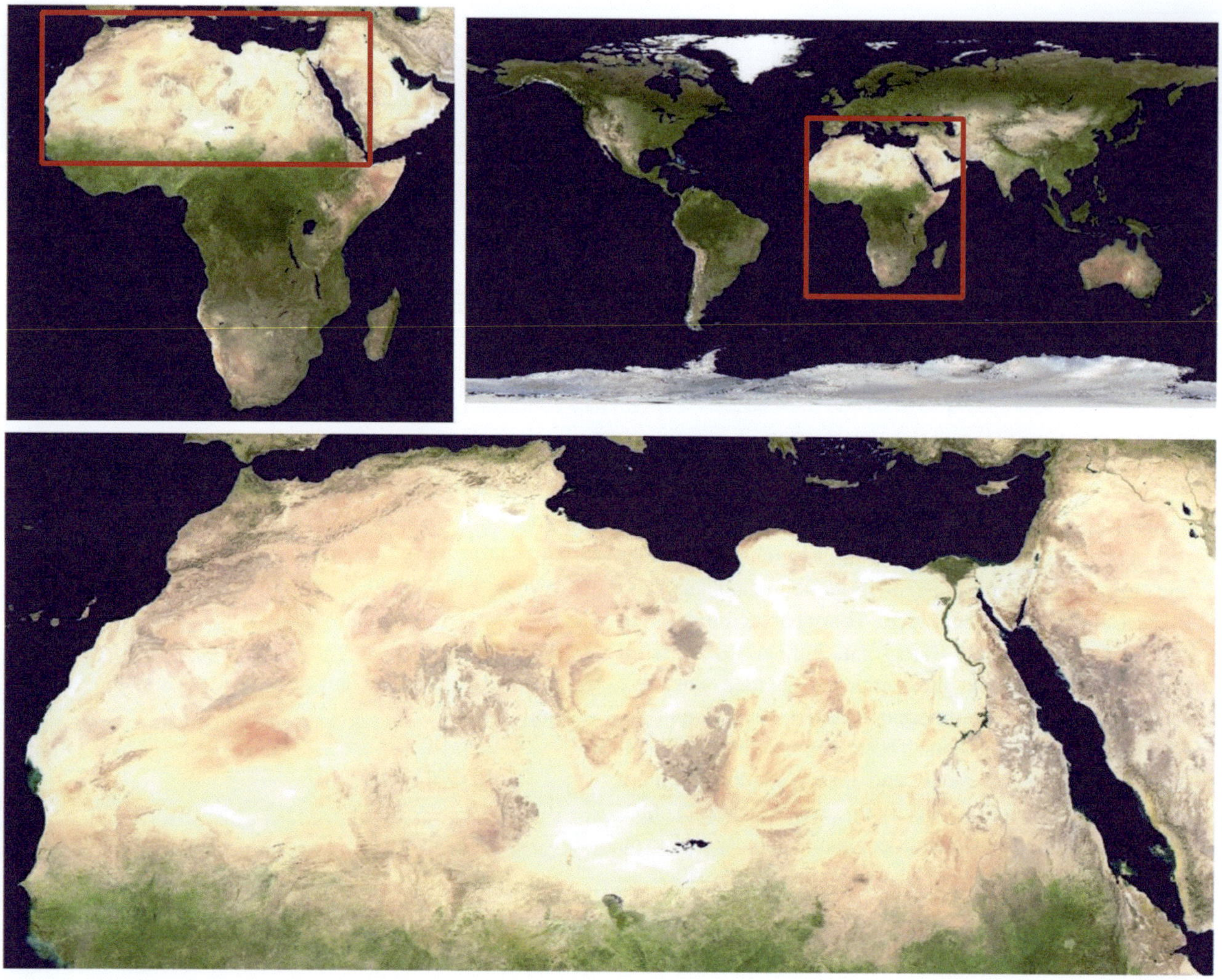

Fig. 3.1 The extend of Sahara Desert in the northern Africa continent. Updated from NASA visible earth, https://visibleearth.nasa.gov/view.php?id=57752

et al. 2010). Based on Hamdi and Abdelhafez (2001), the width of the Nile Valley is about 18 km, whereas the total area coverage of Nile Delta is reported between 19,000 and 22,000 km^2 which represents around 2% of Egypt area, support over 60% of the total fertile land in Egypt, and populated with around half of the Egypt's total population.

Historically, the branches of the Nile in the Delta region were seven, namely, from west to east, Canopic (currently, Rosetta Branch), Bolbitinic, Sebennitic, Fatmetic (currently, Damietta Branch), Mendisy, Tanitic, and Pelusiac Branchs (Zahran and Willis 2009). Now, only two main branches can be recognized (Fig. 3.2): Rosetta branch (235 km length) in the west and Damietta branch (240 km length) to the east (Stanley and Warne 1993; Hamdi and Abdelhafez 2001; Negm et al. 2017).

This chapter focuses on three natural resources in Egypt: land, water, and human resources. General idea about the main geographical locations and geological features in Egypt territory will be discussed. Besides, this chapter provides a general knowledge of the soil map of Egypt with its different units. The geographic information system tools are intensively used in this chapter to provide the results in attractive and easier ways to be understandable by the decision-makers and the scientific community at national and international levels.

3.2 Egyptian Natural Resources

3.2.1 Geographical Location and Land Resources in Egypt

Egypt is geographically located in the northeastern corner of Africa between latitudes of 22° and 32° N and longitudes 25° and 37° E (Fig. 3.3). Based on the location map (Fig. 3.3), two inland countries are bounded west and south borders of Egypt, Libya, and Sudan, respectively. Besides, two water bodies are touched east and north boundaries: Red

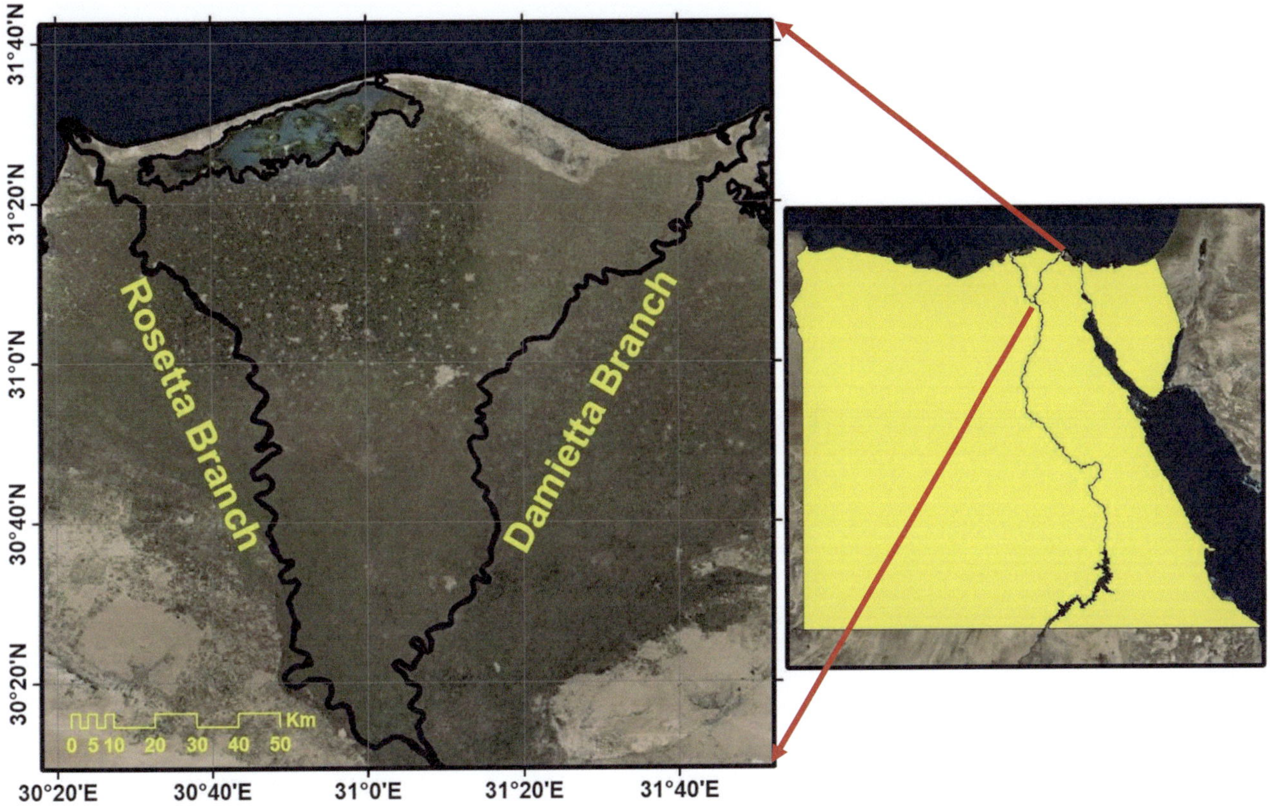

Fig. 3.2 The two main branches of the Nile Delta, Egypt. Created by authors

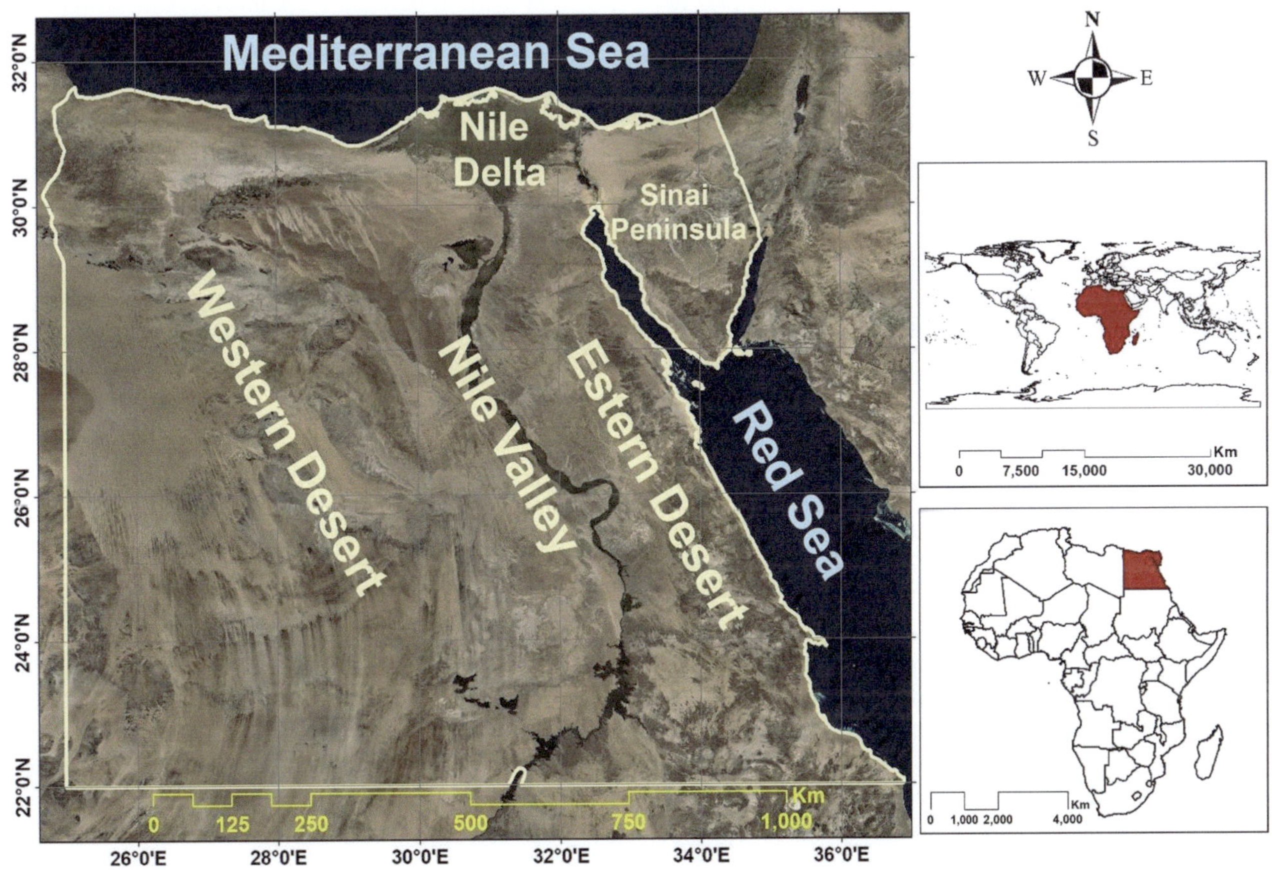

Fig. 3.3 Arab Republic of Egypt general location with the main geographical zones. Created by authors

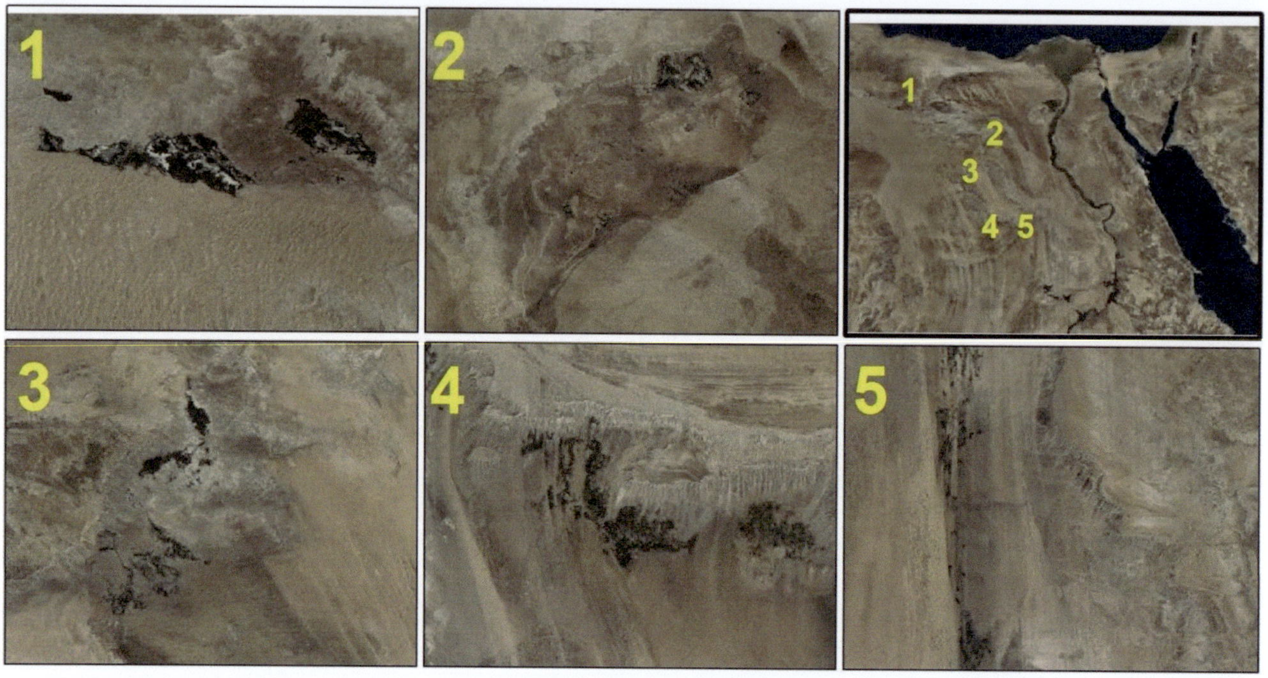

Fig. 3.4 The location of the five oases in the western desert, Egypt, namely, from north to south; (1) Siwa, (2) Bahariya, (3) Farafra, (4) Dakhla, and (5) Kharga oases. Created by authors

and Mediterranean Seas, respectively. The total area of Egypt is around one million Km^2. Geographically, the Nile valley and Delta divided the desert land in Egypt into two main parts: western desert (represents two third of Egypt territory) and eastern desert. Additionally, Sinai Peninsula represents 6% of the Egypt area and belongs to Asia continent which makes Egypt a trans-continental country (Omran 2017).

3.2.1.1 Western Desert

The Western Desert (almost two-thirds of total Egypt area) is characterized by different features. Besides the grazing, as one of the main economic activities in the desert areas, a scattered rained barley crop (as a main annual cultivated crop) has been grown in the northern Mediterranean coastal region of Egypt which is characterized by a seasonal rainfall during winter months. Figs, olives, and almonds are successfully cultivated in that region. Besides, xerophytes are the most abundant natural vegetation species in that coastal area (Zahran and Willis 2009). The Western desert also involved well-known oases, namely, Siwa, Bahariya, Farafra, Dakhla, and Kharga Oases (Fig. 3.4). Besides the oases, three common depressions are highly recognized in the western desert, namely, Fayoum, Wadi El-Natrun, and Qattara depressions. From them, only Fayoum depression is intensively cultivated just like Nile Valley and Delta (Fig. 3.5), whereas the other two depressions are salty areas with less agricultural activities. Rather than those areas, the rest of the western desert is almost barren soils with no observed agricultural activities. Some efforts for reclamation are taken place in different areas in western desert close to the Nile Valley and Delta (Bakr et al. 2010).

3.2.1.2 Eastern Desert

The Egypt's Eastern desert occupies around one-fifth of Egypt area. It characterized by high relief due to the presence of Red Sea Mountains that formed a narrower coastal area compared with the wider coastal land for the Mediterranean region in the north. Due to the high relief and multiple mountains in this region, there are intensive numbers of dry Wadis that are spread over the eastern desert (Fig. 3.6). The coral reefs are the most known water species that occur along the Red Sea coastal area width of 50–100 m wide (Zahran and Willis 2009).

3.2.1.3 Sinai Peninsula

Sinai Peninsula occupies around 6% of Egypt territory and formed as a triangular plateau (Fig. 3.7). Gulf Suez and Aqaba bordered the Peninsula from the west and the east sides, respectively, whereas the Mediterranean and Red Seas bordered its north and south sides, respectively. Wadi El-Arish basin is one of the most well-known geographical features in the northern Sinai (Embabi 2018). Zahran and Willis (2009) stated that due to the difference in the environment within north, central, and south Sinai, hundreds of flora can be recognized in this region.

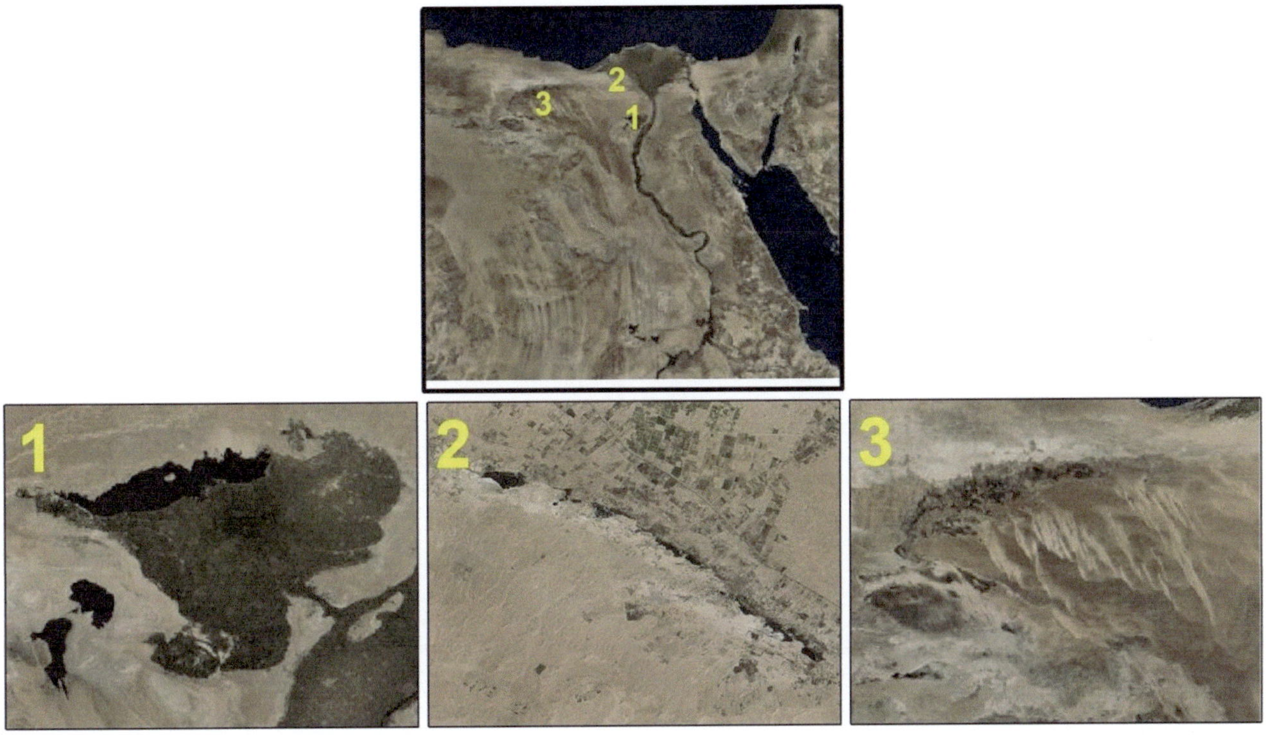

Fig. 3.5 Three main depressions in the western desert, Egypt: (1) Fayoum, (2) Wadi El-Natrun, and (3) Qattara depressions. Created by authors

3.2.1.4 Nile Valley and Delta

The band of the Nile in both of its sides represents that the Nile Valley (also its common name is upper Egypt) has an area around 11,000–12,000 km^2 over a distance of >1000 km (Fig. 3.8) whereas Lower Egypt is a common name of the Nile Delta region (Zahran and Willis 2009). The soils of the Nile Valley and Delta consist of alluvial deposits (Nile mud, besides gravel and sand). Before the construction of Aswan High Dam, the Nile water was carrying the silt particles which gave an annual increase in sediment by a rate of 1 m of mud every 1000 years (Abdel Meguid 2017). Those two areas (Nile Valley and Delta) represent the oldland that are the main cultivated land in Egypt and support over 90% of the agriculture activities in the whole country except the sand areas, salt marshes, sabkha, and swamps features in the northern Nile Delta Mediterranean coast (Fig. 3.8).

The new reclaimed areas around the two Delta Fringes (Fig. 3.8) had taken places over the last six decades till now. Modern irrigation technologies and systems (sprinkler and drip irrigation) have been used to increase the water use efficiency and reduce the water unit requirements per cultivated unit area (ICARDA 2013). The area outside the two fringes of Nile Delta (eastern and western sides) attracts the attention of stakeholders, scientists, and decision-makers as a promising reclaimed area for agriculture development, as those areas are adjacent to the center area of Egypt (Nile Delta). Cropping system in the

Fig. 3.6 The main drainage network in the eastern desert. Created by authors

Fig. 3.7 Sinai Peninsula, Egypt. Created by authors

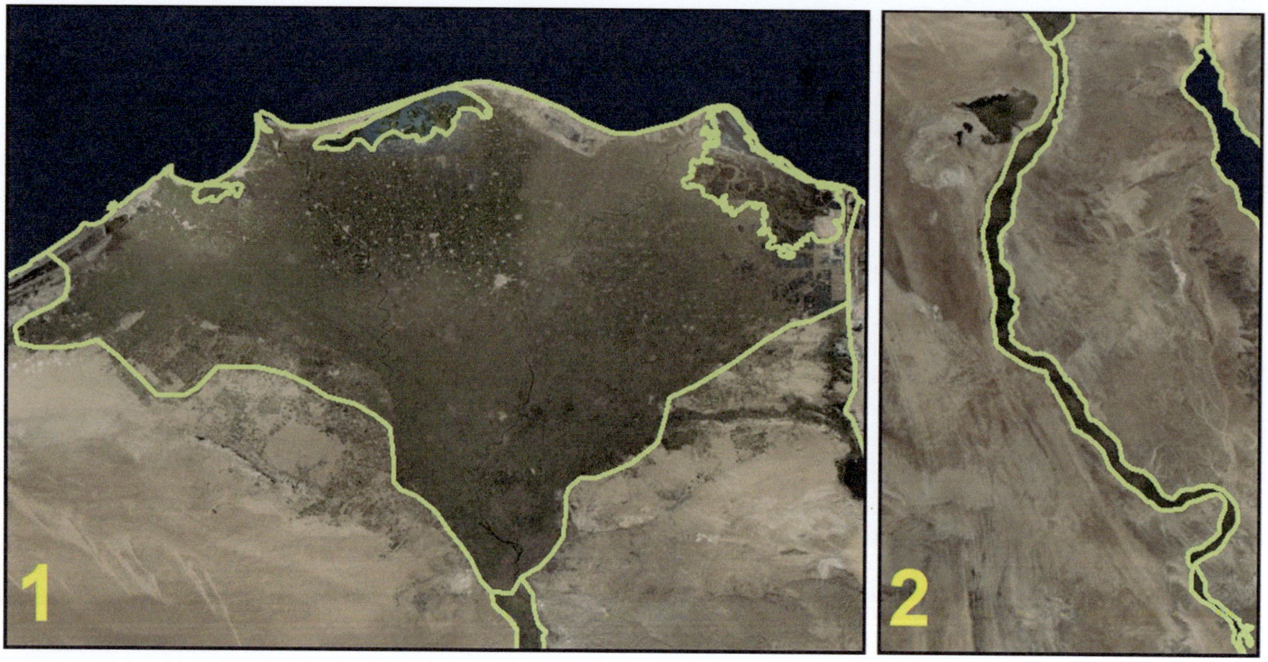

Fig. 3.8 Nile Delta and its extension (1) and the Nile Valley (2), Egypt. Created by authors

oldland has three consecutive cropping seasons: winter, summer, and *nili* seasons, which all depend on the traditional surface irrigation systems from the Nile. Additionally, some areas are cultivated with annual crops, such as sugarcane (mainly in Upper Egypt) and fruit trees. This cropping rotation is exponentially the cropping area to around 13.7 million feddans with crop intensity of 172% (Abdel Meguid 2017).

3.2.2 Egypt Landform

The general climate of Egypt is hot dry summers and warm rainless moderate winters. The rain is limited to the narrow strip of north coastal area with less than 200 mm/year during winter season (Zahran and Willis 2009). The detailed climate information and data are discussed in Chap. 5. Based on the digital elevation model (DEM), the relief of Egypt is ranged

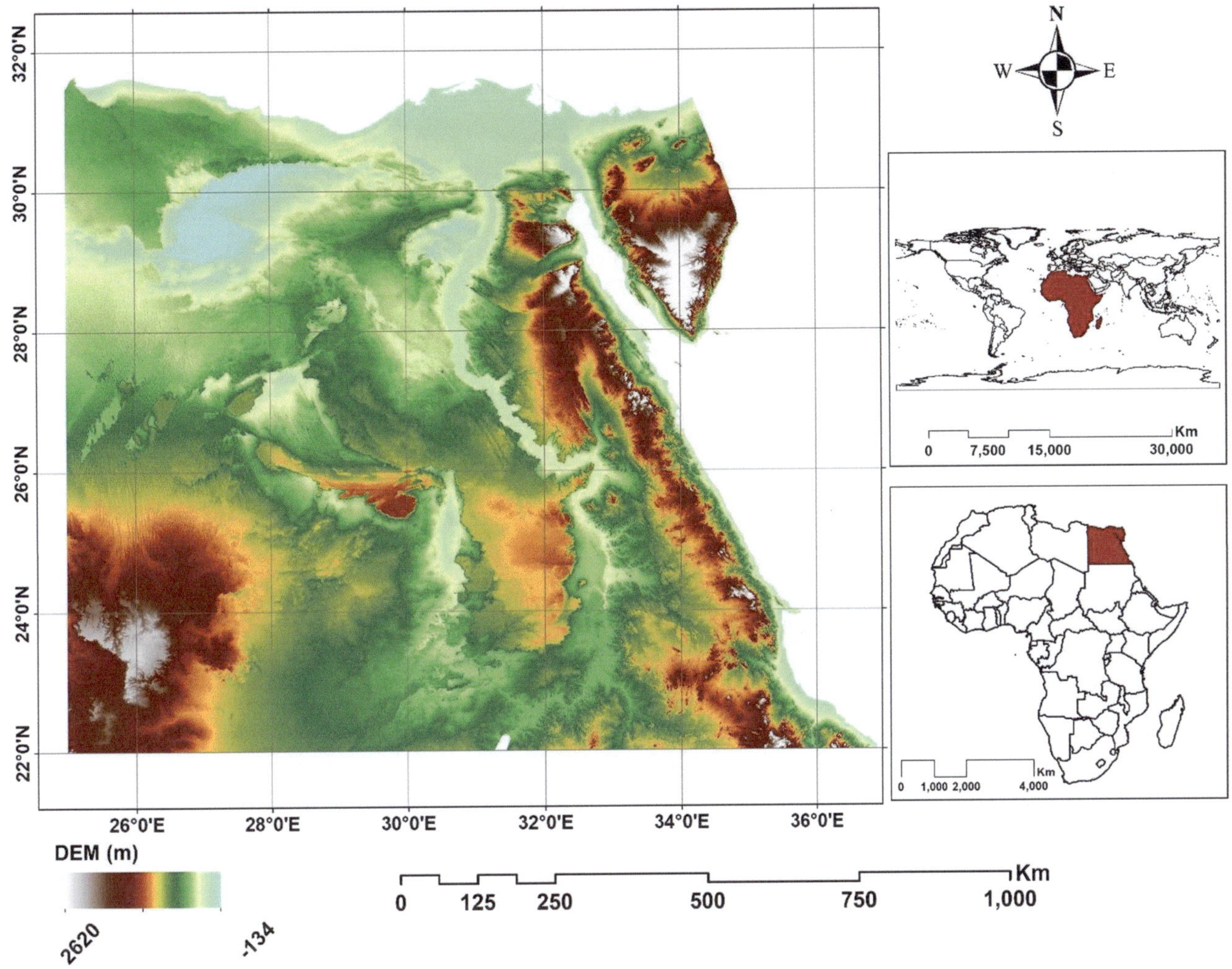

Fig. 3.9 The digital elevation model (DEM) of Egypt. Created by authors

from −133 m below the sea level (lowest point in Qattara Depression) to a maximum of 2620 m above sea level at Mount Catherine in southern of Sinai Peninsula (Fig. 3.9). Other high altitudes can be observed in the Red Sea Mountains along the Red Sea coast in the eastern desert and the Gilf Kebir plateau in the southeastern corner of Egypt (Worldatlas 2017). The final DEM was produced based on the NASA Shuttle Radar Topographic Mission (SRTM) data, 90 m resolution that was downloaded from the US Geological Survey website, in georeferenced tagged image file format (GeoTIFF), using Earth Explorer (U.S. Geological Survey 2017). The Global Mapper 17 software was used to mosaic the different scenes together and produced the final DEM map (Fig. 3.9). Embabi (2018) discussed the main landforms of Egypt territory as

- Nile Delta includes Deltic Plain, sabkhas and stabilized sand dunes, and Nile Valley.
- Eastern Desert with four major landform: (a) El-Galalah Plateaus in the northern eastern desert which involve El-Galalah El-Bahariyah, north, El-Galalah El-Qibliyah, south, and Wadi Araba, in between; (b) dune fields in southeastern corner of eastern desert (El-Hebal dunes, El-Allaqi dune, a Wadi Kraf dune); (c) drainage basin such as Wadi Qena; and (d) Red Sea Mountains that include dyke swarms, intermountain basins, and ring complexes circular mount.
- Sinai Peninsula; Sinai mountains (south Sinai) involve same features like Red Sea Mountains. North Sinai involves north Sinai Sand Sea, different types of dunes, and Wadi El-Arish.
- Western Desert involves carbonate ridges, coastal lagoons, and sabkhas in northwest coast, sandstone plateaus of Gilf Kebir and Abu Ras in southwest, Qattara, Kharga–Dakhla, and Fayoum depressions, Karstified Carbonate Platforms, Great Sand Sea, and Ghard Abu Moharik.

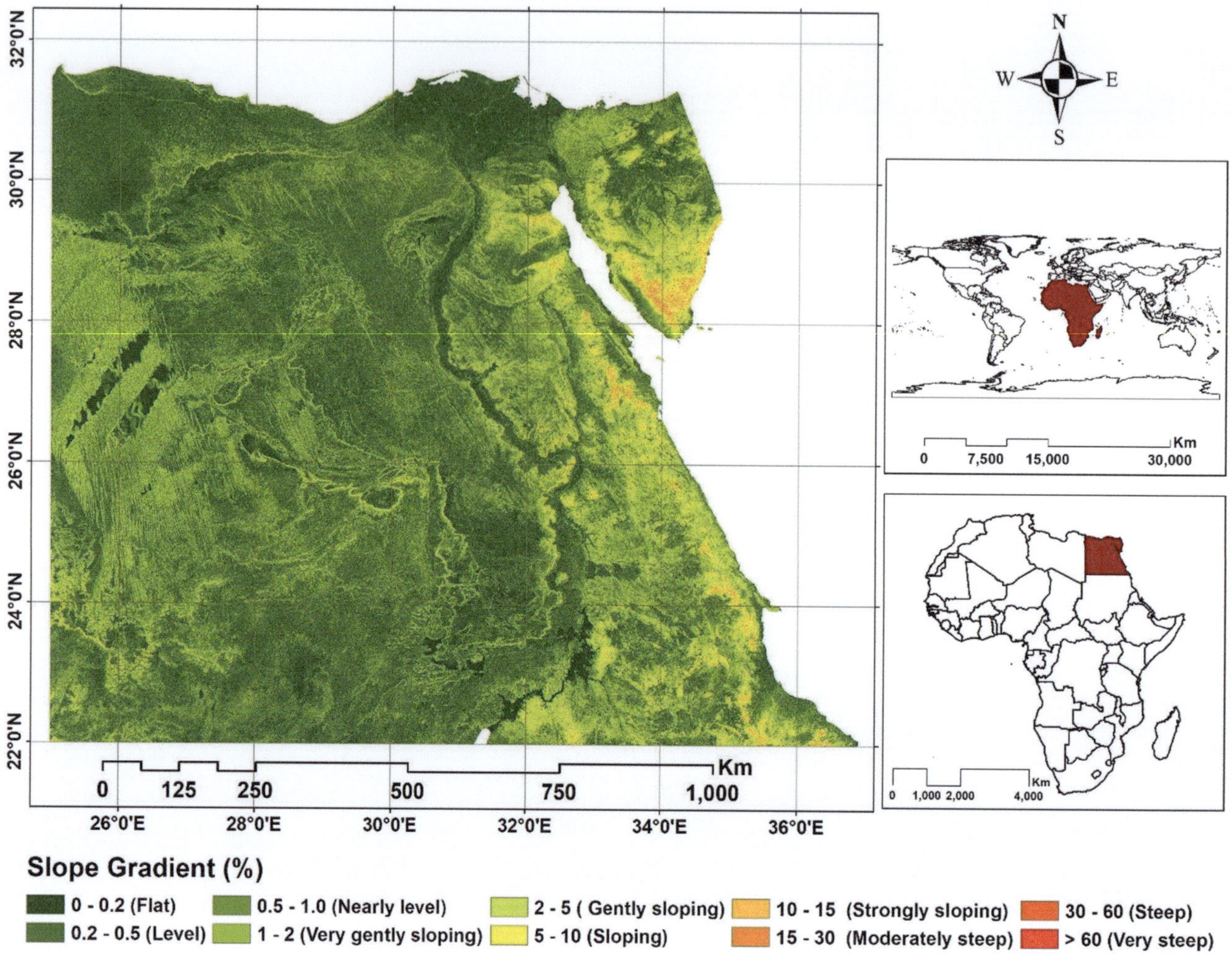

Fig. 3.10 The slope gradient of Egypt extracted from DEM. Created by authors

From the obtained DEM, the slope percentage was calculated for the entire country. The resulted slope map indicated that the flat slope (<2%) can be observed in the Nile valley and Delta regions as well as the northwestern corner of western desert (Fig. 3.10). The highest slope (very steep slope > 60%) is shown in the southern Sinai Peninsula where Mount Catherine locates. Other high slopes can also be noticed along the coastal zone of Red Sea Mountains. Generally, the eastern desert and Sinai Peninsula have higher slope compared with the western desert.

3.2.3 Water Resources in Egypt

The natural water resources in Egypt are very limited as almost all human activities are depending on a fixed share of Nile River of 55.5 BCM annually. Due to the vast increases in the Egyptian population (around 94 million in 2017, CAPMAS 2017a), the annual share per capita decreases to 700 m^3 in 2011, which is already below the water poverty line of 1000 m^3/capita/year and expected to be 500 m^3 by 2030 (Egyptian Ministry of Agriculture and Land Reclamation, MALR 2009). The other natural water resources such as rainfall and groundwater are contributed by significantly smaller amounts which are not sufficient to meet the increasing demand of the population needs. CAPMAS (2016) stated that in 2015, the total water resources of Egypt were around 76 BCM involving the 55.5 BCM of the Nile share and the non-conventional water resources which include mainly recycled agricultural drainage (around 12 BCM), shallow groundwater in the Nile Valley and Delta (about 7 BCM), and wastewater recycling (1.3 BCM) (Elbana et al. 2017). These non-conventional water sources can be used in irrigation as the agricultural

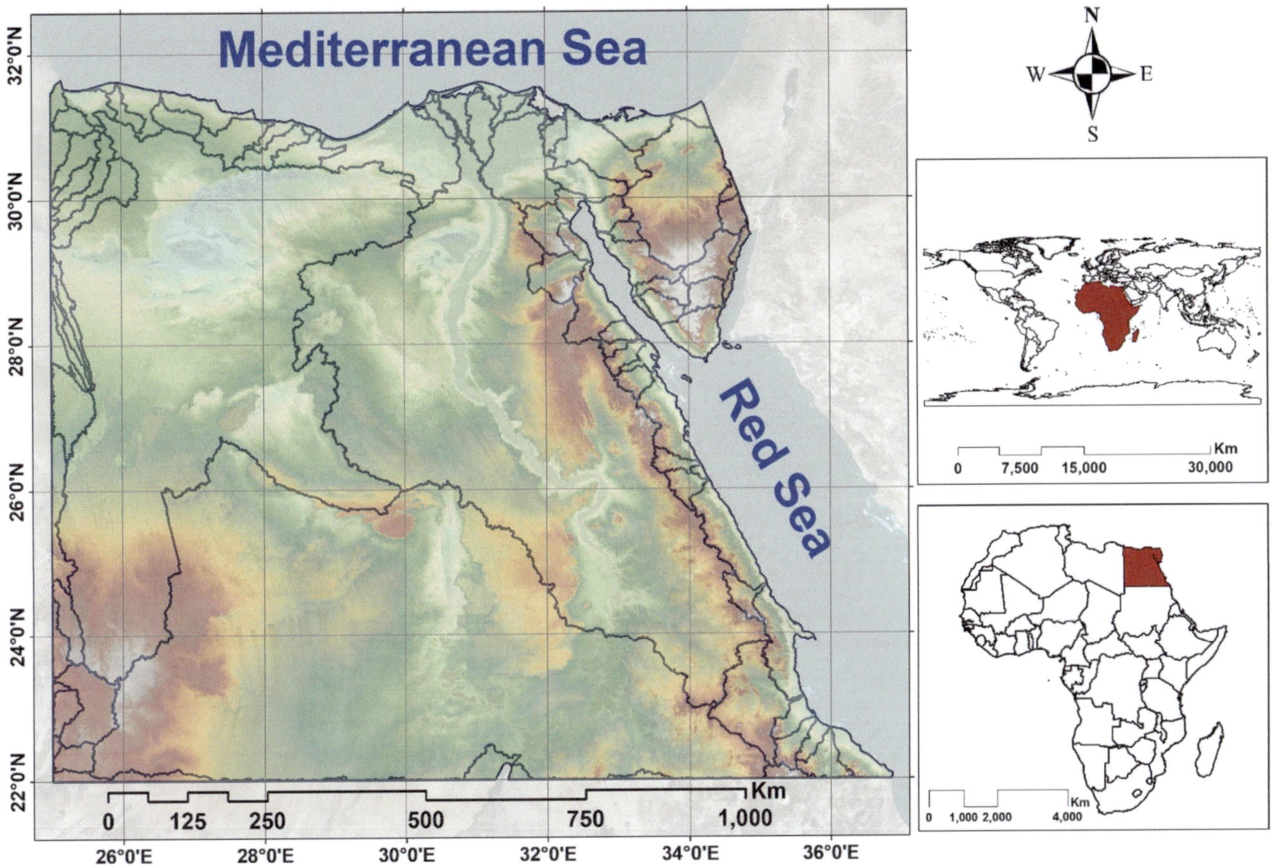

Fig. 3.11 The main basins in Egypt based on hydrological analysis of DEM in Egypt. Created by authors

sector consumes from 80 to 85% of available freshwater in Egypt (Ali and Mahmoud 2004; MWRI 2014). Based on the digital elevation model, the hydrology tools in ArcMap were used to extract the main basins for the entire country. Different basins can be recognized in Egypt as shown in Fig. 3.11.

Additionally, the hydrology tools were also used to extract the natural streams (drainage network) in each basin. Due to the excessive number of basins in Egypt, some basins were selected to study the natural streams in them separately. Those main selected basins were three largest basins (east, west, and south parts of the country), another three basins in Nile Delta, besides one basin in Sinai Peninsula which is mainly Wadi El-Arish basin (Fig. 3.12). Embabi (2018) stated that there are 40 basins in Sinai (Fig. 3.11) among them Wadi El-Arish basin (Fig. 3.12) represents 36% (21,700 km^2) of Sinai Peninsula area.

The Nile River and its branches, as the main source of natural water resource in Egypt, besides the main natural drainage network as well as the main surface water areas for the entire country are presented in Fig. 3.13. As the relief in Sinai Peninsula and eastern desert increases, the drainage network is intensively existed in those areas. Based on Zahran and Willis (2009), wadis are one of the main and important ecosystems of the Egyptian deserts. Wadis represent drainage systems that are used to collect water from the catchment areas and additionally are spread over the northwestern coast (NWC) of Egypt. Embabi (2018) illustrated that the eastern desert involved 21 and 28 basins from Nile (e.g., Wadi Qena) and Red Sea (e.g., Wadi Araba) drainage systems, respectively. This is why eastern desert had extensive drainage networks.

3.2.4 Human Resources and Socioeconomic Sittings in Egypt

At a national level, Egypt involves 27 administrative divisions that are known as governorates (CAPMAS 2017a) (Fig. 3.14; Table 3.1). Five governorates have shared boundaries with

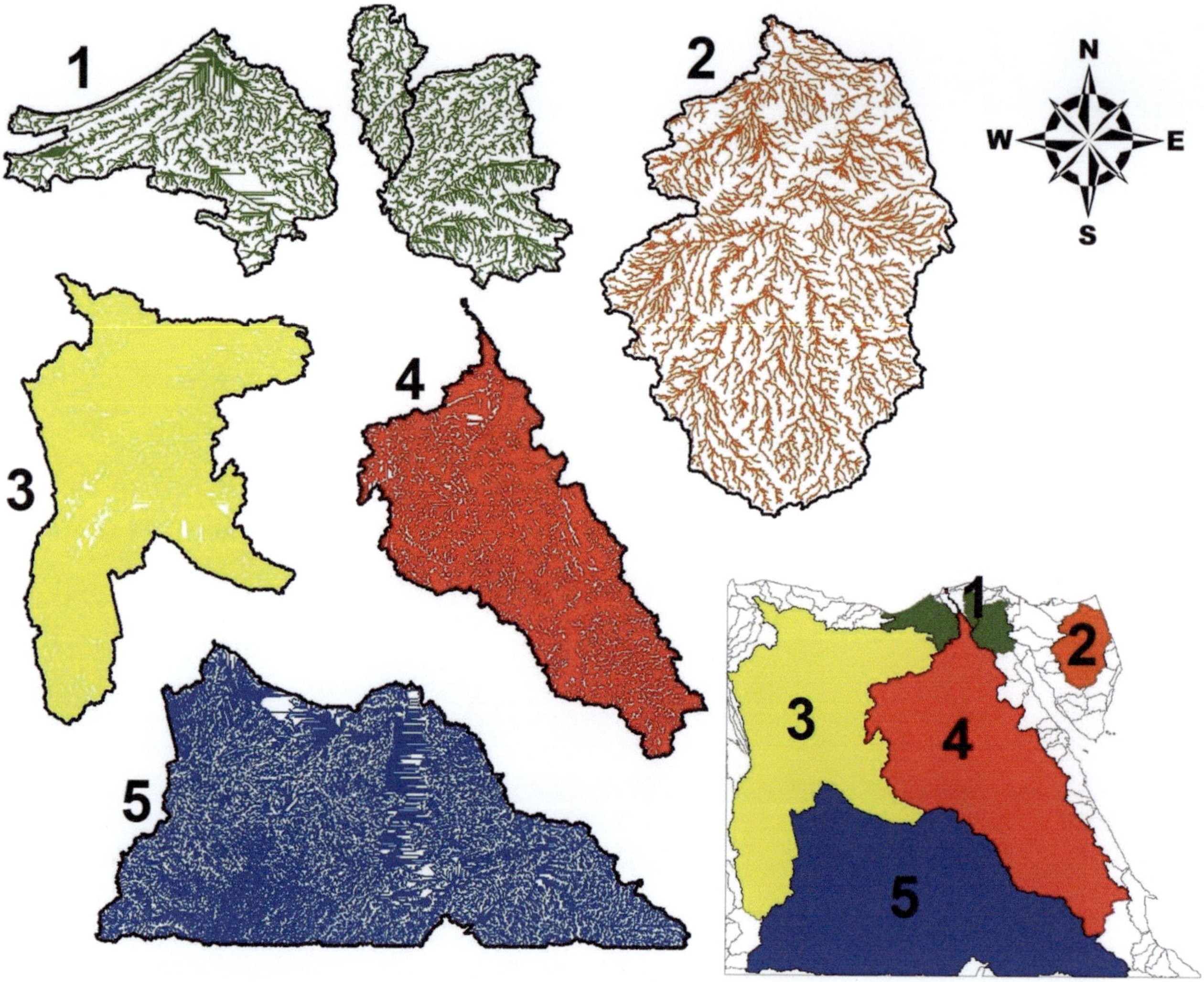

Fig. 3.12 The streams of the chosen main basins in Egypt: (1) east, west, and central Nile Debasins, (2) main basin in Sinai Peninsula, (3) western main basin, (4) eastern main basin, and (5) southern main basin in Egypt territory. Created by authors

foreigner countries (Matrouh, New Valley, Aswan, Red Sea, and North Sinai). The largest governorate in area is New Valley (42% of total Egypt's area) and only 0.25% of the Egypt total population live in, whereas around 25% of population are settled in a Great Cairo (Cairo, Giza, and Kalyoubia governorates) (Table 3.1). The lowest governorate in population is South Sinai with only 102,000 persons living in it. Based on CAPMAS (2017), the Egypt total population is 94,798,827 and the total inland area is 1,004,458 km^2 (Table 3.1).

Those governorates are administratively divided into 364 districts. The Capital "Cairo" involves the highest number of districts of 43 (Table 3.1). Based on the general infrastructure of Egypt (Fig. 3.15), within the 27 governorates, there are around 24 main cities, 122 towns, and 117 villages. There are huge road networks that connect all these cities, towns, and villages together. The estimated total length of these networks is around 27,300 km^2, ranged from highways to smaller width roads between villages (Fig. 3.14) (FAO-Stat 2017) (Fig. 3.15).

Agricultural sector in Egypt is the main source of income to around 60% of Egyptians (Fig. 3.16), as it supports the livelihood to around two-thirds of the Egyptians (ICARDA 2013). Within the last five decades (1960–2016), the population in the rural areas sharply decreased from 62% (during 1960s) to around 56% (during 1975). Recently, the rural population represents around 58% of Egyptian total population (Fig. 3.16). The rural population growth was less than the national population growth by around 1% between 1960 and 1980, whereas after 1980 till now the rural population growth is higher than the national population growth especially after 2010 based on the presented data shown in Fig. 3.16.

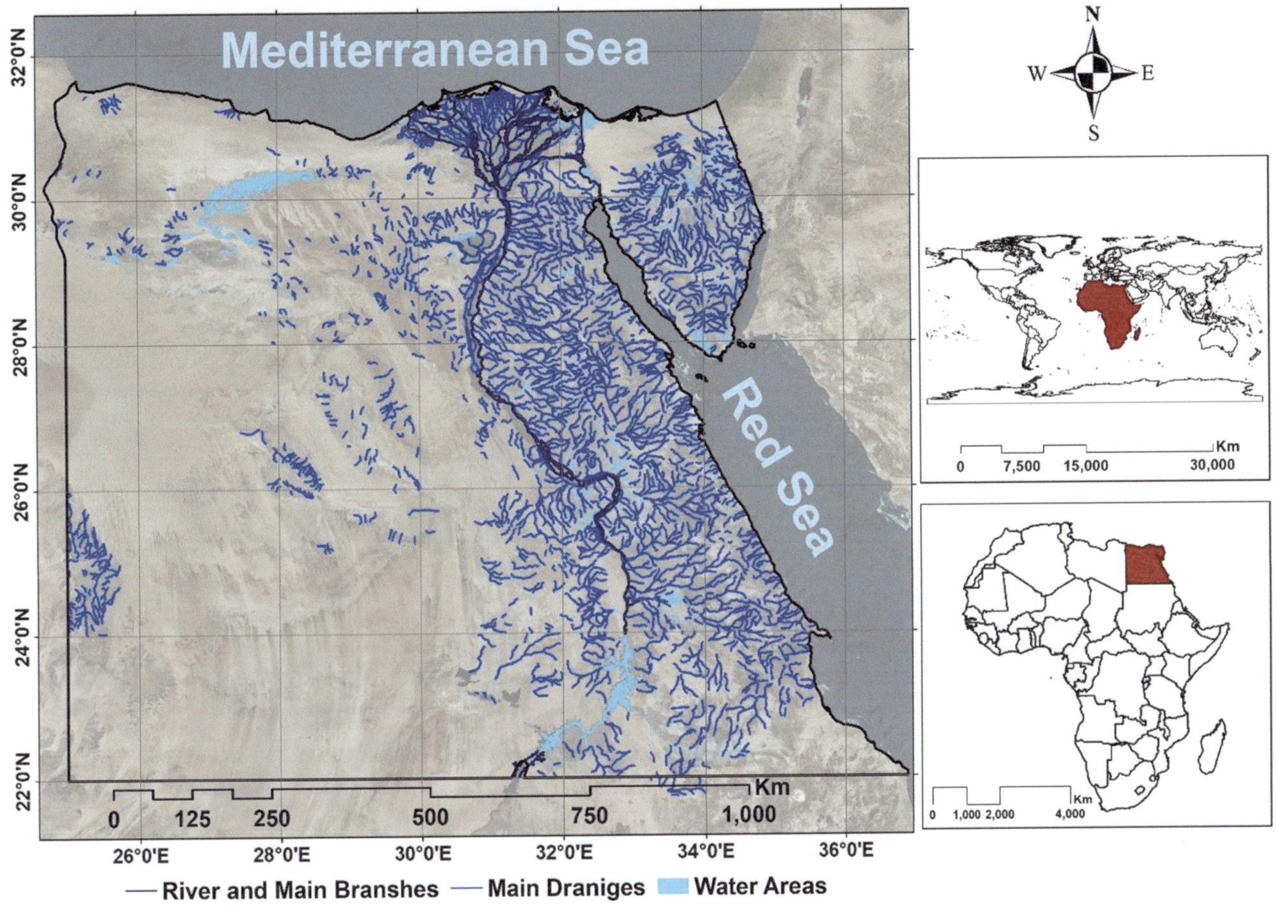

Fig. 3.13 The main drainage network in Egypt. Created by authors

The employment in agricultural sector is decreased dramatically from 55 to 40% between 1970 and 1980, respectively. With continuous decrease in employment in agricultural sector, the most recent value is less than 25% (Fig. 3.17). With these deductions in agricultural employment, the agricultural sector participation in the total gross domestic product (GDP) reduced, respectively, from around 30% to around 10% over the last 55 years (1960–2016) as shown in Fig. 3.17.

3.3 The Geological Features and Soil Map of Egypt

Different land uses of soil are closely related to the geological characteristics and soil properties of the earth surface. Several attempts have been achieved to map the geological features and Egypt soils based on their characteristics. The oldest geological features in Egypt are dated to Precambrian Era where the Archaen rocks cover mainly in Red Sea Mountains, South Sinai Mountains, and Uweinat Mountain (on Gulf Kabir Plateau, southwestern corner) and represent around 10% of Egypt area (Zahran and Willis 2009; Hammad 1975). Then, within the Mesozoic era, the Cretaceous period formation (e.g., Nubian Sandstone) occurred mainly in south Egypt and north Sinai covered around 40% of Egypt surface (Zahran and Willis 2009). The Cenozoic era involving Paleogene (Paleocene, Eocene, and Oligocene), Neogene (Miocene and Pliocene), and the Quaternary (Pleistocene and Holocene) periods have been distributed over the rest of Egypt area in eastern and western deserts. The most recent geological features (Quaternary period) can be observed mainly in recent Nile Alluvium in the Nile Valley and Delta (Hammad 1975).

Based on those geological features, there are different soil units that occur in Egypt territory and can be observed in Fig. 3.18 and Table 3.2. There are 33 different soil units distinguished in Fig. 3.18, as the description of each unit is too long so that the abbreviation for each unit is used as suggested by authors. Table 3.2 shows the area coverage and description for each soil unit. Based on the soil map, the sand dune and sandy areas (Sa1 to Sa8) cover around 41% of Egypt land, besides 16% of shallow or stony loamy sand/sandy loam soils. The Nile alluvium represents only 2.6% of Egypt soils. This later soil type is the main and the highest fertile soils that can be used for agricultural production in Egypt.

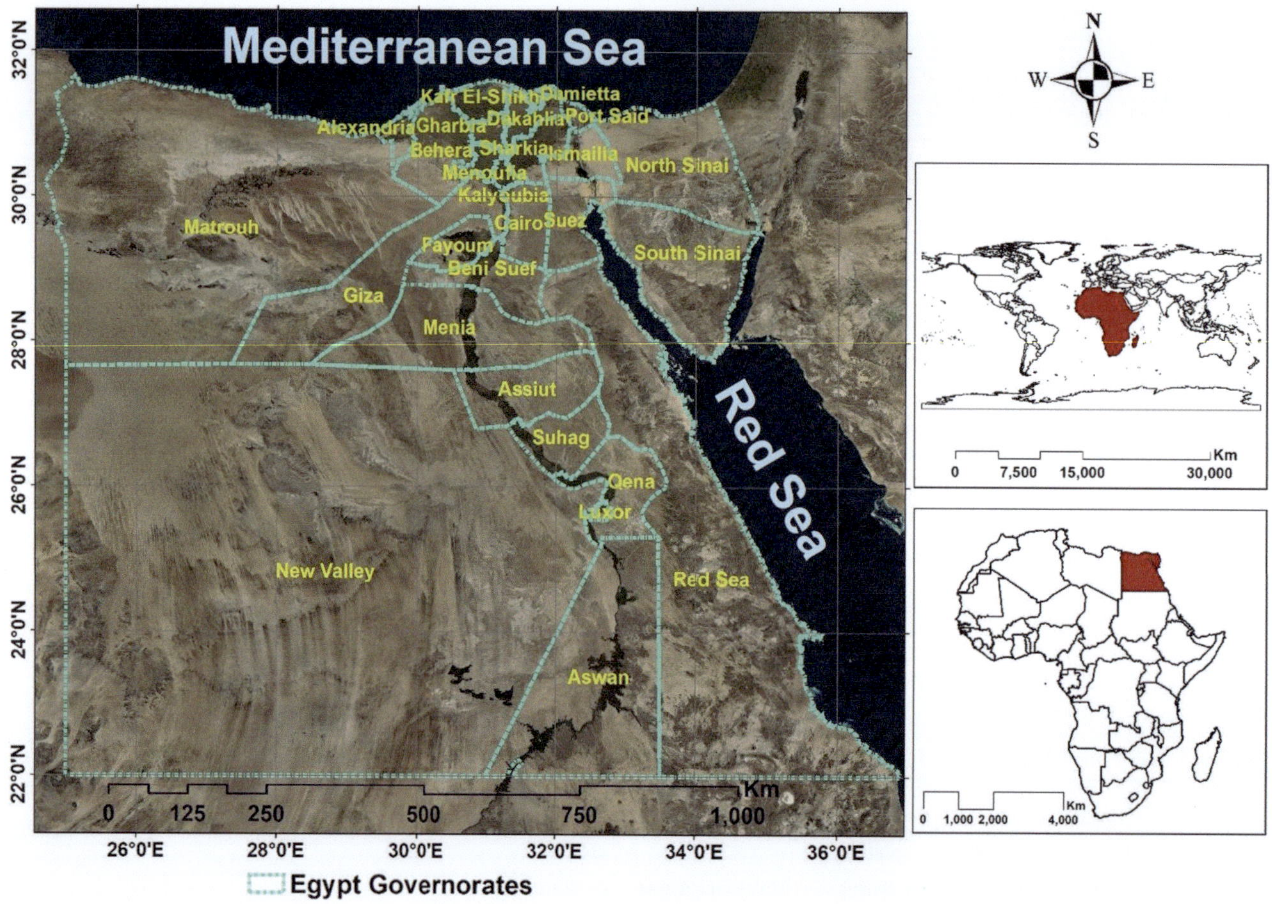

Fig. 3.14 The main Egypt's governorates. Created by authors based on CAMPAS 2017 database

Table 3.1 The area and total population of the Egyptian governorates

No.	Governorates	Number of districts	Area (km^2)	(%)	Total population* (×1000) (%)	
1	New Valley	4	424,073.75	42.22	241.25	0.25
2	Matrouh	8	162,437.75	16.17	425.62	0.45
3	Red Sea	8	121,043.78	12.05	359.89	0.38
4	Aswan	10	68,967.82	6.87	1473.98	1.56
5	Menia	13	32,768.72	3.26	5497.10	5.80
6	South Sinai	8	30,143.55	3.00	102.02	0.11
7	Giza	21	29,245.34	2.91	8632.02	9.11
8	North Sinai	11	27,376.59	2.73	450.33	0.48
9	Qena	14	13,546.84	1.35	3164.28	3.34
10	Assiut	15	12,084.22	1.20	4383.29	4.62
11	Sohag	20	11,021.28	1.10	4967.41	5.24
12	Beni Suef	10	10,752.35	1.07	3154.10	3.33
13	Beheira	19	10,328.80	1.03	6171.61	6.51
14	Suez	6	9198.54	0.92	728.18	0.77
15	Cairo†	43	7773.36	0.77	9539.67	10.06
16	Fayoum	9	6444.09	0.64	3596.95	3.79
17	Ismailia	8	5043.19	0.50	1303.99	1.38
18	Sharqia	22	4497.17	0.45	7163.82	7.56
19	Dakahlia	21	3906.17	0.39	6492.38	6.85
20	Kafr El Shiekh	12	3820.62	0.38	3362.19	3.55
21	Monofiya	12	2440.39	0.24	4301.60	4.54
22	Gharbia	12	1996.20	0.20	4999.63	5.27
23	Alexandria	18	1697.88	0.17	5163.75	5.45
24	Port Said	12	1458.27	0.15	749.37	0.79
25	Kalyoubia	15	1264.76	0.13	5627.42	5.94
26	Damietta	9	862.76	0.09	1496.77	1.58
27	Luxor	4	263.64	0.03	1250.21	1.32
Total		364	1,004,457.81	100	94798.88	100

*Data updated from CAPMAS 2017a, b, †The Egypt Capital

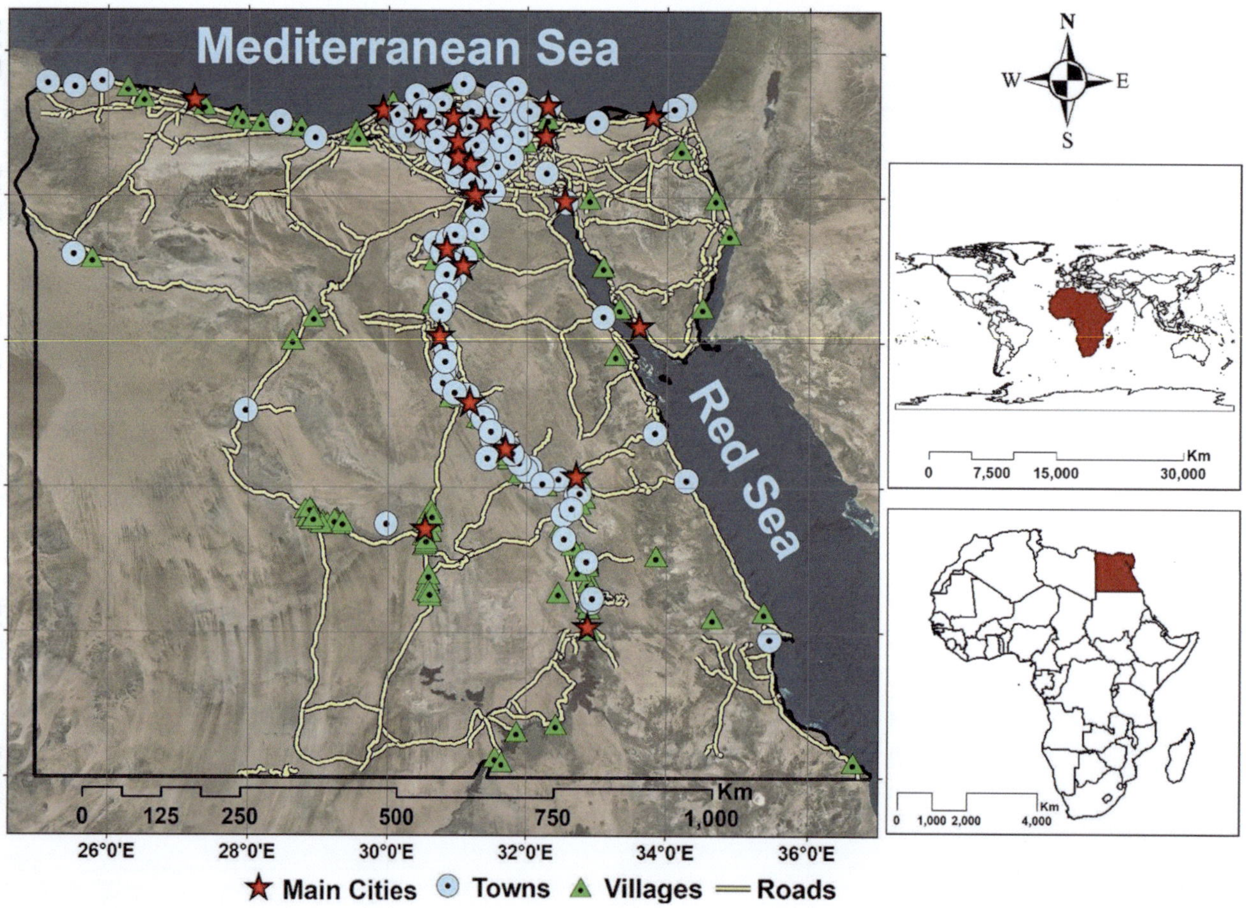

Fig. 3.15 The main infrastructure of Egypt. Created based on FAOStat 2017

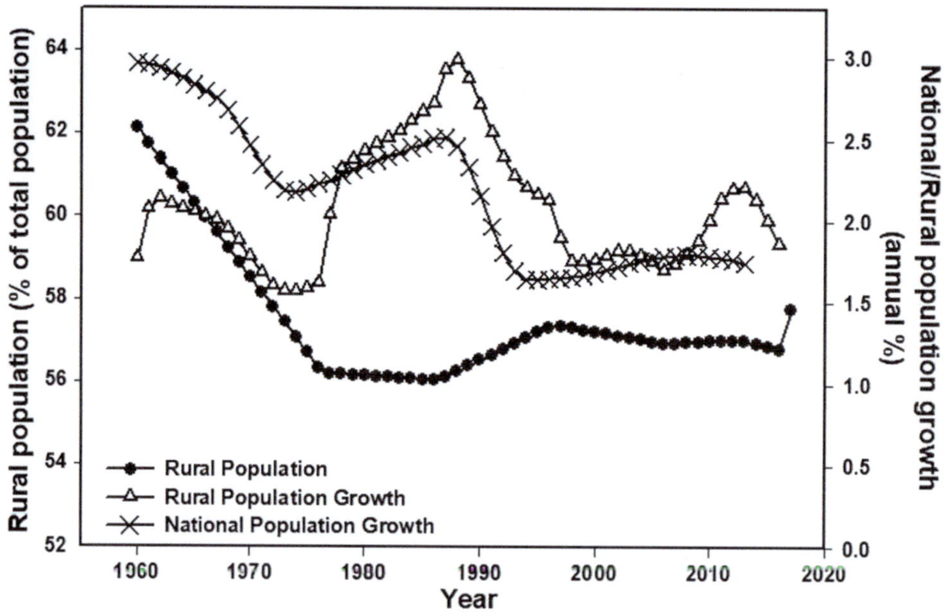

Fig. 3.16 The rural population as a percentage of total population and the rural population growth comparing with the national population growth. Created based on World Bank, World Urbanization Prospect and CAPMAS (2017a, b) database

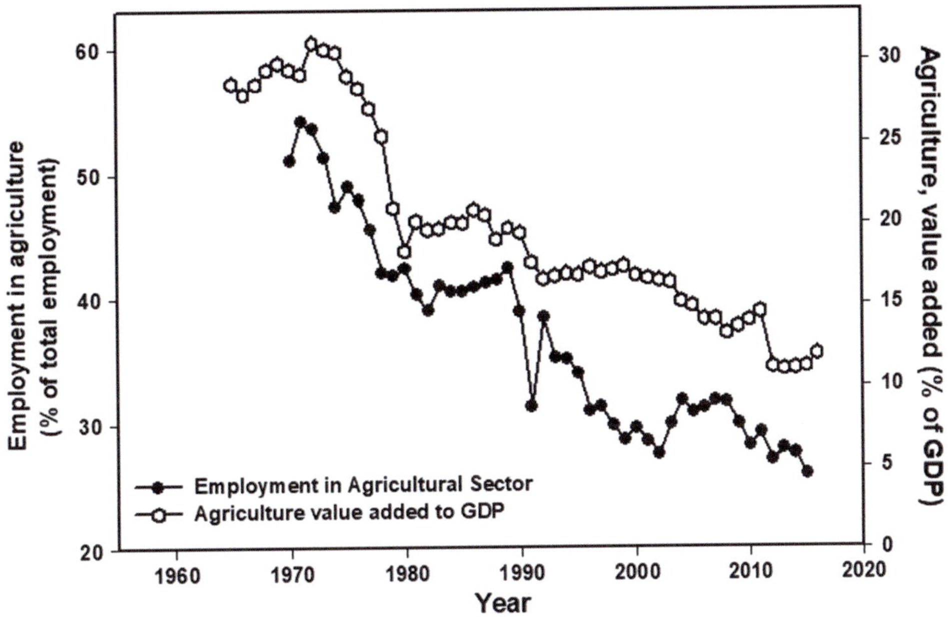

Fig. 3.17 The percentage of employment in agricultural sector compared with the Egypt's total employment and the participation percentage of agricultural sector in gross domestic product (GDP). Created based on World Bank, World Urbanization Prospect and CAPMAS (2017a, b) database

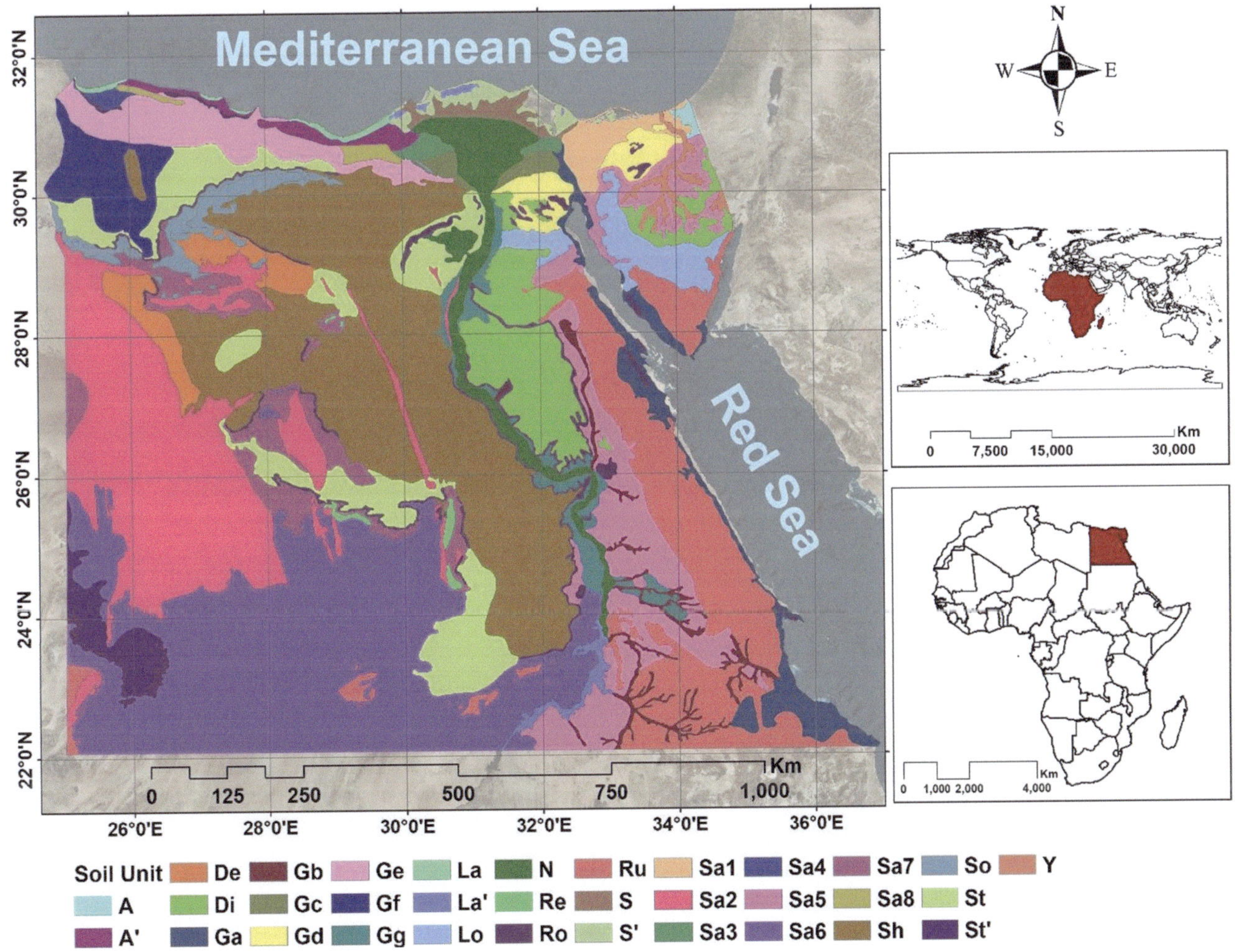

Fig. 3.18 The soil map of Egypt. Created by authors based on http://www.un-spider.org/ and confirmed by Hammad (1975) https://esdac.jrc.ec.europa.eu/images/Eudasm/Africa/images/maps/download/afr_egsma2.jpg

Table 3.2 Egypt soil units based on the legend of Fig. 3.18

ID	Soil unit description	Area (%)
I. Soils developed mainly from limestones		
St	Stoney and loamy sand Lithosols on rough to undulating denuded terrain	5.95
Sh	Shallow or stoney loamy sand - sandy loam soils of Peneplains with	16.01
Gf	Gravelly sand to gravelly loamy sand Lithosols with scattered Solonchaks	1.92
Sa7	Sandy soils with stoney hill remnants of the Piedmont plains	2.99
Ge	Gravelly sand Lithosols with brown loamy soils in scattered patches of the desert plains	2.58
A′	Arid brown loamy soils with remnants of rocky ridges of the coastal plains	0.57
Sa4	Sandy and loamy sand soils with remnants of rocky ridges of the coastal plains	0.14
II. Soils developed mainly from sandstones		
Sa5	Sandy soils and gravelly Lithosols of the desert plains with rocky hills	5.99
A	Arid brown soils with sand dunes	0.09
Sa6	Sandy soils of the Nubia sandstone plains with stoney hill remnants	18.04
St’	Stoney sand Lithosols on rough terrain with rock lands	1.19
Sa8	Sandy to loamy sand soils with Lithosols	0.29
III. Soils developed mainly from a variety of rocks		
Lo	Loamy sand to sandy loam soils with Lithosols	1.56
Gg	Gravels and gravelly sand soils of the alluvial fans, outwash plains, Nile terraces and Piedmont plains	1.33
Gd	Gravelly sand and gravels of a denuded rock land with sand dunes	1.05
IV. Soils Derived Mainly from Shales		
Re	Reddish brown calcareous clayey soils of the desert oases	0.19
V. Solonchaks and Salt-Affected Soil		
So	Solonchaks with rocky hill remnants	0.71
So	Solonchaks with rocky hill remnants	0.70
S	Salt-affected soils of the lower Nile delta areas	0.51
VI. Saltwater marshes		
S′	Salt marshes of the lower delta plain	0.57
La	Lagoons–coastal limestone ridges combinations	0.20
VII. Alluvial soils and alluvial lacustrine complexes		
N	Nile alluvium	2.58
Gb	Gravelly loamy sands of the drainage channels	1.15
Sa3	Sands, clay loams with calcareous crusts and sand dunes of the delta lacustrine complex	0.31
Gc	Gravels and gravelly sand soils of deltaic phase with sand dunes	0.44
Y	Yellowish brown soils of Wadi El-Arish	0.48
VIII. Regosols		
Ga	Gravelly and gravelly sand beaches, sometimes with rock outcrops	2.32
Sa2	Sand dunes and sand sheets of the Western Desert	11.92
Sa1	Sand dunes and sand sheets of the Northern Sinai	0.87
Ru	Rugged rock land mainly of the basement complex	8.94
Ro	Rocky escarpments of different country rocks	2.08
De	Denuded rock land with few sand dunes	1.64
Di	Dissected limestone plateau with lithosols	4.18
La′	Lakes	0.49

3.4 Conclusion

The earth surface and its use are highly dynamic and affected by the soil characteristics as well as the environmental circumstance of every geographical region. Egypt characterizes by several and unique natural features and resources, so this chapter provides an overview of the main natural resources in Egypt, mainly land and water resources as well as related socioeconomic features. The main geographical regions discussed were western desert, eastern desert, Sinai Peninsula, and Nile Valley and Delta. For each region, the general characteristics and landforms were described. Additionally, as the natural drainage network and basin are highly influenced by the geological structure, topology, and relief of the earth surface, the digital elevation model was used to extract the streams and natural basin in Egypt. Besides, human resources and socioeconomic features were generally presented with the main focus on agricultural sector as a key of socioeconomic sector in Egypt and support the livelihood for more than two-thirds of Egyptian population. Based on the most recent population census in 2017, the total population in Egypt is around 95 million who unevenly distributed over 27 governorates. Furthermore, the soil map of Egypt was provided in the last section of this chapter as an introduction to the detailed soil-related studies in the following chapters of this book.

References

Ali H, Mahmoud M (2004) Management of Egypt crop pattern according to reservoir water storage. In: Second regional conference on Arab water: action plans for integrated development, Cairo

Bakr N, Weindorf DC, Bahnassy MH, Marei SM, El-Badawi MM (2010) Monitoring land cover changes in a newly reclaimed area of Egypt using multitemporal Landsat data. Appl Geogr 30(4): 592–605

CAPMAS (2016) Central Agency for Public Mobilization and Statistics. Statistical yearbook. Cairo. Ref. No. 71-01111-2016. http://www.capmas.gov.eg

CAPMAS (2017a) The population of Egypt in 2017. Arabic reference, http://www.capmas.gov.eg/party/party.html

CAPMAS (2017b) Geographical Egypt Data. Arabic Reference. http://geoportal.capmas.gov.eg/arcgis/apps/webappviewer/index.html?id=67fdd9906ae6a7be42bf70dc&locale=ar

Elbana TA, Bakr N, Elbana M (2017) Reuse of treated wastewater in Egypt: challenges and opportunities. In: Negm AM (ed) Unconventional water resources and agriculture in Egypt: The handbook of environmental chemistry. Springer, Berlin, Heidelberg. https://doi.org/10.1007/698_2017_46

Embabi NS (2018) Landscapes and landforms of Egypt: landforms and evolution. In: Migoń P, Wroclaw P (eds) World geomorphological landscapes. Springer, Switzerland

FAOStat (2017) Egypt: country indicators. http://www.fao.org/faostat/en/#country/59

Hamdi H, Abdelhafez S (2001) Agriculture and soil survey in Egypt. In: Zdruli P, Steduto P, Lacirignola C, Montanarella L (eds) Soil resources of Southern and Eastern Mediterranean countries. http://om.ciheam.org/om/pdf/b34/01002089.pdf

Hammad MA (1975) Geological map of Egypt. Appendix 1. Soil Survey Papers no. 11

ICARDA (2013) Water and agriculture in Egypt. Technical paper based on the Egypt. Australia ICARDA Workshop on on-farm water-use efficiency

MALR (2009) Egypt sustainable agricultural development strategy towards 2030. Egyptian Ministry of Agriculture and Land Reclamation, Cairo

MWRI (2014) Water scarcity in Egypt: the urgent need for regional cooperation among the Nile Basin Countries. Egyptian Ministry of Water Resources and Irrigation, Cairo

Meguid AM (2017) Key features of the Egypt's water and agricultural resources. In: Negm AM (ed) Conventional water resources and agriculture in Egypt: The handbook of environmental chemistry. Springer, Berlin, Heidelberg. https://doi.org/10.1007/698_2017_41

Negm AM, Saavedra O, El-Adawy A (2017) Nile Delta biography: challenges and opportunities In: Negm AM (ed),The Nile Delta, Barceló D, Kostianoy AG (Series eds) The Handbook of environmental chemistry 55, pp 3–18

Omran ESE (2017) Land and groundwater assessment for agricultural development in the Sinai Peninsula, Egypt. In: Negm AM (ed) The handbook of environmental chemistry. Springer, Berlin, Heidelberg. https://doi.org/10.1007/698_2017_43

Stanley DJ, Warne AG (1993) Nile Delta: recent geological evolution and human impact. Science 5108(628):634–7046. https://doi.org/10.1126/science.260.5108.628

U.S. Geological Survey (2017) EarthExplorer. http://earthexplorer.usgs.gov/

Worldatlas (2017) https://www.worldatlas.com/webimage/countrys/africa/eg.htm

Zahran MA, Willis AJ (2009) The vegetation of Egypt. In: Werger MJA (ed) Plant and vegetation, vol 2, 2nd edn. Springer, Berlin, Heidelberg. ISBN 978-1-40208755-4

4 Land Use/Land Cover and Vegetation Status

Noura Bakr and Mohamed H. Bahnassy

Abstract

With an area around one million square kilometers, five land use/land cover (LULC) classes are distinguished in Egypt, namely, agricultural land, barren land, urban areas, natural vegetation (aquatic and terrestrial), and water bodies. In this chapter, each LULC class is discussed, and their changes are briefly introduced. The LULC in Egypt is not enormously varied as over 90% of Egypt soils are desert, and other economic activities and settlement existed in around 4 to 5%. Although the agricultural land represents around 4% of the total area, the agricultural sector is a crucial player in the Egyptian economy. The old cultivated land (Oldlands) are observed in the Nile Valley and Delta, whereas, the Newlands represent the horizontal expansion in the desert, mainly around the fringes of the Nile Delta. Wheat, rice, barley, and maize are the main cultivated cereal crops that are recognized in Egypt. Both cereal crops yield and their cultivation extend are almost doubled within the last five decades. Barren lands are observed in three locations, western desert, eastern desert, and the Sinai Peninsula. Egypt oases and depressions locate in the western desert, the Red Sea Mountains are in the eastern desert, and the highest spot in Egypt (St. Catherine Mountain) locates in the Sinai Peninsula. The urban sprawl is one of the main anthropogenic challenges that causes an observed change in LULC in Egypt. Thus, the government adopts several policies to relocate Egyptian outside the Nile Valley and Delta regions by constructing new communities in the desert and conserved the Oldlands from being lost due to the urban sprawl. Besides Nile River and its branches, other water bodies such as the five northern lakes, Nasser Lake, and Toshka depression are recognized. The northern lakes are subjected to change as a result of both environmental and anthropogenic impacts.

Keywords

Land use/Land cover • Agriculture • Cereal crops
Change detection

4.1 Introduction

Di Gregorio and Jansen (2000) introduced the definition of land cover as "the observed (bio) physical cover on the earth's surface." Whereas the land use describes "the arrangements, activities and inputs people undertake in a certain land cover type to produce, change or maintain it." Under the same context, Prakasam (2010) and Rajeswari et al. (2014) stated that land cover refers to the biophysical cover of the earth's surface which naturally occurred, such as vegetation and water bodies, or anthropogenic effects such as built-up land. Whereas land use refers to the specific utilization of each land cover, for example, the built-up land cover class represents several land use types such as industrial, residential, and/or commercial area. Thus, the land use/land cover (LULC) patterns are mainly related to the human usage of the land in terms of natural and socioeconomic development through space and time (Hua 2017). Thus, the changes in LULC are determined by the interactions of environmental and socioeconomic factors. The study of LULC and their changes is highly required and extremely important for land use management and planning for sustainable development of land resources (Esmail et al. 2016), urban sprawl (e.g., Megahed et al. 2015; Kamh et al. 2012; Shalaby and Gad 2010), national level reclamation efforts (Bakr et al. 2010), water resources management (Schilling et al. 2008), and socioeconomic studies.

Change detection is the process of identifying differences in the state of an object by observing it over a period of time

N. Bakr (✉)
Soils and Water Use Departmant, National Research Centre (NRC), Cairo, Egypt
e-mail: nourabakr@yahoo.com

M. H. Bahnassy
Soil and Water Sciences Department, Faculty of Agriculture, Alexandria University, Alexandria, Egypt

H. El-Ramady et al. (eds.), *The Soils of Egypt*, World Soils Book Series,
https://doi.org/10.1007/978-3-319-95516-2_4

(Singh 1989). Two primary categories of change detection exist. One category focuses on detection of detailed change trajectories, called "from-to". Post-classification comparison (PCC) is a common example of this approach. The second category focuses on the detection of binary change and non-change features, such as vegetation index differencing (Lu et al. 2004). Change detection can be performed by unsupervised or supervised approaches (Singh 1989). A supervised technique, also called PCC, requires the availability of ground truth points to derive a training set containing information about the spectral signatures of the changes that occurred in the considered area between two dates. An unsupervised technique performs change detection without any additional information besides the raw images considered; however, it is also fraught with some critical limitations (Bruzzone and Prieto 2002). As these basic approaches have limited utility independently, a hybrid classification method was used in order to obtain both high change accuracy and efficiency (Schowengerdt 2007).

Cultivation is one of the main anthropogenic driving forces that lead to the change in the ecosystem. Urbanization is another important demographic pressure on the land especially with the vast increase in the population on the earth (Rajeswari et al. 2014). Thus, the study of LULC and monitoring the environmental changes are essentially required to meet the increasing demands of population needs and simultaneously achieving more sustainable land use and/or management planning.

The LULC in Egypt is not extremely varied as around 95–96% of Egypt soils are just desert and all other economic activities and settlement are only in around 5% of Egypt territory (Shalaby and Tateishi 2007). Egypt faces several environmental and anthropogenic challenges in both water and land resources due to: (1) since 1959, Egypt receives a fix share of water from the Nile River equal to 55.5 BCM yr^{-1} (billion cubic meter per year) regardless of the exponential increases in its population, (2) scarcity of precipitation which limited to the narrow coastal zones at the northern region, (3) the sea water intrusion in the Nile Delta region, (4) coastal erosion and sea level rise, and (5) the new development plans in southern Nile basin countries such as Grand Renaissance Dam in Ethiopia. Moreover, there are multiple anthropogenic challenges such as: (1) the urbanization and industrialization that are taking place within the Nile valley and Delta regions are associated with a huge amount of municipal effluents and industrial wastes that are causing vast deterioration for both land and water resources in those areas, (2) Urban sprawl and exponential increase in population and (3) the changes in LULC that are taking place in that region (Negm et al. 2017).

This chapter focuses on the LULC in Egypt and their changes over time. Additionally, more attention is given to agricultural land use as the agricultural sector is a key player in Egypt's economy. After presenting the overview of the current status of LULC in Egypt, a comprehensive review about the changes that have been taken place in different areas within the country, is discussed. This chapter is supported with lots of maps using geographic information system tools to provide the results in easier ways to be understandable by the decision-makers for better land use planning and by the scientific community for general overview about the status of LULC in Egypt.

4.2 Main Land Use/Land Cover Classes in Egypt

Five main LULC classes can be distinguished in Egypt; agricultural land, barren land, urban areas, water class and natural vegetation (Fig. 4.1). The agricultural land can be observed in the Nile Valley and Delta regions which represent the old cultivated land in Egypt (*Oldland*). The eastern and western desert as well as Sinai Peninsula represent the barren desert land. The water class is assigned mainly for the Nile River and Egypt's lakes. The natural vegetation could be differentiated based on their occurrence either as terrestrial or aquatic natural vegetation.

4.2.1 Agricultural Land

Agriculture is the main LULC in Egypt and a key player sector of the Egyptian economy. Agriculture supports the livelihood of over half of the Egyptian population besides approximately 35% of the total employment is assigned for agricultural sector (CAPMAS 2017). Egypt's agricultural land can be divided into two main categories; *Oldlands* and *Newlands* (Fig. 4.2). The *Oldlands* refer to the highly fertile lands that have been intensively cultivated for long time in the Nile Valley and Delta which depends mainly on the Nile as the main water source for irrigation. The *Newlands*, which are considered as the horizontal expansion for the agriculture sector, are the desert lands that have been reclaimed relatively recently (post-1980) or in the process of being reclaimed now (Bakr et al. 2009). The *Newlands* are less fertile, more fragile, and face more restricted access to high-quality water for irrigation, but over time and with good management practices, their productivity could be improved (Bakr et al. 2010). As a part of agricultural development strategies, horizontal and vertical expansion programs were adopted to increase the cultivated land (by reclamation) and crop production/productivity for the same unit of land, respectively (Noaman 2017). Examples of the reclamations efforts in the western Nile Delta include El-Tahrir, Maryout, and Nubariah projects. Whereas El-Salam Canal is an example of reclamation project in the eastern Delta and northern Sinai (Fig. 4.2) which aimed to

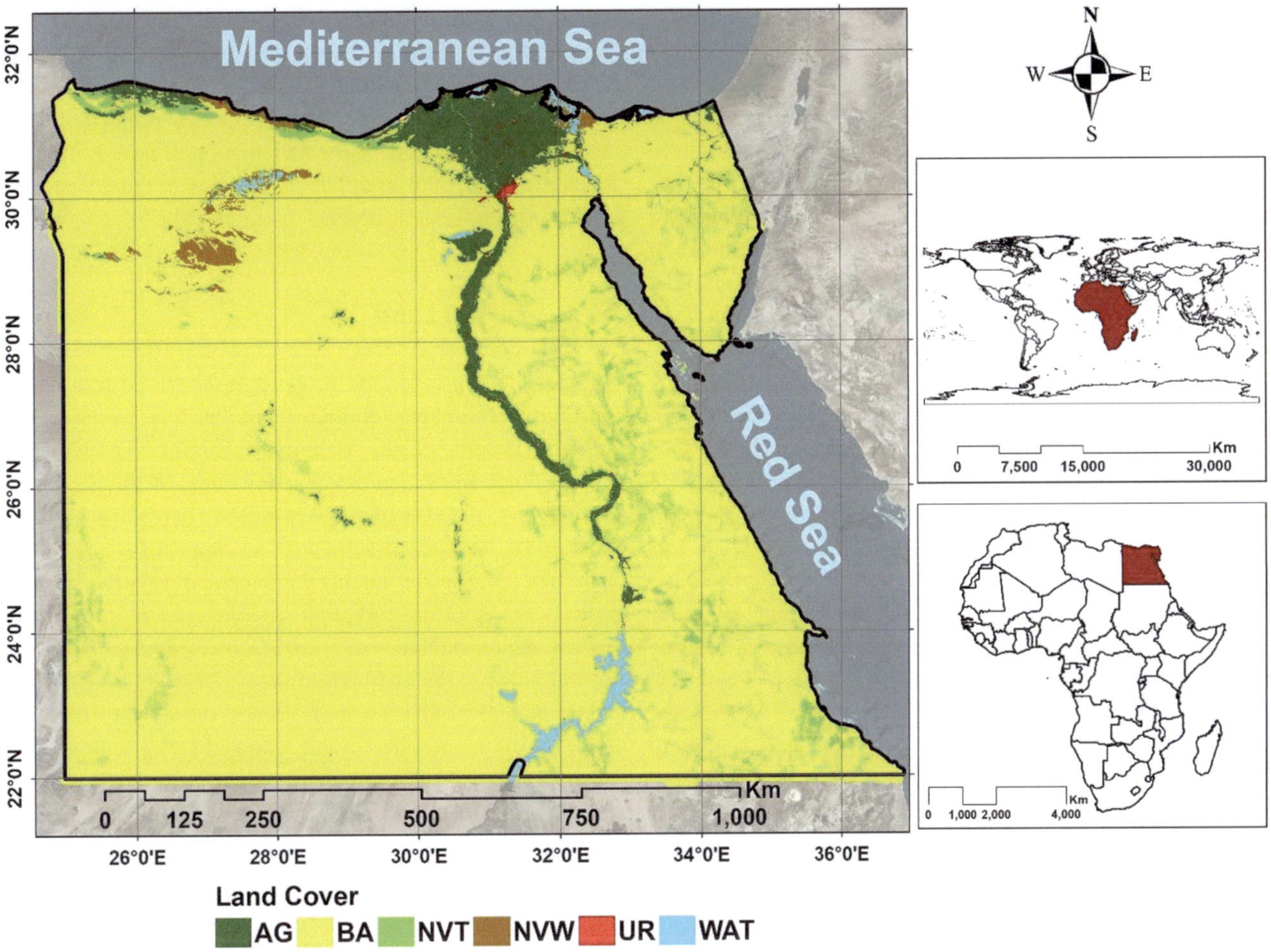

Fig. 4.1 The main LULC classes in Egypt; AG: Agriculture, BA: barren, NVT: terrestrial natural vegetation, NVW: aquatic natural vegetation, UR: urban and WAT: water classes. Created by authors based on http://www.un-spider.org/

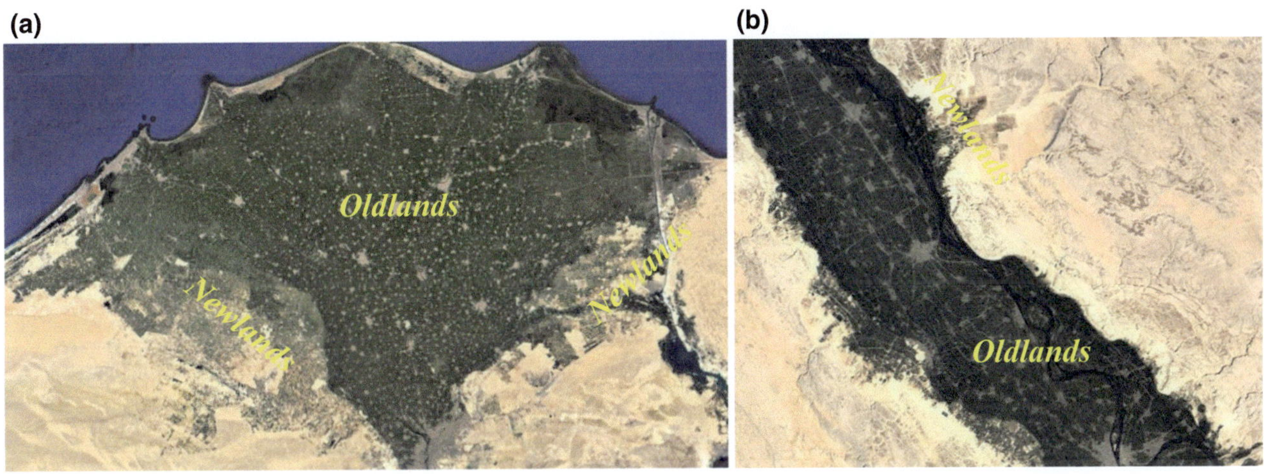

Fig. 4.2 *Oldlands* in Nile Delta (**a**) and Valley (**b**) and the *Newland* extension around their margins. Created by authors based on *Google Earth*

reclaim a total of 462,000 hectares of desert land using 1:1 mixture of Nile water and treated wastewater (Hereher 2013).

The agricultural practices in Egypt have been revolved dramatically as a result of Aswan High Dam construction in 1960s. Before the High Dam establishment, only one irrigation season existed which was after the Nile flooding in summer every year. However, with the existence of the High Dam, the irrigation was regulated, no flooding occurred. The water is saved or stored in Lake Nasser and then was utilized to reclaimed *Newlands* (Abu Zeid 1993).

Based on 2014 the available data survey, the total agricultural land in Egypt was around 9,270,000 feddans, 3.9 Mha, (fed = 4200 m^2 = 0.42 ha) which represented 3.5% of the total Egyptian territory, from which around 6,095,000 feddans (2.6 Mha) were assigned as *Oldland* and 3,175,000 feddans (1.3 Mha) were allocated for the newly reclaimed area (Negm et al. 2017).

Recently, the Egyptian government adopted different national projects to reclaimed new desert land as a kind of horizontal expansion of the agricultural sector. These are done to overcome the exponential population increases, achieve self-sufficiency of food production, and start new communities to redistribute the population over a larger area. Most areas of potential reclamation included Nile Delta margins, the Western Desert, and North Sinai (Hereher 2013). The central western desert involved Bahariya oasis reclamation efforts (Fig. 4.3). In south western desert, two main projects are running (Fig. 4.4); Toshka depression (irrigated from Lake Nasser) and east Oweinat (irrigated from the Nubian Sandstone Aquifer, underground water). Additionally, one of the mega national projects that is recently proposed aims at the reclamation of new 1.5 million feddan (630,000 ha) in different locations in the desert land, mainly in the western desert.

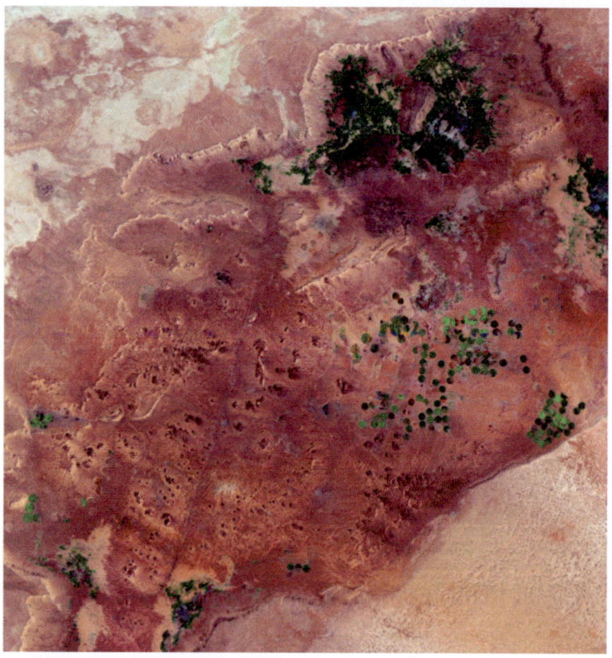

Fig. 4.3 New reclaimed areas (green spots) in Bahariya oasis (Western Desert), Egypt. Created by authors

4.2.2 Barren Land

It represents the bare land in western desert, eastern desert, and Sinai Peninsula. Rather than the five oases in the western desert; Siwa, Bahariya, Farafra, Dakhla, and Kharga, and three depressions; Fayoum, Wadi El-Natrun, and Qattara, the rest of western desert (two third of Egypt territory) is bare desert soils with no support for any human activities. The government implemented several strategic policies to relocate Egyptian people outside the limited narrow region in Nile Valley and Delta by constructing new communities in the huge desert areas (Hereher 2013). These governmental policies would; reduce the density of population in the main cities in the Nile and Delta regions, conserve the *Oldlands* from being lost due to the urban sprawl and urbanization stress over that areas, and create new job opportunities which reduced the high rate of unemployment in Egypt.

The main desertic landforms in the Western Desert from north to south are included: Marmarika plateau, Sand Sea, Gilf Kebir, and Owienat Mountain besides the sand dunes that spread out on the desert surface. The main features in the Eastern Desert are Red Sea chain of Mountains and Elba Mountains. The highest spot in Egypt is St. Catherine Mountain in the Sinai Peninsula (Fig. 4.5).

4.2.3 Urban Areas

They are the residential or built-up land for settlement in Egypt territory. Egypt involves 27 governorates the lowest governorates in the population are South Sinai, New Valley, Red Sea, Matrouh, and North Sinai with a total summation of all around 1.7% for all population where their area coverage is more than 76% of Egypt area (CAPMAS, 2017). Whereas, around 25% of Egyptian live in Greater Cairo (Cairo, Giza, and Kalyoubia governorates) which has an area of 3.8% of Egypt area coverage, considering the hug desert extend of Giza governorate to Bahariya oasis. This confirms the higher density population in Nile Valley and Delta regions. Figure 4.6 shows the urban expansion over part of the Greater Cairo and some new communities close to it.

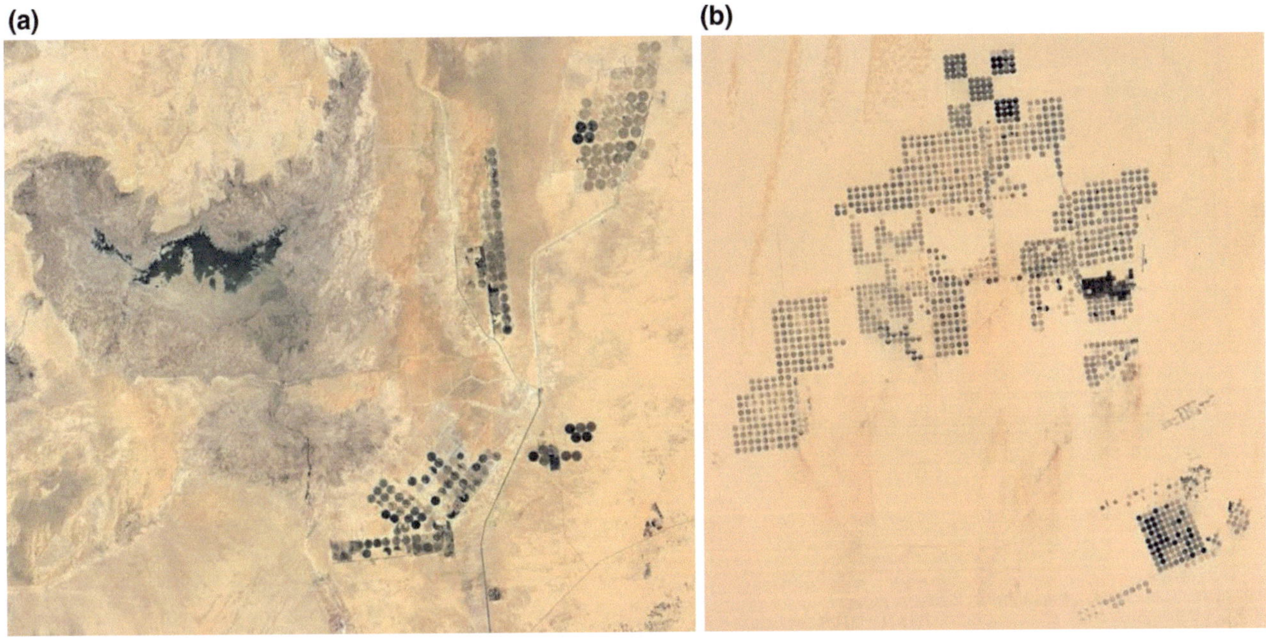

Fig. 4.4 Toshka depression (a) and east Oweinat (b) projects in the southern Western Desert), Egypt. Created by authors based on *Google Earth*

4.2.4 Water Class

Besides the Nile River and its two branches; Damietta to the east and Rosetta to the west, other water bodies can be observed in Egypt. Specifically, the northern lakes (from west to east are: Mariout, Idko, Brullus, Manzala, and Bardawell). These lakes receive the excess irrigation water and municipal wastewater from the surrounded regions (Fig. 4.7). In the western desert, there are several water bodies could be distinguished mainly: Qarun lake and Wadi El Rayan in Fayoum governorate (Fig. 4.7). They also fed with excess drainage water from the Fayoum depression. Additional water bodies in the western desert are Qattara Depression (North West), Nasser Lake and Toshka depression (South and South West, respectively) (Fig. 4.7). Whereas in the eastern desert, Ismailia irrigation canal and the Great Bitter Lake (between the Red Sea and the Suez Canal) could be observed in Fig. 4.7.

4.2.4.1 Natural Vegetation

Two types of natural vegetation could be distinguished in Egypt: aquatic and terrestrial natural vegetation. The aquatic natural vegetation refers to the natural vegetation that exists in the wetlands or on the top of the surface water. Whereas, the vegetation that naturally grows on the soil surface is known as terrestrial natural vegetation. The description of general kinds of aquatic and terrestrial natural vegetation is illustrated in Table 4.1 based on FAOStat 2017.

Generally, four floral provinces can be recognized in Egypt (Zahran and Willis 2009); Mediterranean (northern coast), North African-Indian Desert (dominant), Central Asiatic (Sinai Peninsula and some parts of Eastern desert), and African Forest and Steppe (Elba mountain). The north western coastal belt of Egypt is the richest part of its floristic composition as around 50% of the total Egyptian flora is existed in that area due to the rainfall. The Xerophytes are the dominated flora in this region and cover around 90% of that area. Aquatic and terrestrial Halophytes exist in and around Egypt lakes, respectively.

4.3 Detailed Land Use/Land Cover Map of Egypt

Based on the main LULC classes (Fig. 4.1), further subclasses (total of 48) can be categorized (Fig. 4.8). The area coverage for each class was calculated and presented in Table 4.2. Although, the agricultural sector is the main LULC class in Egypt, as previously mentioned, the area of agricultural land is still very low as it represents around 4% from the total area of Egypt including both irrigated and rainfed cultivated areas. Almost the same area coverage (around 4%) is assigned for natural vegetation with its both types (terrestrial and aquatic). The water bodies cover an area of around 1% of Egypt soil whereas only 0.2% is used as urban or built-up land that mainly exists within the agricultural land. The dominant LULC class is the barren land that covers over 90% of Egypt area involving; loose sand, sand dune, bare rock, bare soil, and salt crust areas (Fig. 4.8; Table 4.2).

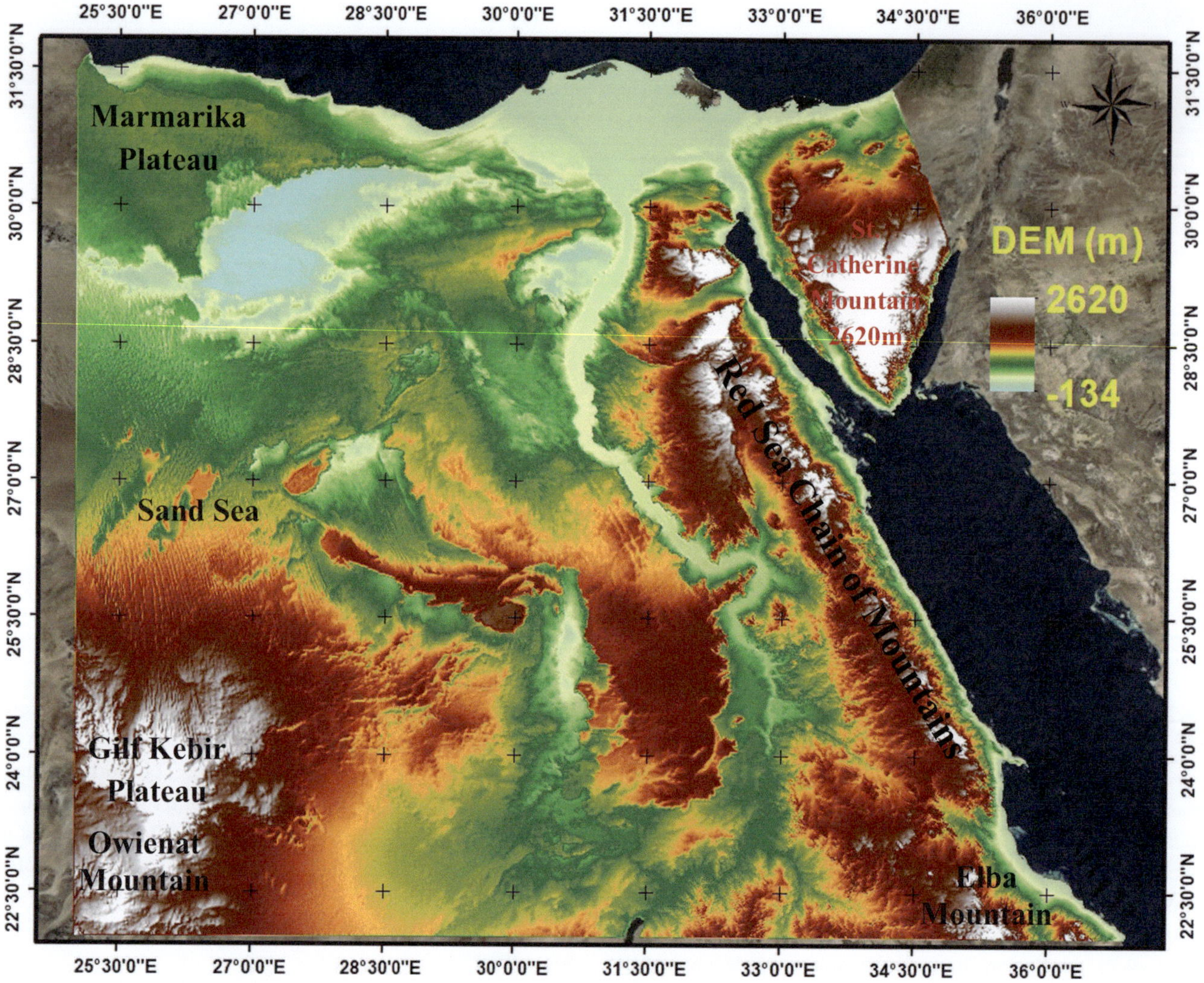

Fig. 4.5 The main desertic features in the bare desert soils in Egypt. Created by authors

4.3.1 Cropland

Over 55 years (1960–2015), the agricultural land has been increased from around 2.5% to the assigned percentage of around 4%. The observed increase was shown after 1990s due to the successful governmental reclamation efforts to increase the cultivated area in Egypt by reclaimed the desert land (Bakr, 2009; 2010). With this increases in the cultivated lands, a respective increase in the permanent cropland area from around 0.05% in 1960s to around 1.1% by 2015 was recognized (Fig. 4.9). An exponential increase in the permanent cropland area occurred in 2005 as an increase by twofold (from around 0.45 to around 1.1%) was observed.

4.3.2 Cereal Crops

Within the different cultivated crops, the cereal crops could be considered as the most crop type that has been intensively used worldwide. Wheat, rice, barley, and maize are the main cereal crops that can be recognized in Egypt's cultivated lands. Over the last five decades, the cultivated cereal croplands have been doubled from around 1.5 Mha to about 3.2 Mha (FAOStat 2017; Fig. 4.10). Besides the horizontal expansion in the cultivated land for cereal crops, the vertical expansion using highly productive new varieties led to exponential increases in the cereals yield from 3 to around 7.5 ton per hectare (FAOStat 2017; Fig. 4.10).

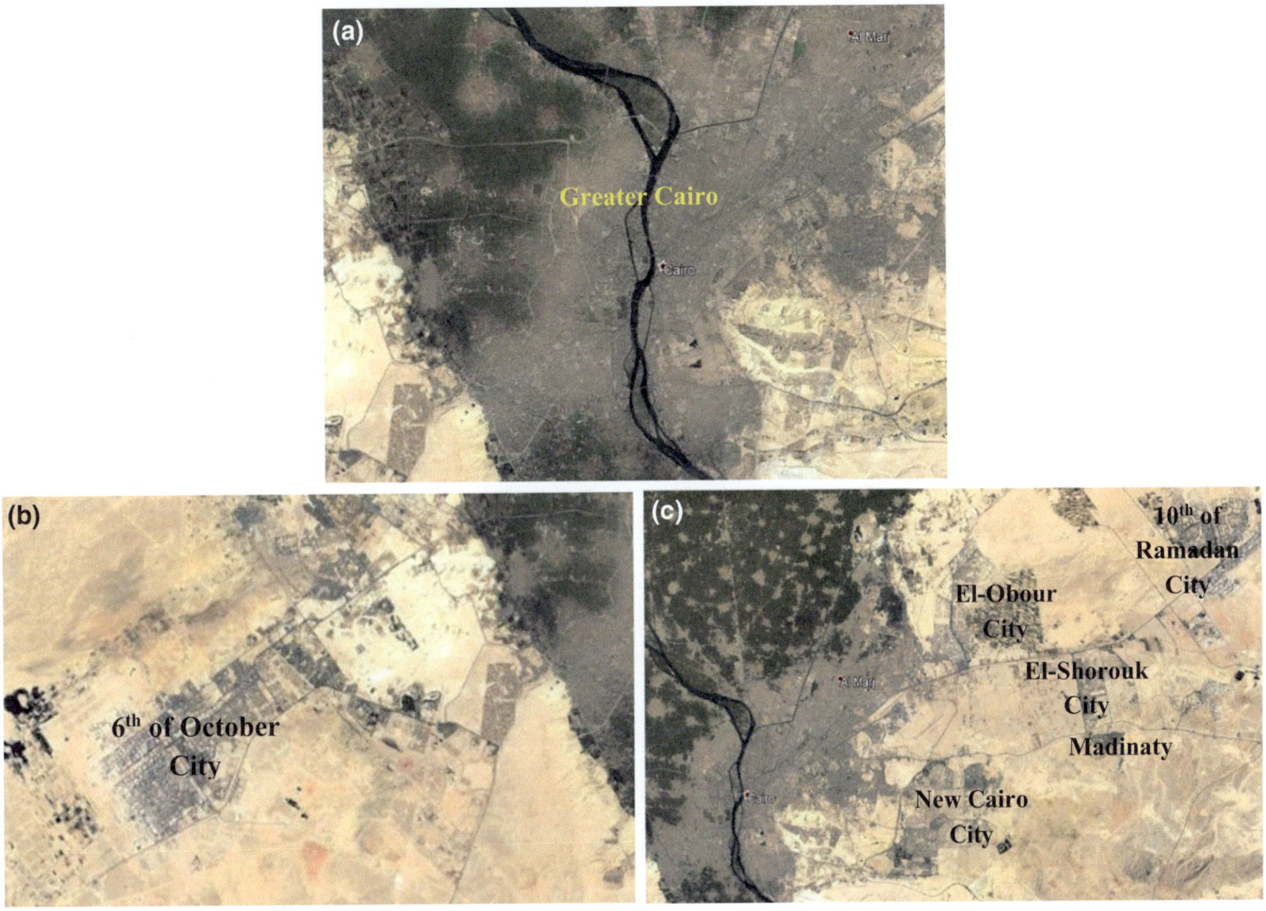

Fig. 4.6 Great Cairo urban expansion (**a**), 6th of October city west of Cairo (**b**), and some new cities east of Cairo governorate (**c**). Created by authors based on *Google Earth*

Wheat, as the main strategic cereal crop, is the essential component of the Egyptian's daily food and the most consumed cereal crop in Egypt. Wheat is a winter crop that has been intensively cultivated in around 47% of the total cropped area in Egypt, mainly in the Nile valley and Delta. However, Egypt is the world's biggest wheat importer, as it represents 20% of all agricultural imports to fulfill the increasing consumption of wheat by the increased population (McGill et al. 2015). Based on FAO (2015) and USDA monthly reports, and published statistics by (MALR 2009) in Egypt, there was an increase in wheat production by fourfold from around 2 Mton (during 1960s to 1980s) to about 8 Mton currently. This increase is supported by increasing in harvested area (from 0.4 to 1.2 Mha) as well as the breading efforts to offer new varieties that can resist the environmental stresses.

4.4 Land Use/Land Cover Change Detection

Mapping of the distribution of vegetation on the earth surface and the change of its status over time have attracted the attention of scientists several decades ago as forests, grassland, and agricultural land are the main LULC classes. With the extent of such classes over large areas, satellite images can be an excellent tool to study and detect the current utilization of the landscape. Currit (2005) stated that the archived satellite images are also an information treasure to monitor and determine the observed change in the LULC in the same area at different times (multi-temporal LULC change detection). One of the most useful and commonly used satellite images worldwide is Landsat. Since the first generation (Landsat-1, 1972) till now, the Landsat provides highly efficient remotely sensed images for all types of users globally (Williams et al. 2006). Geoinformatic technologies, such as geographic information system (GIS), remote sensing (RS), and modeling, can provide excellent tools to offer the database needed for the proceeding development plan over a large area at a long period of time. Remote sensing is cost-effective and time-efficient tool compared to the intensive fieldwork. Geographic information system (GIS) is a powerful set of tools for detecting, investigating, and interpreting land resources as they are capable to study soils at spatial and temporal domain with a cost-effective manner. Timely and accurate identification of the Earth's surface

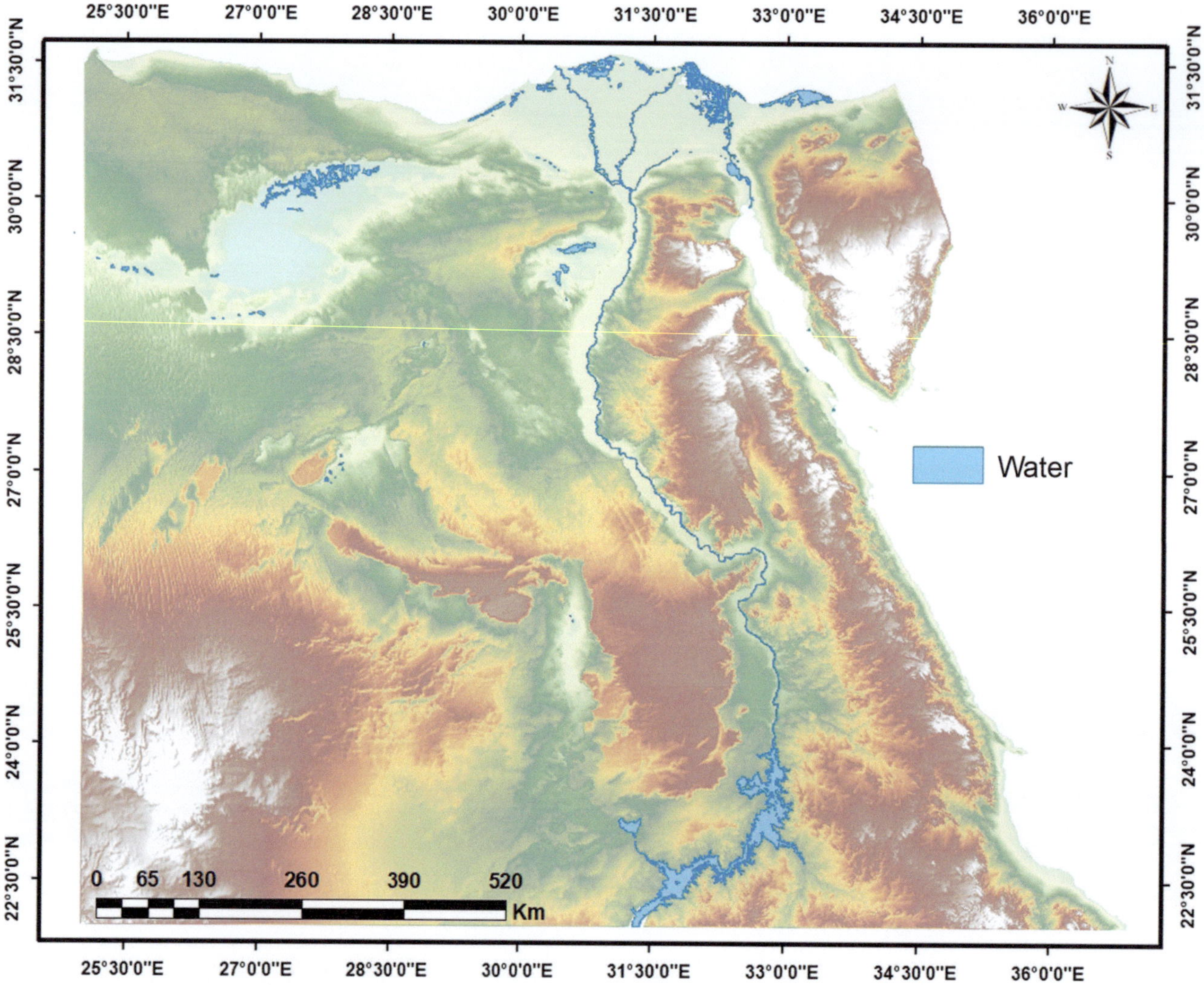

Fig. 4.7 The main water resources in Egypt (created based on FAOStat 2017)

features provides better understanding of the interactions between human and natural phenomena to better manage and use the natural resources (Lu et al. 2004).

Based on Singh (1989), the change detection process can be defined as the identifying of the differences in the state of an object or phenomenon by observing it at different times. Two primary categories of change detection can be distinguished: first, category focuses on detection of detailed change trajectories (called "from-to"), whereas second, category focuses on the detection of binary change and non-change features. Post-classification comparison (PCC) and vegetation index differencing are common examples of the first and second categories, respectively (Lu et al. 2004).

The change detection methods can be grouped into two categories: (1) pixel-based methods and (2) object-based methods (Hussain et al. 2013). Pixel-based change detection methods are suitable for medium and coarse resolution images whereas object-based imaged analysis techniques are considered more suitable for very high-resolution image data (Rajeswari et al. 2014). The change detection methods are grouped into seven categories: (1) algebra, (2) transformation, (3) classification, (4) advanced models, (5) Geographical Information System (GIS) approaches, (6) visual analysis, and (7) other approaches (Lu et al. 2004). In his intensive review, Singh (1989) listed eight different digital change detection techniques, namely, univariate image differencing, image regression, image rationing, vegetation index differencing, principal components analysis (PCA), post-classification comparison, direct multi-date classification, and change vector analysis. From them, the image classification and vegetation index differencing (Lu et al. 2005) have been intensively used in several studies worldwide. A supervised classification is achieved by deriving training sets containing information about the spectral signatures of each LULC then comparing the resulted classified image between two dates. An unsupervised technique

Table 4.1 Description of the main kinds of terrestrial and aquatic natural vegetation in Egypt

	Description
Terrestrial natural vegetation	Sparse herbaceous vegetation. The height of shrubs varies from 0.3 to 5 m. Shrubs density varies from 4 to 15%
	Sparse shrubs with sparse herbaceous layer. The height of shrubs varies from 0.3 to 5 m. Shrubs density varies from 1 to 15%
	Continuous shrubs with herbaceous layers. The height of shrubs varies from 0.3 to 5 m. Shrubs density varies from 15 to 40%
	Fragmented shrubs. The height of shrubs is lower than 0.5 m. Shrubs density varies from 15 to 40%
Aquatic natural vegetation	Short forbs on waterlogged soil—salt crust. The height of herbaceous varies from 0.03 to 0.3 m. Forbs density varies from 4 to more than 15%
	Short forbs under reclamation on waterlogged soil—salt crust. The height of herbaceous varies from 0.03 to 0.3 m. Forbs density varies from 4 to 15%
	Grassland on permanently flooded land. The height of herbaceous varies from 0.3 to 0.8 m. Herbaceous density is more than 65%. Flooded land for more than 4 months/year
	Grassland on permanently brackish area. The height of herbaceous varies from 0.3 to 0.8 m. Herbaceous density is more than 65%. Flooded land is more than 4 months/year
	Grassland on temporarily brackish area. The height of herbaceous varies from 0.3 to 0.8 m. Herbaceous density varies from 15 to 40%. Flooded land from 2 to 4 months/year
	Broadleaved evergreen shrubs on a permanently brackish area. The height of shrubs varies from 0.5 to 5 m. Shrubs density is more than 65%. Flooded land for more than 4 months/year

performs classification based on the clustering of the natural groupings of pixels in image data (Bruzzone and Prieto 2002). The recent classification method, hybrid classification, was used to obtain both high change accuracy and efficiency (Schowengerdt 2007). Normalized difference vegetation index (NDVI) is widely used to monitor vegetation dynamics and plants' health (Sader 1987; Pax Lenney et al. 1996).

In Egypt, two main types of significant agricultural related changes can be determined; first, the soil reclamation that has taken place in the desert lands to be transformed into agricultural land along the barren lands especially for the areas near to the fringes of the Nile Delta. Second, the urban sprawl that it is proceeding on the old agricultural land in Nile Valley and Delta.

Thus, the change detection studies for Egypt's *Oldlands* have been mainly focused on urban sprawl or extended of the main cities and land degradation of the old and highly fertile cultivated lands in Nile Valley and Delta regions (Elhag et al. 2013). Whereas, for detecting the change in *Newlands* most of the achieved studies were focused on the transformation of barren land into agriculture land. Figures 4.11 and 4.12 show that for 200 years (1818 to 2018), there is a remarkable change in the Nile Delta and Valley regions, respectively. Figure 4.12 shows a significant change in the El-Fayoum depression over the last two centuries.

4.4.1 LULC Change Detection in *Oldlands*

Yin et al. (2005) studied the extended of built-up land and population distribution pattern in Greater Cairo between 1989 and 1999 using high spatial resolution images such as IRS, CORONA, and IKONOS. Their results indicated that the built-up area in Greater Cairo increased from 24 to 32% between 1986 and 1999, respectively. Besides, the population density also increased from 7158 to 9074 person/km^2 between 1986 and 1999, respectively. Additionally, Megahed et al. (2015) analyzed the urban growth modeling of Greater Cairo using Landsat images of 1984, 2005, and 2014. In their research, they utilized FRAGSTATS software to calculate spatial metrics for urban classes between 1984 and 2014 besides the Land Change Modeler to model the major transitions of vegetated and desert lands to urban areas and predict the future scenarios up to 2025. Their results indicated that 13 and 12% of the vegetation were lost due to urban areas between 1984–2003 and 2003–2014, respectively. While 3 and 5% of desert areas became urban between the two respective periods. For 2025 map, 14 and 4% from vegetation and desert are estimated to have transferred to urban, respectively.

Shalaby and Gad (2010) studied the urban sprawl impacts on the fertile agricultural in Qalubiya governorate within the three different dates (1993, 2001, and 2009). Their results showed that the built-up lands within the governorate

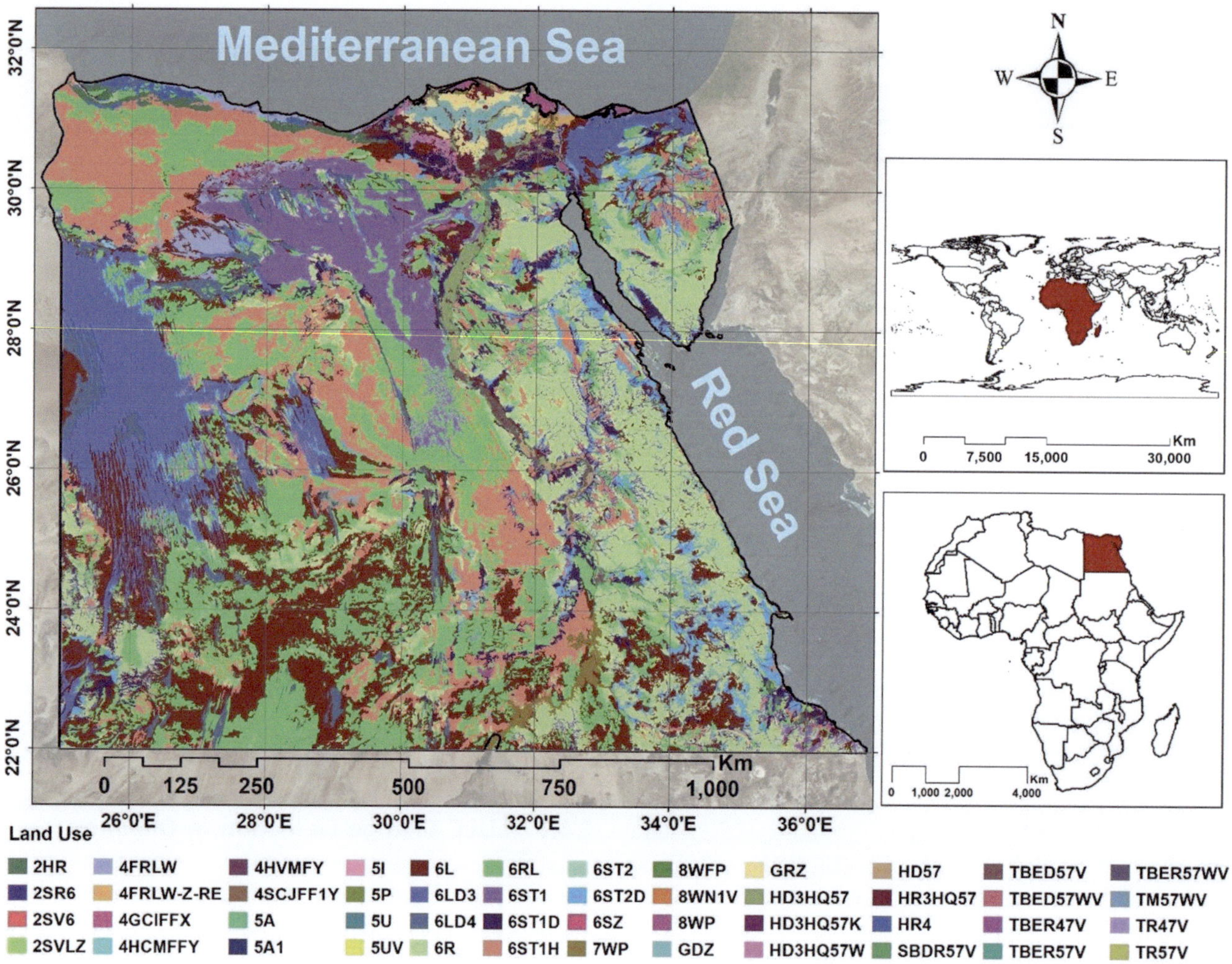

Fig. 4.8 The land use classes (LUC) of Egypt soils. Created by authors based on FAOStat 2017 and http://www.un-spider.org/

increased from 90, 134, and 262 km^2 for the three respective dates, where the urbanization mainly occurs on the most fertile soils of Qalubiya. Under the same concept, Shalaby and Moghanm (2015) evaluate the hazard of urban sprawl and its impact on agriculture land in northern Nile Delta region. They used Landsat images in three different dates (1984, 1992 and 2006). Their results indicated that the urban areas increased from 297, 406, and 986 km^2 in 1984, 1992 and 2006, respectively. Additionally, this vast increasing was over the most fertile soils, where most of the urban expansion was taken place between 1992 and 2006.

Elhag et al. (2013) evaluate two supervised classifiers (maximum likelihood and support vector machine) to detect the change in LULC in the Nile Delta region over time series of 1984, 2000, and 2005. They differentiated five LULC classes; agricultural land, urban area, desert, fish farms, and surface water. Their results indicated that a decrease by around 14% in the desert and increase by 7, 6, and 3%, in agricultural land, urban area, and fish farms, respectively, were noted. Esmail et al. (2016) monitored the changes in land use around Damietta promontory over the period of 1987, 1999, 2003, and 2015. The resulted indicated that the agriculture land decreased from 16 to 13% between 1987 and 2015, respectively, while the urban areas increased from 9.7 to 11.2% for the respective dates. Ezzeldin et al. (2016) studied the change in the northeastern part of the Nile Delta from Damietta Branch to Suez Canal from west and east, respectively, and from El-Manzala Lake to the Cairo-Suez Desert Road from north to south, respectively. Five dates were used; 1984, 1990, 1998, 2006, and 2015. They applied a hybrid classification to distinguish six land cover classes: agricultural, urban, desert, fallow, water, and marshland. Their results revealed that the agricultural and fallow lands classes increased by 9% between 1984 and 2015 whereas the urbanization increased by 5% mainly on agricultural land.

Table 4.2 Egypt's main and detailed LULC classes based on the legend in Fig. 4.8

Main LULC	Detailed LULC	Description	Area (%)
AG	GDZ	Graminoids—large to medium fields—rice	0.512
	GRZ	Graminoids—small fields—rice	1.012
	HD3HQ57	Irrigated herbaceous crop, large to medium fields	0.624
	HD3HQ57K	Irrigated (Sprinkler) herbaceous crop, large to medium fields	0.018
	HD3HQ57W	Irrigated (Drip) herbaceous crop, large to medium fields—pulses & vegetables	0.142
	HD57	Irrigated herbaceous crop, large to medium fields—sugarcane	0.195
	HR3HQ57	Irrigated herbaceous crop, small fields—cotton, wheat, maize, clover	0.900
	HR4	Rainfed herbaceous crop, small fields	0.356
	SBDR57V	Irrigated shrub crop, small fields—grapes	0.000
	TBED57V	Irrigated orchard, large to medium fields	0.329
	TBED57WV	Irrigated tree crop, large to medium fields	0.038
	TBER47V	Rainfed orchard, small fields	0.018
	TBER57V	Irrigated orchard, small fields	0.097
	TBER57WV	Irrigated tree crop, small fields—olive	0.018
	TM57WV	Irrigated orchard, medium fields—apple, fig	0.023
	TR47V	Rainfed tree crop, small fields	0.026
	TR57V	Irrigated tree crop, small fields	0.075
	Total		4.381
BA	6L	Loose and shifting sands	16.713
	6LD3	Longitudinal dunes	7.727
	6LD4	Dunes (undifferentiated)	2.640
	6R	Bare rock	15.935
	6RL	Bare rock with a thin sand layer	20.579
	6ST1	Bare soil stony	4.645
	6ST1D	Bare soil stony (deep soil)	3.237
	6ST1H	Bare soil stony (shallow soil)	15.721
	6ST2	Bare soil very stony	0.025
	6ST2D	Bare soil very stony (deep soil)	3.352
	6SZ	Salt crusts	0.001
	Total		90.574
NVT	2HR	Sparse herbaceous	0.567
	2SR6	Low sparse shrubs with herbaceous	1.834
	2SV6	Very open shrubs with closed to open herbaceous	0.003
	2SVLZ	Very open low shrubs—fragmented	0.184
	Total		2.588
NVW	4FRLW	Sparse low forbs on waterlogged soil—salt crust	0.938
	4FRLW-Z-RE	Sparse low forbs under reclamation on waterlogged soil—salt crust	0.092
	4GCIFFX	Closed high grassland on permanently flooded land—saline water	0.072
	4HCMFFY	Closed medium herbaceous on permanently flooded land—brackish water	0.017
	4HVMFY	Very open medium herbaceous on temporarily flooded land—brackish water	0.046
	4SCJFF1Y	Closed shrubs (broadleaved evergreen) on permanently flooded land—brackish water	0.004
	Total		1.168

(continued)

Table 4.2 (continued)

Main LULC	Detailed LULC	Description	Area (%)
UR	5A	Airport	0.007
	5A1	Archeological area	0.000
	5I	Industrial area—general	0.002
	5P	Port	0.001
	5U	Urban Areas (general)	0.227
	5UV	Vegetated Urban Areas	0.001
	Total		0.238
WAT	7WP	Artificial Lakes or Reservoirs	0.607
	8WFP	River	0.069
	8WN1V	Inland water non-perennial with scattered vegetation	0.013
	8WP	Natural Lakes	0.362
	Total		1.051

AG Agriculture, *BA* barren, *NVT* natural terrestrial vegetation, *NVW* natural aquatic vegetation, *UR* urban and *WAT* water classes

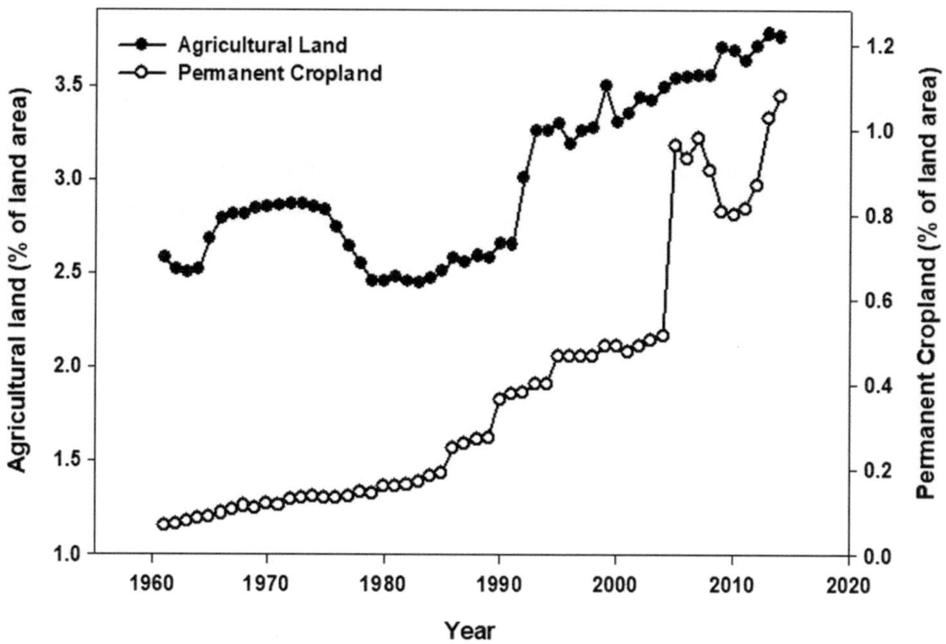

Fig. 4.9 The percentage of agricultural land from the total land area of Egypt and the permanent cropland area compared to the total area. Created based on FAOStat and World Bank 2017 database

For the Nile Valley region, Esam et al. (2012) studied the change in agriculture and urban areas, and River Nile over the period from 1987 to 2009 in west Tahta Region, Sohag Governorate, Upper Egypt. Their result indicated that the average annual change in urban area was around 101 ha/year whereas the average annual change for agriculture area was around 81 ha/year. Additionally, Mahmoud and Divigalpitiya (2017) studied the urban growth and its effect on the old cultivated land in Asyut region besides expecting the future scenario till 2030. They use three satellite images to study LULC changes of the study area from 1990, 2003, and 2015. They utilized a framework of the Markov-cellular automata and a multi-criteria evaluation to predict the future scenario. Their results showed that the net growth rate of built-up areas

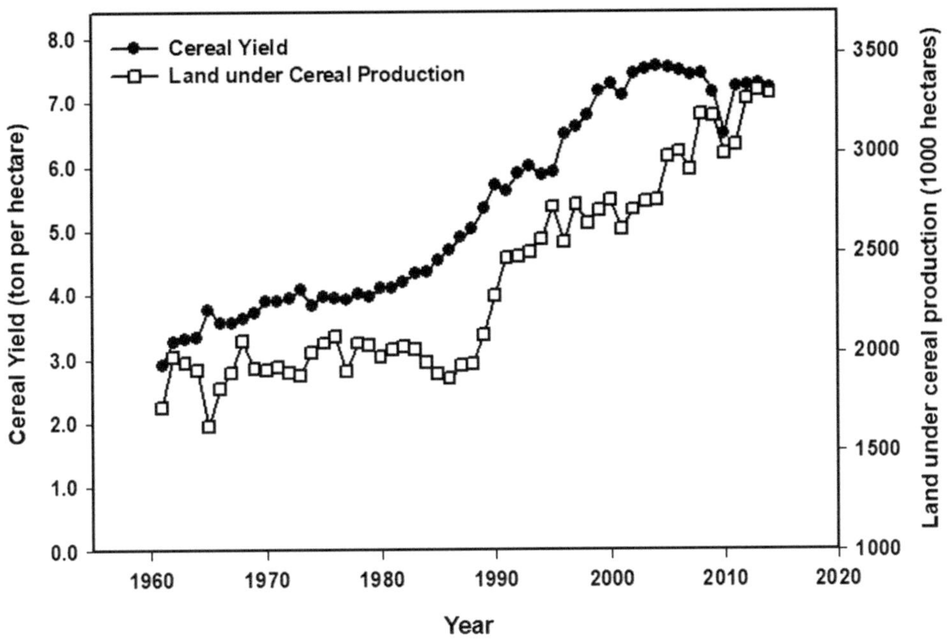

Fig. 4.10 The cereal yield and the land under cereal cultivation in Egypt. Created based on FAOStat 2017 database

is expected to be around 42% with around 18% loss in agricultural lands due to urban sprawl. Faid and Abdulaziz (2012) used multitemporal Landsat images of the Kom Ombo area to map LULC and monitor their changes during 1988, 1999, and 2008. They also applied post-classification change detection analysis and the results revealed that the agricultural development increased by 39% at the same time the built-up land was also increased.

Kamh et al. (2012) integrated the GIS and socioeconomic to detect the urban expansion directions and land use conversions that have occurred in the Hurghada area within time periods of 1987, 2000, and 2005. The results indicated that the annual urban growth rate increased was around 20, 28, and 22% for the assigned periods of 1987–2000, 2000–2005, and 1987–2005, respectively, as a result of urban expansion over that area.

4.4.2 LULC Change Detection in *Newlands*

Western Nile Delta exhibits the attention of several scientists because it is considered as a promising area for reclamation by the governmental authority as it has wider extent compared with the eastern Nile delta region. Sadek (1993) mapped the horizontal expansion of agricultural lands around delta fringes from east and west using satellite imagery to monitor the reclamation process in that region. Pax Lenney et al. (1996) used multi-temporal normalized difference vegetation index (NVDI) that was derived from ten Landsat TM images for a time period from 1984 to 1993 to assess the status of agricultural lands in western Nile Delta and coastal regions, in Egypt. Shalaby and Tateishi (2007) used maximum likelihood supervised classification and post-classification comparison change detection techniques to map land cover changes on the northwestern coast of Egypt using Landsat images for the time series of 1987 and 2001, respectively. Bakr et al. (2010) monitored the changes in land cover in the Bustan 3 area, as a newly reclaimed area in the western Nile Delta. They utilized five multi-temporal Landsat images from 1984 to 2008. They applied two techniques to classify the satellite images that cover the study area; hybrid classification approach and NVDI. The results revealed that from 1984 to 2008, the Bustan 3 area transformed from 100% barren land to 79% agricultural land which reflects the successful land reclamation efforts. Abd El-Kawy et al. (2011) applied the supervised classification on four Landsat images collected on (1984, 1999, 2005, and 2009) for the new reclaimed areas in the western Nile Delta. The results showed that approximately 28, 14, and 9% of the barren land was changed to agricultural land in the periods between 1984–1999, 1999–2005, and 2005–2009, respectively.

Zaki et al. (2011) monitored the LULC changes in the new urbanized zone in Northeast Cairo using Landsat images for a period from 1990 to 2003. Their final results indicated that the urban areas increased from 4 to 11% between 1990 and 2003, respectively, also a respective

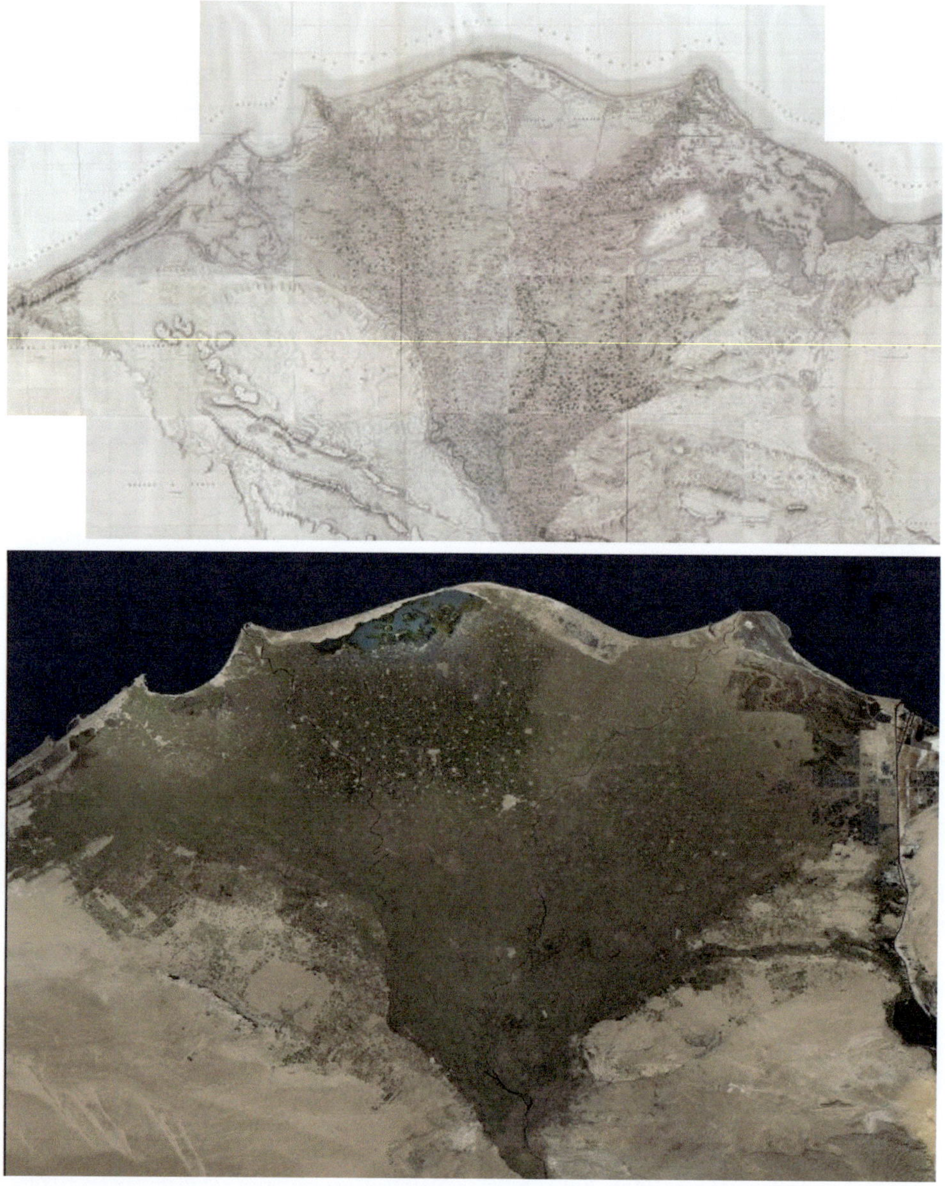

Fig. 4.11 The Nile Delta, Egypt. Up, the Nile Delta in 1818 (updated from: http://rumsey.geogarage.com/maps/mosaicegypt29_48.html). Down, the Nile Delta in 2017. Created by authors

increase in agriculture land from 8 to 13% has taken place in this area due to the successful reclamation efforts of eastern desert lands. Abdulaziz et al. (2009) used a hybrid classification approach on Landsat images from 1984, 1990, and 2003 to monitor land cover change in eastern Nile Delta. Their results showed that the agricultural land increased by 14% with also increase in the urbanization due to urban sprawl into old cultivated land at the fringes of urban centers in the Nile Delta region.

Badreldin and Goossens (2014) monitored the multi-temporal LULC maps for years 1999, 2001, 2005, and 2010 in El-Arish City, North Sinai governorate. The vegetated lands increased from 1 to 8% between 1999 and 2010, respectively. Remarkable increases were detected for the cultivated land from 5 to 24% between 1999 and 2001, respectively, then decreased from 25 to 18 between 2001 and 2005, respectively. Whereas, the bare soil decreased by more than 20% between 1999 and 2005.

Fig. 4.12 The Nile Valley, Egypt. Left, the Nile Valley in 1818 (updated from: http://rumsey.geogarage.com/maps/2mosaicegypt17_28.html, http://rumsey.geogarage.com/maps/mosaicegypt7_16.html). Down, the Nile Valley in 2017. Created by authors

4.5 Conclusion

The earth surface is highly dynamic and affected by the soil characteristics as well as the environmental circumstance of every geographical region on the earth. The study of LULC at the nationwide is a key for proper and more sustainable land use planning based on the information about the previous and current LULC of the proposed areas. The outstanding ability of new technologies such as remote sensing and GIS provides more efficient and accurate information for the current and past LULC for a specific region, besides, predicts the expected change in the future. Egypt faces several challenges: among them the progressive increase in population and the water scarcity represent the foremost. Five main LULC classes can be recognized in Egypt land: agricultural lands (representing the *Oldland* in Nile Valley and Delta and *Newlands* for the new reclaimed areas), bare soils (the dominant LULC class), urban areas (built-up lands), natural vegetation (terrestrial and aquatic), and water bodies (mainly Nile River and Egypt's lakes). With the importance of agriculture as one of the main national income and socioeconomic activity in Egypt, the sustainable development of current cultivated land and the new reclaimed desert lands has become a strategic priority for the government. Thus, it is highly required to study the current LULC, their changes over time, and estimating changes in the future, for more sustainable land use planning.

References

Abd El-Kawy OR, Rød JK, Ismail HA, Suliman AS (2011) Land use and land cover change detection in the western Nile delta of Egypt using remote sensing data. Appl Geogr 31:483–494

Abdulaziz MA, Hurtado JM, Al-Douri R (2009) Application of multitemporal Landsat data to monitor land cover changes in the Eastern Nile Delta region, Egypt. Int J Remote Sens 30(11):2977–2996

Abu Zeid M (1993) Egypt's water resources management and policies. In: Faris M, Khan M (eds) Sustainable agriculture of Egypt. Lynne Rienner Publishers Inc, Boulders, Colorado, pp 71–79

Badreldin N, Goossens R (2014) Monitoring land use/land cover change using multi-temporal Landsat satellite images in an arid environment: a case study of El-Arish, Egypt. Arab J Geosci 7:1671–1681. https://doi.org/10.1007/s12517-013-0916-3

Bakr N, Bahnassy MH, El-Badawi MM, Ageeb GW, Weindorf DC (2009) Land capability evaluation in newly reclaimed areas: a case study in Bustan 3 area, Egypt. Soil Surv Horiz 51(3):90–95

Bakr N, Weindorf DC, Bahnassy MH, Marei SM, El-Badawi MM (2010) Monitoring land cover changes in a newly reclaimed area of Egypt using multitemporal Landsat data. Appl Geogr 30(4):592–605

Bruzzone L, Prieto DF (2002) An adaptive semiparametric and context-based approach to unsupervised change detection in multitemporal remote-sensing images. IEEE Trans Image Process 11:452–466

CAPMAS (2017) The population of Egypt in 2017. Arabic reference. http://www.capmas.gov.eg/party/party.html

Currit N (2005) Development of a remotely sensed, historical land-cover change database for rural Chihuahua, Mexico. Int J Appl Earth Obs Geoinf 7:232–247

Di Gregorio A, Jansen LJM (2000) Land cover classification system (LCCS): classification concepts and user manual. For Software Version 1.0. FAO, Rome. ISBN 92-5-104216-0

Elhag M, Psilovikos A, Sakellariou-Makrantonaki M (2013) Land use changes and its impacts on water resources in Nile Delta region using remote sensing techniques. Environ Dev Sustain 15:1189–1204

Esam I, Abdalla F, Erich N (2012) Land use and land cover changes of West Tahta Region, Sohag Governorate, Upper Egypt. J Geogr Inf Syst 4:483–493. https://doi.org/10.4236/jgis.2012.46053

Esmail M, Masria A, Negm A (2016) Monitoring land use/land cover changes around Damietta promontory, Egypt, using RS/GIS. Procedia Eng 154:936–942

Ezzeldin MM, El-Alfy KS, Abdel-Gawad HA, Abd-Elmaboud ME (2016) Land use changes in the Eastern Nile Delta Region; Egypt using multi-temporal remote sensing techniques. Int J Sci Eng Res 7 (12):78–98

Faid AM, Abdulaziz AM (2012) Monitoring land-use change-associated land development using multitemporal Landsat data and geoinformatics in Kom Ombo area, South Egypt. Int J Remote Sens 33 (22):7024–7046. https://doi.org/10.1080/01431161.2012.697207

FAO (2015) Egypt: wheat sector review. FAO Investment Center, Report N. 21, FAO/EBRD Cooperation

FAOStat (2017) Egypt: Country indicators. http://www.fao.org/faostat/en/#country/59

Hereher ME (2013) The status of Egypt's agricultural lands using MODIS Aqua data. Egypt J Remote Sens Space Sci 16:83–89

Hua AK (2017) Land use land cover changes in detection of water quality: a study based on remote sensing and multivariate statistics. J Environ Public Health 7515130:12. https://doi.org/10.1155/2017/7515130

Hussain M, Chen D, Cheng A, Wei H, Stanley D (2013) Change detection from remotely sensed images: From pixel-based to object-based approaches. ISPRS J Photogrammetry Remote Sens 80:91–106

Kamh S, Ashmawy M, Kilias A, Christaras B (2012) Evaluating urban land cover change in the Hurghada area Egypt, by using GIS and remote sensing. Int J Remote Sens 33(1):41–68. https://doi.org/10.1080/01431161.2010.550331

Lu D, Mausel P, Brondizio E, Moran E (2004) Change detection techniques. Int J Remote Sens 25:2365–2407

Lu DS, Mausel P, Batistella M, Moran E (2005) Land-cover binary change detection methods for use in the moist tropical region of the Amazon: a comparative study. Int J Remote Sens 26:101–114

Mahmoud H, Divigalpitiya P (2015) Modeling future land use and land-cover change in the Asyut Region using Markov Chains and cellular automata. In: Bisello A, Vettorato D, Stephens R, Elisei P (eds) Smart and sustainable planning for cities and regions. SSPCR 2015. Green Energy and Technology. Springer, Cham

MALR (2009) Egypt sustainable agricultural development strategy towards 2030. Egyptian Ministry of Agriculture and Land Reclamation, Cairo

McGill J, Prikhodko D, Sterk B, Talks P (2015) Egypt wheat sector review. FAO/EBRD Cooperation, FAO, Rome

Megahed Y, Cabral P, Silva J, Caetano M (2015) Land cover mapping analysis and urban growth modelling using remote sensing techniques in greater Cairo Region-Egypt. ISPRS Int J Geo-Inf 4:1750–1769. https://doi.org/10.3390/ijgi4031750

Negm AM, Saavedra O, El-Adawy A (2017) Nile Delta biography: challenges and opportunities. In: Negm AM (ed) The Nile Delta, Barceló D, Kostianoy AG (Series eds) The handbook of environmental chemistry 55, pp 3–18

Noaman MN (2017) Country profile. In: Satoh M, Aboulroos S (eds) Irrigated agriculture in Egypt: past, present and future. Springer, Switzerland

Pax Lenney M, Woodcock CE, Collins JB, Hamdi H (1996) The status of agricultural lands in Egypt: the use of multi temporal NDVI features derived from Landsat TM. Remote Sens Environ 56:8–20

Prakasam C (2010) Land use and land cover change detection through remote sensing approach: a case study of Kodaikanal taluk, Tamil nadu. Int J Geomatics Geosci 1(2):150–158

Rajeswari AV, Saritha S, Kumar GS (2014) Classification based land use/land cover change detection through Landsat images. In: 2014 International conference on data science & engineering (ICDSE), 26–28 Aug 2014, Kochi, India. https://doi.org/10.1109/icdse.2014.6974644

Sadek SHA (1993) Use of Landsat imagery for monitoring agricultural expansion of East and West Nile Delta, Egypt. Egypt J Soil Sci 33:23–34

Sader SA (1987) Digital image classification approach for estimating forest clearing and regrowth rates and trends. In: International Geoscience and Remote Sensing Symposium, Ann Arbor, Michigan, 18–21 May, pp 209–213

Schilling KE, Jha MK, Zhang Y, Gassman PW, Wolter CF (2008) Impact of land use and land cover change on the water balance of a large agricultural watershed: historical effects and future directions. Water Resour Res 44(7):1–12

Schowengerdt RA (2007) Remote sensing: models and methods for image processing, 3rd edn. Elsevier Inc., USA, p 515p

Shalaby A, Gad A (2010) Urban Sprawl impact assessment on the fertile agricultural land of Egypt using remote sensing and digital soil database, Case study: Qalubiya Governorate. In: US-Egypt workshop on space technology and geo-information for sustainable development, Cairo, Egypt 14–17 June

Shalaby A, Tateishi R (2007) Remote sensing and GIS for mapping and monitoring land cover and land-use changes in the Northwestern coastal zone of Egypt. Appl Geogr 27(1):28–41

Shalaby A, Moghanm FS (2015) Assessment of urban Sprawl on agricultural soil of Northern Nile Delta of Egypt using RS and GIS. Chin Geogra Sci 25(3):274–282. https://doi.org/10.1007/s11769-015-0748-z

Singh A (1989) Digital change detection techniques using remotely-sensed data. Int J Remote Sens 10:989–1003

Williams DL, Goward S, Arvidson T (2006) Landsat: yesterday, today, and tomorrow. Photogram Eng Remote Sens 72:1171–1178

Yin ZT, Stewart DJ, Bullard S, MacLachlan JT (2005) Changes in urban built-up surface and population distribution patterns during 1986–1999: a case study of Cairo, Egypt. Comput Environ Urban Syst 29:595–616

Zahran MA, Willis AJ (2009) The vegetation of Egypt. In: Werger MJA (ed) Plant and vegetation, vol 2, 2nd edn. Springer Science+Business Media, Switzerland, p 451. ISBN 978-1-40208755-4

Zaki R, Zaki A, Ahmed S (2011) Land use and land cover changes in Arid Region: the case New Urbanized Zone, Northeast Cairo, Egypt. J Geogr Inf Syst 3:173–194. https://doi.org/10.4236/jgis.2011.33015

Climate

5

Samia El-Marsafawy, Noura Bakr, Tamer Elbana, and Hassan El-Ramady

Abstract

There is no doubt that climate change nowadays is considered one of the hottest spots worldwide. These changes in climate penetrated all fields in our life including agriculture, soil, water, air, etc. and climate has a deciding role in all agricultural systems. The distribution of crops in different geographical regions and the selection of cropping patterns mainly depend on climate. All plant growth stages including seed germination and its growth, flowering, fruiting, and ripening are also mainly controlled by different climatic elements (e.g., moisture, temperature and day length). Therefore, the cultivation and management of cultivated crops require a proper weather and suitable climatic conditions. Due to the mismanagement of soil and burning fossil fuels, emissions of greenhouse gases have alarmingly increased. In Egypt, climate changes have three serious dimensions representing in the threat of sinking of the Nile Delta, water crisis and the decrease in crop production. Therefore, this chapter will focus on climate and its parameter variations in Egypt as well as expected effects from these changes mainly on water crisis and the production of crops.

Keywords

Egypt • Climate • Water crisis • Greenhouse gases Emissions • Mitigation

S. El-Marsafawy
Soils, Water and Environment Research Institute, Agricultural Research Center, Giza, Egypt

N. Bakr · T. Elbana
Biological and Agricultural Division, Soils and Water Use Department, National Research Centre, Dokki, Giza, Egypt

H. El-Ramady (✉)
Soil and Water Department, Faculty of Agriculture, Kafrelsheikh University, Kafrelsheikh, Egypt
e-mail: ramady2000@gmail.com

H. El-Ramady et al. (eds.), *The Soils of Egypt*, World Soils Book Series, https://doi.org/10.1007/978-3-319-95516-2_5

5.1 Introduction

Climate is the group of changes in climatic elements including temperature, humidity, atmospheric pressure, wind, precipitations, etc. for a long time (Chen et al. 2017), whereas the change in climate mainly focuses on the change in temperature, the greenhouse gases (CO_2, CH_4, and N_2O in the atmosphere), sea level rise, etc. (Omambia et al. 2017). These changes in climate penetrate all sides in our life including the growth and production of crops (food and feed), the health of humans, the management of water and land resources, the activities of human societies, the energy and its resources, etc. (Wesseh and Lin 2017). Concerning the concentration of CO_2, CH_4, and N_2O in the atmosphere, it is reported that these concentrations have been increased by 37, 156, and 19% in the atmosphere, respectively, over 1750 values (Osman 2013). One global debate more is the rising in global temperature; it has been raised from 0.75 °C to by 1.8–4 °C in 1850 to the twenty-first century, respectively (Osman 2013). These changes also include the rise in sea level in several places worldwide and the melting of arctic ice, the extreme events such as floods and droughts, the global warming, etc. (Figure 5.1; Hope et al. 2017). Therefore, climate changes have direct and indirect effects on different agroecosystems including the soil, water, air, and plants (Tataw et al. 2016; Masud et al. 2017).

Concerning the international climate negotiations, globally, the UN Framework Convention on Climate Change (UNFCCC) established four phases till the Paris Agreement in 2015. The first phase started from the adoption of the UNFCCC in 1992 up to the adoption of the Berlin Mandate in 1995. The second phase describes the negotiations that led to the adoption of the Kyoto Protocol and the creation of binding emission reduction targets for developed countries. The third phase includes the implementation of the Kyoto Protocol. This protocol had hardly begun in 2005 when new efforts and partnerships emerged with the goal to bring all major emitters to the table. The fourth phase or Paris

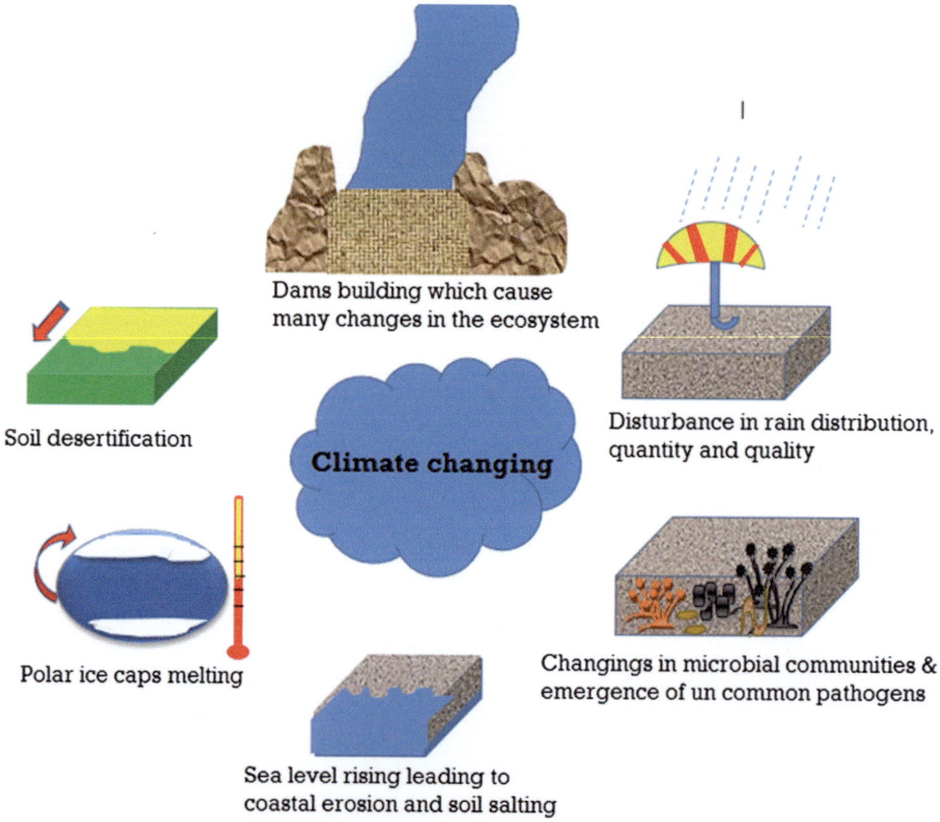

Fig. 5.1 Change in climate is a crucial issue. This change leads to a lot of features such as melting of ice in the pole, raising the level of sea, change in the activities of soil microbial communities, and disturbance in rain distribution, and others. Drawn by Dr. Tamer Elsakhawy, SWERI, Sakha, Kafr El-Sheikh

Agreement was characterized by the UNFCCC as a more horizontal, inclusive and balanced approach. This Agreement has a common framework, in which the individual countries are requested to define their action plans or their contributions taking into account the overall goal of this Agreement as well as their own capacities (Streck and Gay 2017).

Numerous investigations have been published regarding particular climatic drivers related to soil processes (e.g., Coyle et al. 2017; Durán et al. 2017). Effects of climate change on soil chemical, biological and physical health indicators were also reviewed by several authors (e.g., Allen et al. 2011; Gao et al. 2017). Soil organic carbon, greenhouse gas emissions and erosion were also well studied as short-term responses to climate changes (e.g., Edenhofer et al. 2014; Frank et al. 2015; Yigini and Panagos 2016; Bojko and Kabala 2017; Muñoz-Rojas et al. 2017; Olaya-Abril et al. 2017; Soleimani et al. 2017). The direct influence of precipitation and air temperature on soil temperature regimes and its hydrology also has been determined in the agricultural context (Trnka et al. 2013; Mahmoud 2017). There is a lack in the comprehensive understanding of how crop–soil systems respond to changes in climate and land use, although progress has been made in understanding the impact of human activities on soil functions (Yigini and Panagos 2016; Bojko and Kabala 2017; Mäkinen et al. 2017; Soleimani et al. 2017).

Soil–climate combination is a key role in projections of agricultural land use, and climate-induced change in land use is essential for climate change adaptation (Bojko and Kabala 2017; Mäkinen et al. 2017; Lozano-García et al. 2017; Soleimani et al. 2017). Soil types mainly vary in their agronomic characteristics such as soil clay content, water-holding capacity, thermal conductivity, soil organic content, etc. Soil geography and its magnitude of crop responses to climate are highly sensitive to this soil type (Rötter et al. 2013; Lisson et al. 2016; Lozano-García et al. 2017). There are few systematic empirical studies on crop yield responses to agro-climatic variables (e.g., Kahiluoto et al. 2014; Mäkinen et al. 2015, 2016), and nearly there is practically no quantitative empirical information regarding the dependence on soil type of the yield response to agro-climatic variables (Mäkinen et al. 2017). Therefore, the soil–climate combination is very critical in designing and assessing adaptation of cropping including breeding goals and cultivation techniques (Mäkinen et al. 2017).

It could be divided Egypt into four regions according to geological zones including climate of the Nile Delta, the Western desert, the Eastern desert, and Sinai (Elbasiouny et al. 2017). The effects of climate changes in Egypt are represented by three serious problems including (1) the rise in sea level, which may threaten the coastal and Nile Delta region, (2) the rise in temperature, which would enforce

Egypt to change her agricultural strategy, and (3) the water crisis (El Quosy 2013; McCarl et al. 2015; Ahmed 2017; Mahmoud 2017). Therefore, a great threat might be expected facing Egypt from the climate changes towards the agricultural sector, food security, and water scarcity as well as its crisis. Thus, an urgent plan should be scheduled by Egypt to avoid different troubles might expect resulting from these changes in climate of Egypt (Froehlich and Al-Saidi 2017).

Regarding the climate changes in Egypt, an increasing concern from all specialists has been recorded for handling this serious problem and its mitigation. Several publications have also been issued regarding the changing climate of Egypt including the sustainability of Egyptian agriculture under this change (e.g., El-Ramady et al. 2013; Froehlich and Al-Saidi 2017; Mahmoud 2017), impact of changing climate on crop production (e.g., El Massah and Omran 2015; El Afandi 2017), mitigation the changing of climate in Egypt (e.g., Hinkel et al. 2012; El Quosy 2013; Ahmed 2017), impacts of changing climate on soil organic carbon in Egypt (e.g., Muñoz-Rojas et al. 2017), climate changes and water resources in Egypt (e.g., Elshinnawy 2008; El Quosy 2013; Younos and Grady 2013; Mahmoud 2017) and using of remote sensing in following and mitigation of the changing in climate in different zones of Egypt (e.g., Hassaan 2013; Elbasiouny et al. 2017; Hassan and Omran 2017).

Therefore, this chapter will highlight the climate of Egypt including her different geological or climatic zones and the expected problems would result from these changes in climate. The water crisis in Egypt and its dimensions related to climate changes also will be focused on it mainly.

5.2 Climate of Egypt and Its Parameter Variations

Climate is one of the main soil formation factors, which is considerably affecting the development of soil profile. The climate of Egypt is generally characterized by the arid and semiarid condition that has low precipitation during winter season. For quantifying the main Egyptian climatic parameters, 34 years (1/1/1983–31/12/2016) of meteorological data was collected from NASA **Prediction of Worldwide Energy Resource (POWER)—Climatology Resource for Agroclimatology** website (http://power.larc.nasa.gov/). The data was tabulated and analyzed for characterizing the recently Egyptian climate. Specifically, the data was gathered for 41 different locations that cover all Egyptian territory (Fig. 5.2). The collected meteorological data includes mean, minimum, and maximum air temperature (°C) as well as relative humidity (%) at 2 m above the surface of the earth. Additionally, data of the wind speed (m/s) at 10 m above the surface of the earth was collected too. The variation of averaged air temperature, relative humidity and wind speed of the collected data for January (representing winter months) and August (representing summer months) over 34 years (1983–2016) is presented in Table 5.1. The meteorological data reveals that air temperature varied between 6.3 and 24.6 °C with an average mean temperature of 13.9 °C during January. The respective minimum and maximum temperatures for August are 20.7 and 39.9 °C with an average mean temperature of 29.7 °C. The relative humidity varied between 35.7 and 54.3% during January months whereas it fluctuated between 19.7 and 67.8% in August. The data of wind speed for January and August reveals that the maximum wind speed is generally low with value less than 5.5 m/s (Table 5.1).

Concerning the climate of Egypt, Egypt is a nation which has a very famous civilization that flourished more than 5000 years ago due to her distinguished position among three continents including Africa, Asia, and Europe. Egypt has a special weather due to its location, which calls Mediterranean for the Egyptian coastal regions in winter and arid for the deserts, which have higher temperature more than 40 °C sometimes in summer. The main Egyptian cities located in the deserts (i.e., Luxor, Aswan, Asyut, Siwa or Sohag) are hotter in their temperature in summer; whereas the higher places in elevations (e.g., some mountainous located in Sinai like Saint Catherine) are cooler in their temperatures. It could explain the climate of Egypt through the geological zones of Egypt including the Nile valley and delta (about 4% of the total area), the Eastern desert (22%), the Western desert (68%), and the Sinai Peninsula (6%).

5.2.1 Changes in Precipitation

Precipitation (mainly rain or rainfall) in Egypt is generally low and only exists over the narrow north coastal area of Egypt. Whereas, the rest of Egypt territory (more than 95%) is mainly desert. The rainfall varies from 0 mm/year in the desert to around 200 mm/year at the northern coastal area (Abdel-Shafy et al. 2010). Flash flooding and incident heavy rain are observed in some areas in Egypt. For the Mediterranean basin, during 1973–2010, the majority of extreme events with rainfall more than 200 mm/day happened in October (Mariani and Parisi 2014). There are extremes of rainfall events in Egypt revealed that the maximum of the total annual precipitation of 401 mm was recorded in Marsa Matruh (the northern coastal governorate) during 1994 (Gado 2017). Recently, during October 2016 on the Gulf of Suez at Ras Gharib city, a heavy rainfall event with strong storms caused a flash flood disaster at the oil production area in this city (Elnazer et al. 2017). On the other hand, rainfed agriculture is a common practice in the northern coastal area especially in Marsa Matruh and Northern Sinai where the

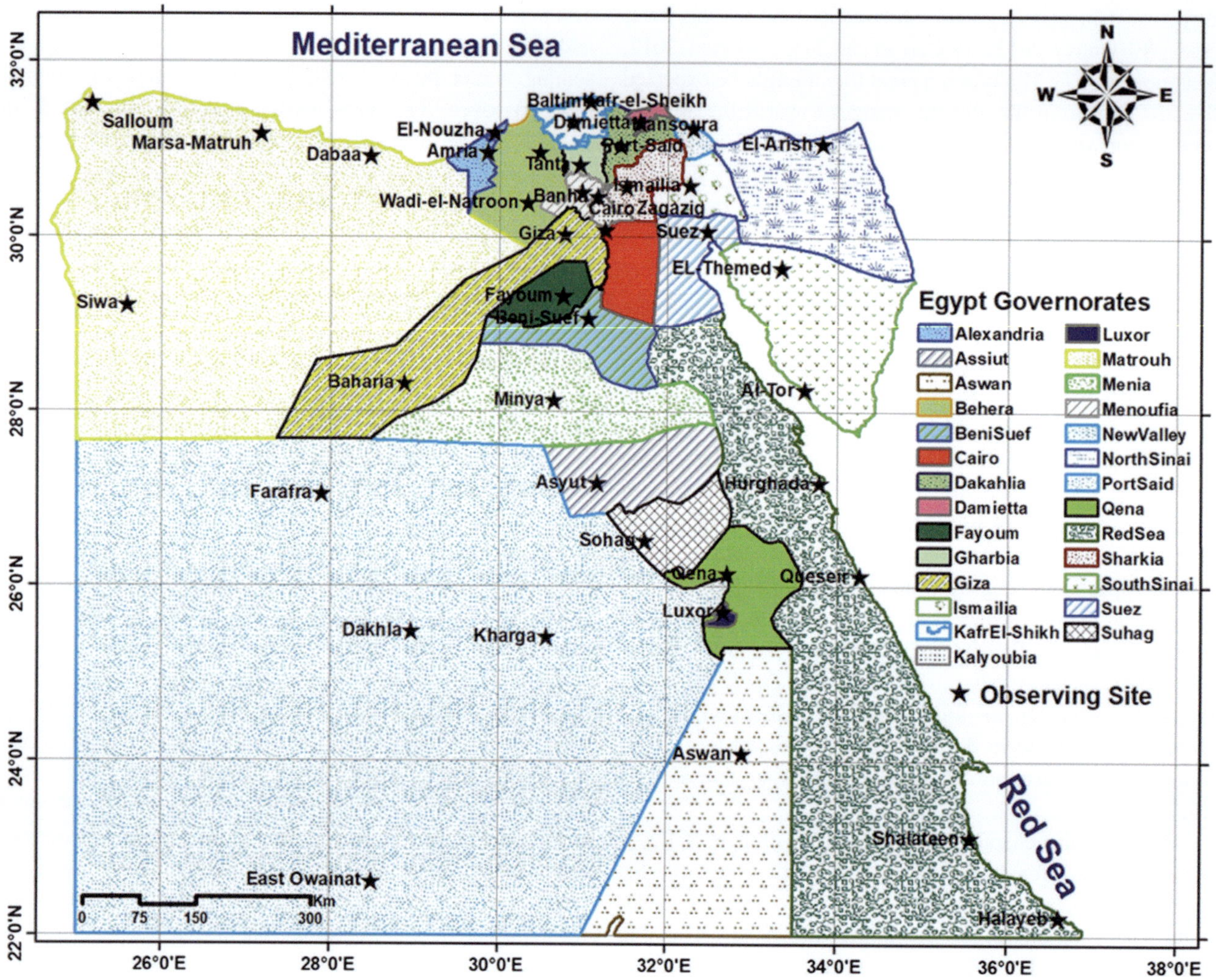

Fig. 5.2 Different locations of the main meteorological stations for collection climatic parameters data over the Egypt governorates. Created by Dr. Bakr, NRC

farmers planted seeds after the occasion of the heavy rainfall in such area. Based on the availability of the rainfall data (1/1/1997–31/12/2014) for the 41 locations obtained from the NASA POWER website, Fig. 5.2 shows that the annual average precipitation ranged from 1.5 mm to 168 mm for Dakhla oasis in the south and Marsa Matruh in the north, respectively. However, the average monthly rainfall on Egypt varied in January between 0.4 and 43.6 mm with the mean of 17.9 mm. In order to show the spatial distribution of the climatic data for the 41 locations, data were interpolated to produce a map with a range values. There are different interpolation techniques that can be used for meteorological data, however, in this study, the ordinary Kriging geo-statistics method had been used as it is considered the best and more accurate interpolation techniques for meteorological data set (Sluiter 2009). ArcMap 10.4 software was used to produce the required maps. Figure 3.3 shows the interpolated map of the average total annual rain over Egypt during 1997–2014. It is stated that the distribution of rain over Egypt was relatively high on the coastal areas and rapidly decreases in land due to Mediterranean depressions and upper troughs effects.

The climatic data shows that more than 60% of the rainfall occasioned during December, January, and February on Egypt. The spatial variability of the monthly average rainfall during December, January, February, and March over Egypt is territory (Fig. 5.3). The highest rainfall is spatially distributed in the coastal area during December and January with rain of 20–44 mm. The decrease in rainfall during February to be less than 20 mm is observed and the precipitation is dropped to less than 10 mm during March. Generally, the rain events are observed only in lower Egypt (above latitude 28 °N northward) this is the area that is characterized by the Mediterranean climate in Egypt. Whereas, the upper Egypt (below latitude 28 °N southward) is completely dry area with almost no rain.

Table 5.1 Averages of 34 years (1983–2016) temperature (°C), relative humidity (%), and wind speed (m/s) during January and August

Location	January					August				
	Temperature (°C)			Relative humidity (%)	Wind speed (m/s)	Temperature (°C)			Relative humidity (%)	Wind speed (m/s)
	Mean	Min.	Max.			Mean	Min.	Max.		
Amria	13.1	8.3	19.4	58	3.7	28.9	22.6	36.4	47	3.9
Aswan	15.1	9.0	22.2	37	3.7	33.0	26.0	39.9	19	4.1
Asyut	12.9	7.1	19.9	45	3.3	30.9	24.0	37.5	27	4.3
Bahariya Oasis	12.6	7.3	19.4	50	3.4	30.4	23.8	37.1	32	4.0
Baltim	14.8	11.7	18.7	65	4.6	28.1	24.1	33.1	60	4.2
Banha	13.0	7.8	19.8	57	3.3	29.3	22.5	37.1	43	3.4
Beni-Suef	12.5	7.1	19.4	53	3.1	29.6	22.6	36.7	36	3.8
Cairo	13.0	7.8	19.8	57	3.3	29.3	22.5	37.1	43	3.4
Dabaa	13.4	8.8	19.4	57	3.8	29.4	23.2	36.6	46	4.2
Dakhla Oasis	12.7	6.7	19.9	42	3.5	31.1	24.0	37.9	23	4.2
Damanhour	13.0	8.0	19.7	58	3.5	29.1	22.4	36.8	45	3.7
Damietta	14.8	11.7	18.7	65	4.6	28.1	24.1	33.1	60	4.2
East Owainat	13.8	7.6	21.1	35	3.8	32.5	25.3	39.5	19	4.2
El-Arish	14.6	11.9	17.7	65	4.4	26.7	23.9	30.4	67	3.9
El-Nouzha	15.3	13.1	18.0	66	5.3	27.2	24.7	30.4	67	4.7
El-Themed	11.2	6.3	17.3	55	3.2	27.5	20.7	34.7	39	4.2
El-Tor	12.9	7.8	19.0	49	3.5	29.4	22.7	36.1	33	4.9
Farafra Oasis	12.7	7.2	19.6	47	3.3	31.0	24.6	37.4	28	3.9
Fayoum	12.9	7.7	19.7	53	3.3	30.0	23.2	37.1	37	4.0
Giza	13.0	8.0	19.7	58	3.5	29.1	22.4	36.8	45	3.7
Halayeb	21.3	18.7	24.6	59	5.1	32.9	29.4	37.4	46	5.3
Hurghada	13.8	8.3	20.7	44	3.7	31.2	24.0	38.4	27	4.9
Ismailia	12.4	7.2	19.3	57	3.3	28.2	21.4	36.0	47	3.5
Kafr El-Sheikh	14.9	12.1	18.5	66	4.8	27.7	24.2	32.2	63	4.4
Kharga Oasis	13.5	7.6	20.6	40	3.6	31.8	24.8	38.5	22	4.4
Luxor	14.3	8.2	21.6	39	3.6	32.4	25.1	39.5	21	4.2
Mansoura	14.8	11.7	18.7	65	4.6	28.1	24.1	33.1	60	4.2
Marsa Matruh	14.8	12.4	17.8	64	5.4	27.5	24.4	31.4	63	5.1
Minya	13.0	7.4	20.1	49	3.2	30.8	24.0	37.6	31	4.1
Port Said	14.8	12.0	18.3	65	4.7	27.4	24.1	31.7	64	4.0
Qena	13.8	7.7	21.1	41	3.4	31.8	24.6	39.0	23	4.3
Qucseir	18.7	16.0	21.9	51	4.7	32.1	28.5	35.8	43	4.9
Salloum	13.9	10.8	17.7	62	5.3	27.6	23.6	32.4	58	5.1
Shalateen	19.3	15.2	24.2	52	4.8	33.4	28.1	39.1	34	4.8
Shebin El-Kom	13.0	7.8	19.8	57	3.3	29.3	22.5	37.1	43	3.4
Siwa Oasis	12.5	7.4	19.1	53	3.5	30.4	23.6	37.2	34	3.9
Sohag	13.1	7.3	20.1	43	3.4	31.3	24.4	38.0	24	4.3

(continued)

Table 5.1 (continued)

Location	January					August				
	Temperature (°C)			Relative humidity (%)	Wind speed (m/s)	Temperature (°C)			Relative humidity (%)	Wind speed (m/s)
	Mean	Min.	Max.			Mean	Min.	Max.		
Suez	12.4	7.2	19.3	57	3.3	28.2	21.4	36.0	47	3.5
Tanta	13.0	8.0	19.7	58	3.5	29.1	22.4	36.8	45	3.7
Wadi El-Natrun	13.0	8.0	19.7	58	3.5	29.1	22.4	36.8	45	3.7
Zagazig	13.0	7.8	19.8	57	3.3	29.3	22.5	37.1	43	3.4

Source Data collected and analyzed from the following web site: http://power.larc.nasa.gov

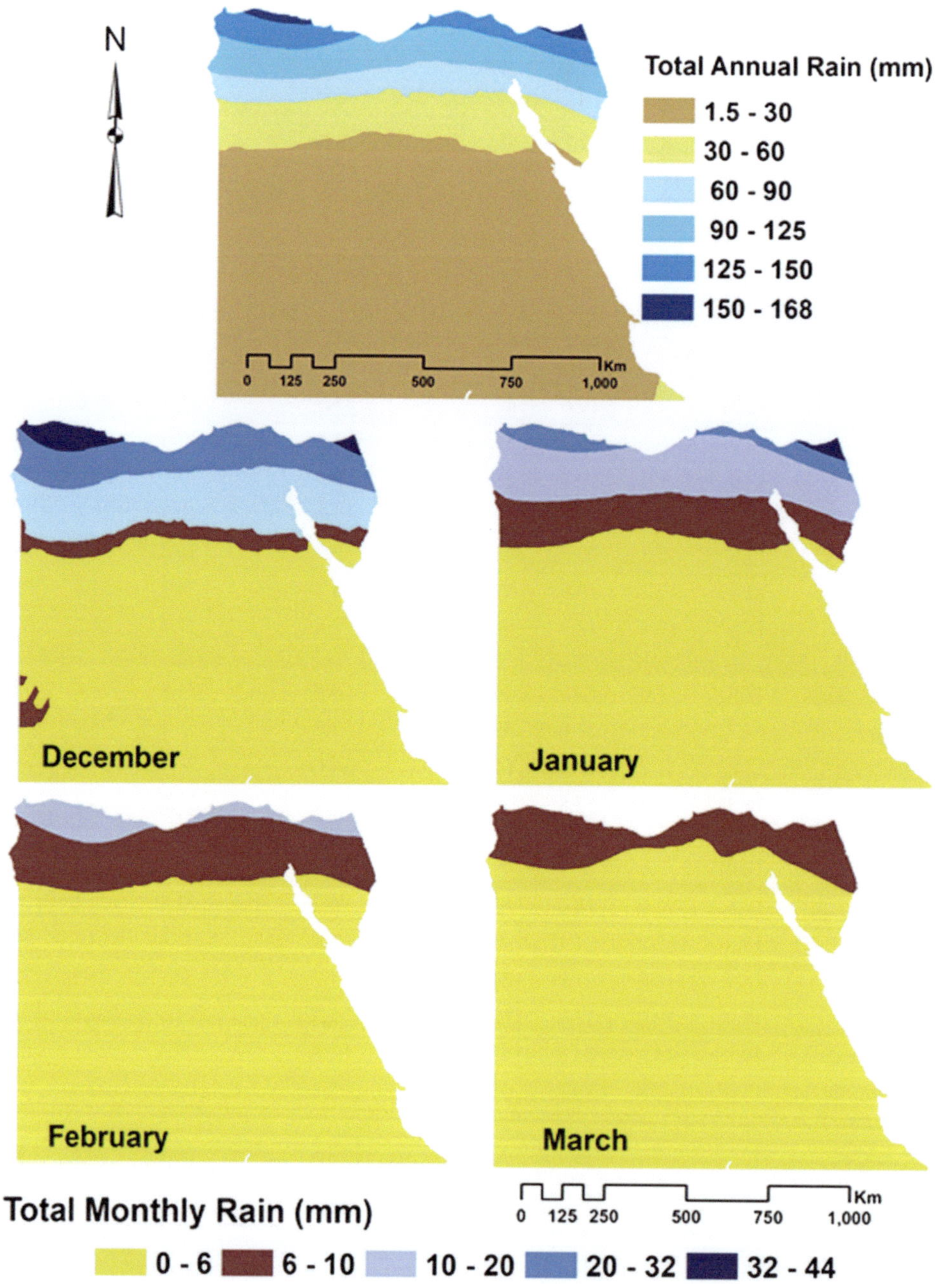

Fig. 5.3 Average of annual and monthly precipitation (mm) during 1997–2014 in Egypt using ArcMap 10.4 software. NASA POWER

Data also shows the averages monthly precipitation for four northern costal locations from west to east (Marsa Matruh, Alexandria, Baltim, and El-Arish) comparing with the averages monthly precipitation rain in an inland area like Cairo (Fig. 5.4). The data indicates that the highest monthly rain was observed in January for El-Arish (43.6 mm), followed by Marsa Matruh (39.6 mm), Alexandria (38.0 mm), Baltim (33.0 mm), and then Cairo (24.4 mm). The precipitation in some Egyptian oases (i.e., Siwa, Baharia, Farafra, Dakhla, and Kharga) in the western desert is extremely low (Fig. 5.4). These oases exhibited the average of annual precipitations of 52.5, 19.0, 11.8, 1.5, and 4.5 mm, respectively, during 1997–2014. It could be concluded that the monthly average of the rain during (1997–2014) for these Egyptian western desert oases is negligible and less than 5 mm except for Siwa (14 mm) during December and January, as it is the closest oasis to the north coastal region.

5.2.2 Changes in Temperature

Egypt is bordered by the Mediterranean Sea to the north which has a significant influence on the climate of northern coastal areas. Accordingly, the climate of the northern region of Egypt is classified as Mediterranean climate. Specifically, Fig. 5.5 shows the averages of 34 years variation of monthly air temperature (°C) for Alexandria governorate in the north (El-Nouzha site) and Aswan governorate in the south. The data reveals that there was a relatively higher temperature and wider differences between the minimum and the maximum air temperature in Aswan compared to the respective values of Alexandria governorate. Such large diurnal and seasonal variation in the thermal state of air affects the energy fluxes between atmosphere–water–soil–plant interfaces. The wide variation of air temperature during the day is commonly observed in arid zone especially for the inland

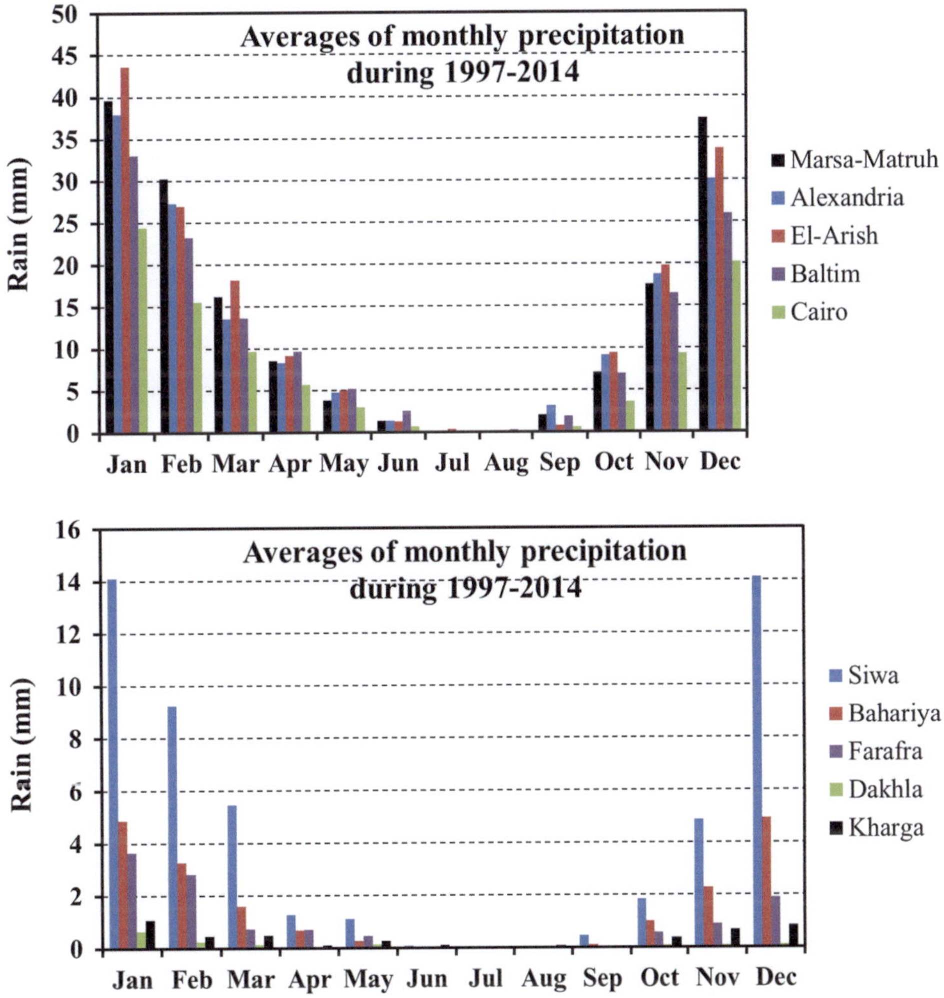

Fig. 5.4 Averages of the rain during (1997–2014) for different locations in the main Egyptian coastal Mediterranean cities compared to the average monthly rain in Cairo (top figure), whereas the bottom figure represents the averages of the monthly rain during the same period for some oases located in the western desert, Egypt. NASA POWER

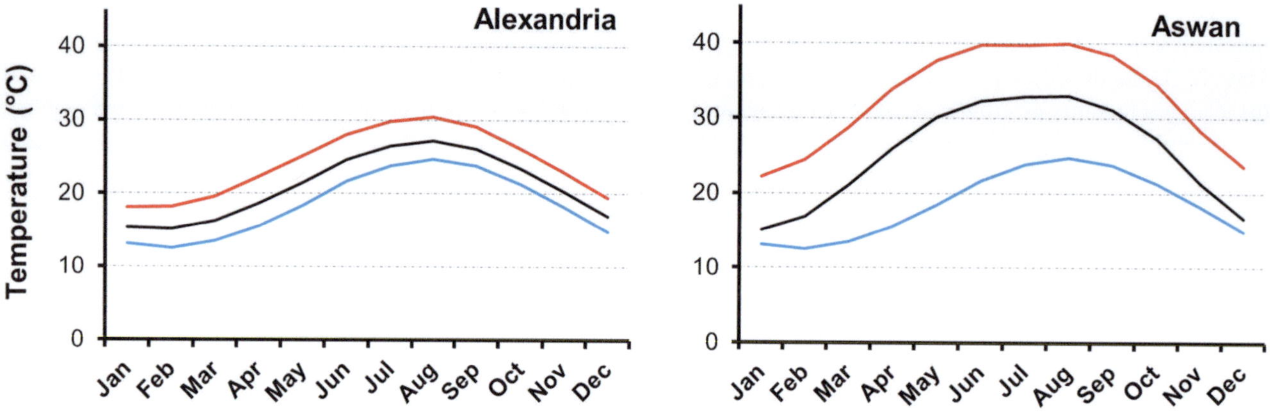

Fig. 5.5 Averages of air temperature (°C) monthly variation for Alexandria and Aswan governorates; red, black, and blue lines for maximum, mean, and minimum air temperature, respectively. NASA POWER

areas (El-Ramady et al. 2013). This fluctuation of soil temperature consequently affects the physiological processes of the cultivated plants and the biological activities in soil. Changes of soil thermal states can affect the soil chemical reactions such as the dissolution and solubility of soil compounds too.

The Egypt's climate is widely different from Mediterranean climate in the northern coastal area to hot desertic climate in the Upper Egypt region. In Egypt, the winter extends from December to February while summer ranges from June to August. The spatial variability of temperature could be presented in this subsection in Egypt during three decades (1983–2012) with 10-year time interval for January (winter month) and August (summer month).

The average maximum temperature results reveal that during January of the first decade (1983–1992), the lowest maximum temperature of 17–18 °C is observed in the north western coast and Sinai Peninsula (Fig. 5.6). The same pattern can be observed for the third decade (2003–2012) as the maximum temperature increased downward to the southeast direction (Fig. 3.6). During 1993–2002 (second decade), a reverse pattern existed since the northern region of Egypt was exposed to highest temperature in January (22–24 °C). For August, maximum temperature of over 40 °C can be observed in the southeast region during the third decade (2003–2012). Maximum temperature up to 38 °C is observed in Upper Egypt area during the first and second decades (1983–2002). In conclusion, the distribution of maximum temperature is significantly varied over the three studied decades.

The average minimum temperature pattern was almost constant over the three decades (1983 2012) in January (Fig. 5.7). The lowest minimum temperature during January (5–8 °C) occurred as transect from Sinai in the northeast to the southwest corner of Egypt. The Mediterranean coastal area had an average minimum temperature of 10–12 °C. The highest minimum temperature can be observed in Halayeb and Shalateen triangle in the southeast corner with a significant increase in its area in the third decade (Fig. 5.7).

In August, the distribution of minimum temperate has a completely different pattern within the three decades. The dominant minimum temperature of 24–26 °C was recorded during second and third decade (1993–2012) whereas the lowest minimum temperature in August (20–22 °C) can be observed in Sinai with significant decrease in its distribution within the three studied decades (Fig. 5.7). The lowest average mean temperature of 10–12 °C is observed mainly in Sinai in January during the first two decades (1983–2003). This mean value completely disappeared during the third decade (Fig. 5.8). The dominant January mean is 12–14 °C while the highest mean (up to 20 °C) is observed within the eastern desert region (Fig. 5.8). During August, the northern Mediterranean region of Egypt exhibits the lowest mean temperature of 26–28 °C (Fig. 5.8) during the three decades. The highest mean temperature of 34–35 is observed in Halayeb and Shalateen triangle during only the third decade (2003–2012).

5.2.3 Changes in Relative Humidity

Humidity is one of the most important climatic elements, which has a vital role in crop production. The Nile Valley has an extremely arid climate including low relative humidity, high temperature, high evaporation, and negligible rainfall, whereas the climate of coastal belt of the delta is an extension of the Western Mediterranean Coast (Abdel Meguid 2017). The averages of 34 years monthly variation of relative humidity (%) for Alexandria governorate in the north (El-Nouzha site) and Aswan governorate in the south

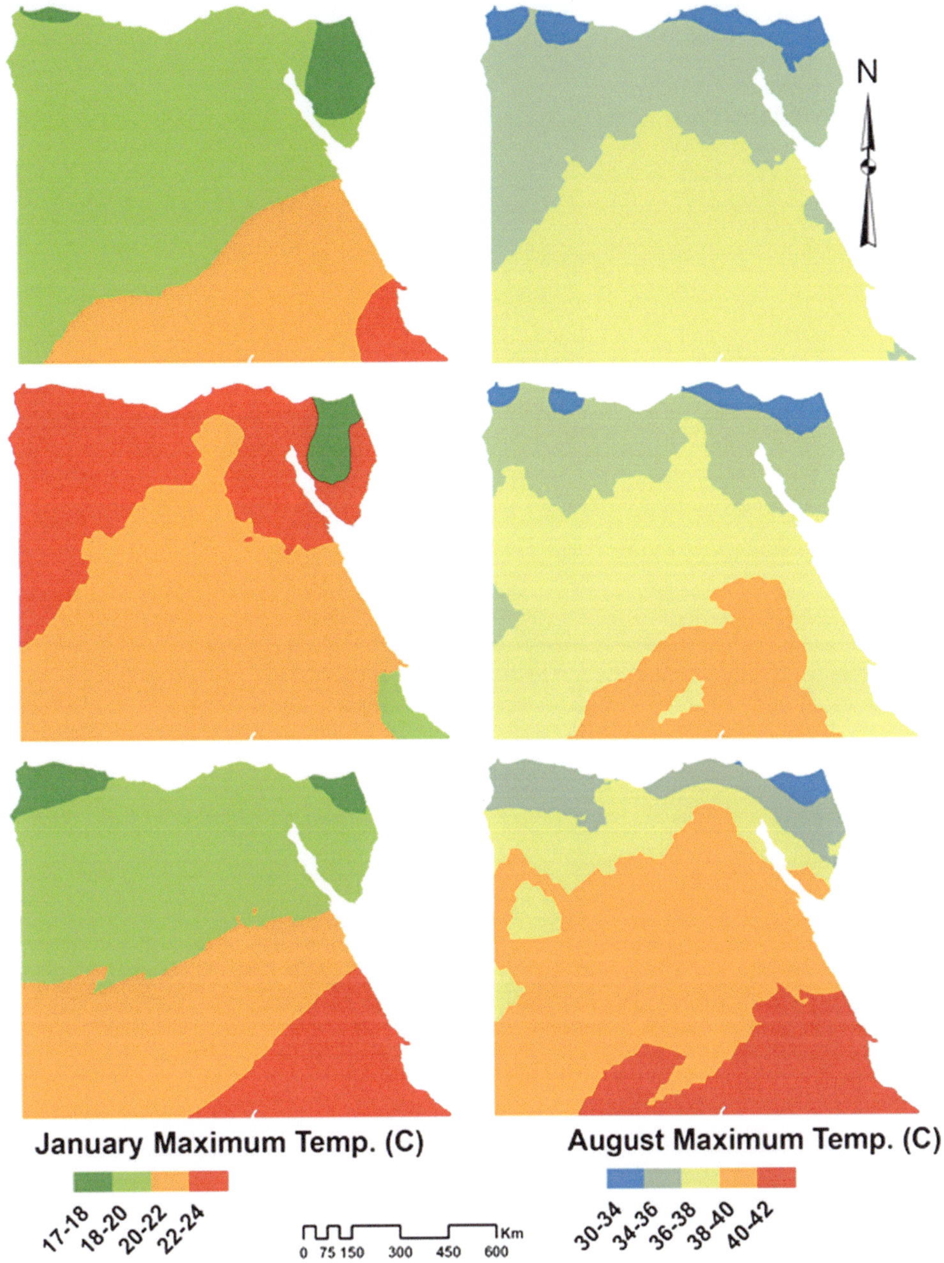

Fig. 5.6 Average of January and August maximum air temperature during 1983–1992 (top), 1993–2002 (middle) and 2003–2012 (bottom). NASA POWER

of Egypt are presented in Fig. 5.9. The data reveals that the average relative humidities in Alexandria and Aswan are found to be 66.1 ± 1.2% and 25.2 ± 8.2%, respectively. Such a relatively constant humidity in Alexandria is attributed to the substantial influence of the Mediterranean Sea on coastal climate.

The relative humidity values range between 32 and 67% during January whereas in August, its range was between 18 and 70% (Fig. 5.10). The relative humidity pattern was the same over the three decades as the maximum values occurred in the north and gradually decreased southward. With the Mediterranean climate in the north Egypt, the relative humidity had its maximum in that area while the minimum values (32 and 18% in January and August, respectively) can be observed in the south border of Egypt territory.

5.2.4 Changes in Wind Speed

Due to the huge areas of desert in Egypt, the speed and direction of wind are very important climatic elements. There are two different wind directions in Egypt including the coastal areas (at the north of Egypt, which mainly is coming in the summer season from northwest to northeast direction) and the south of Egypt, where the wind comes

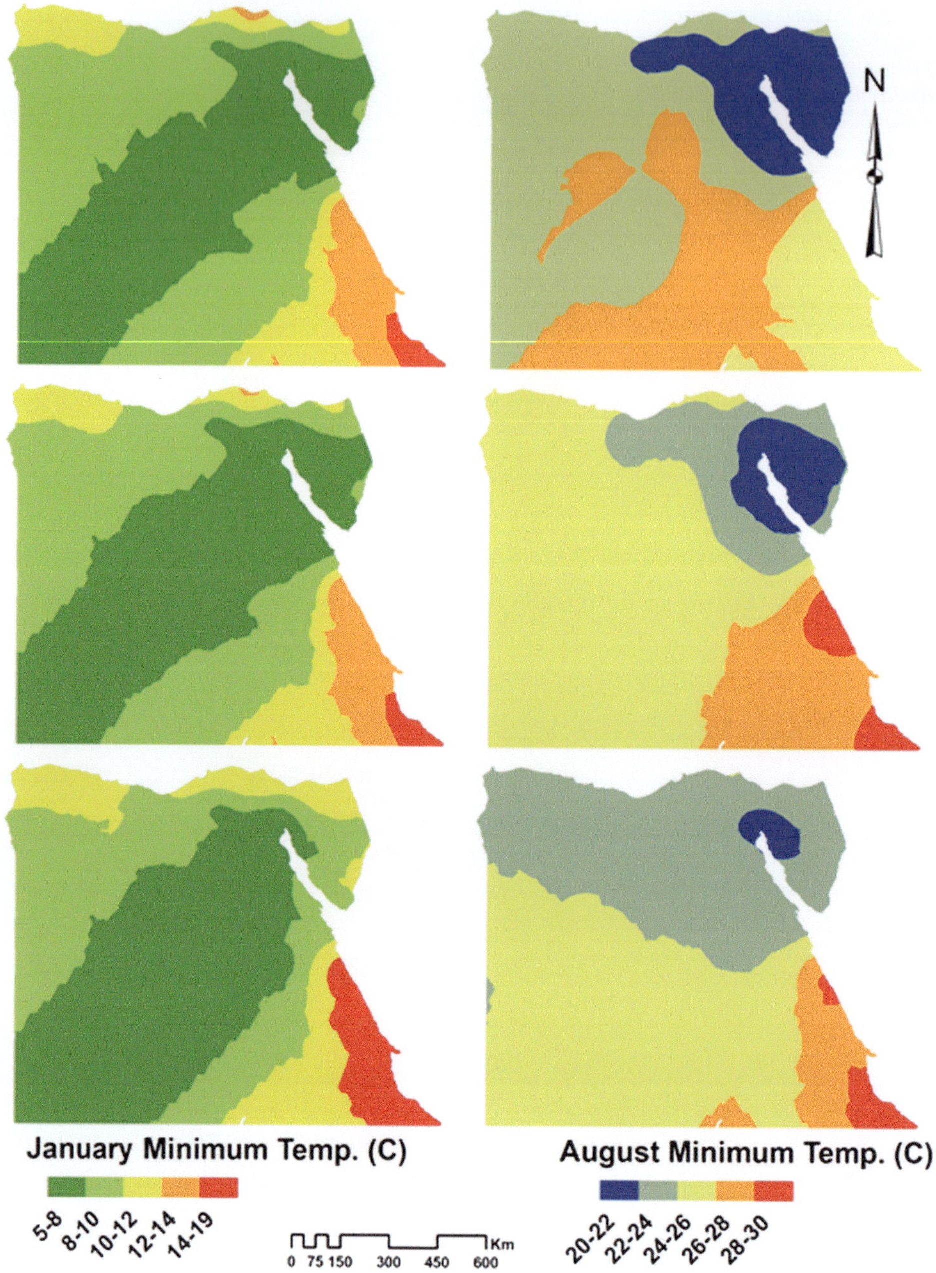

Fig. 5.7 Average of January and August minimum air temperature during 1983–1992 (top), 1993–2002 (middle), and 2003–2012 (bottom). NASA POWER

from southwest to northeast direction (Abdel Meguid 2017). The average wind speeds for Alexandria and Aswan are found to be as low as 4.85 ± 0.34 m/s and 3.93 ± 0.27 m/s, respectively (Fig. 5.11). It is found that the wind speed had its highest value of 5–6 m/s in January over the southeast and northwest corners during the first decade (1983–1992), while there was a reduction in wind speed to 4–5 m/s as a maximum during the second and third decade (1993–2012). The dominant wind speed (3–4 m/s) occurred during the three studied decades (Fig. 5.12). The wind speed in August over the three decades had the same range with different distribution as 4–5 m/s was the dominant wind speed over 30 years.

It could be concluded that Egypt is particularly vulnerable to the change in climate. The Nile River is the main water source and the agricultural sector also depends on water of Nile. Due to a very long coastline of Egypt, it is undergoing intensifying development and erosion. In general, the climate of Egypt is very arid and the annual precipitation also is very low or nearly zero mm per year except the coastlines, which may be less than 200 mm per year. Egypt is the world's 30th—largest country, where its location extended from latitude 22–32 °N and longitudes from 25 to 35 °E awarding this country a long gradient of average temperatures and precipitations (El-Shaer et al. 1996). The change in climate in Egypt could appear in many fields in Egypt including the impact of

Fig. 5.8 Average of January and August mean air temperature during 1983–1992 (top), 1993–2002 (middle) and 2003–2012 (bottom). NASA POWER

global warming on the agricultural sector, the increasing of evapotranspiration and crops water requirements, the expected reduction in yield of crops, distribution rate of plant disease and pests, the drought stress and decrease precipitation, the sea level rising and changing land use (Fig. 5.13), seawater intrusion and secondary salinization (Fig. 5.14), and finally extreme events (Fig. 5.15; Mahmoud 2017).

It could be concluded that Egypt is characterized by the arid climate, where there are high evaporation rates (1500–2400 mm/year) and very low rainfall (5–200 mm/year). However, this climate of Egypt around the year is favorable and ideal for a wide variety of crops. This climate also made it possible to adopt intensive cropping systems and thus permitted the production of more than one crop per year in most of the cultivated areas in Egypt. Although the Delta has a hot desert climate like the rest of Egypt, its northern part is wet. The Delta temperatures are the hottest in July and August, with maximum average temperature of 34 °C, whereas winter temperatures are normally in the range from 9 °C at night to 19 °C during the day (Negm et al. 2017).

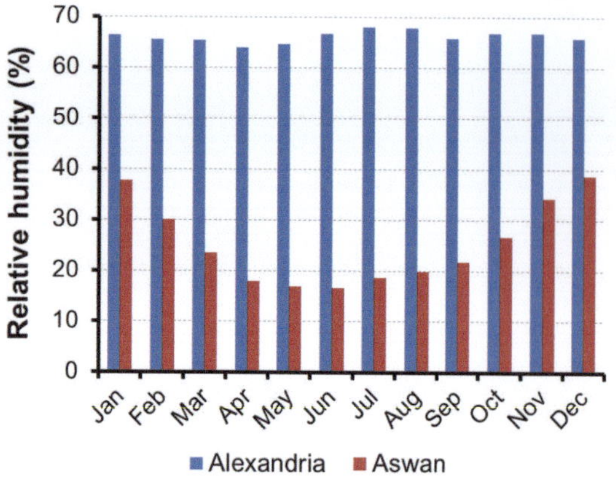

Fig. 5.9 Averages of the monthly relative humidity (%) for Alexandria (blue bars) and Aswan governorates (red bars). NASA POWER

5.3 Climate Changes and Agriculture

5.3.1 Climate Changes and Agricultural Soil

Climate change could be defined as the change in climate variables or elements (e.g., temperature, precipitation, humidity, wind, and pressure) for a long period or for decades or an even longer timescale (Najjar et al. 2017). It is acceptable to define the climate as the study of weather or climatic variables for a long time (usually over a 30-year interval). The climate and its changes have been investigated from many fields including agriculture, water, soil, energy, etc. The topic of climate change is considered nowadays as one of the hottest issues worldwide and studies about it are in an ascending level. This concern could be noticed from Table 5.1, wherein the last months from 2017 till now some thousands have been published about the climate change and soil or water or agriculture.

Concerning climate change and its effects on soils, it is reported that different soil and water resources are at risk globally from loss of access, degradation, and scarcity, which will lead to increase in the pressure on the global and environmental security. So, it is stated by the President of the World Bank (Jim Yong Kim 2014) that "*fights over water and food are going to be the most significant direct impacts of climate change in the next 5–10 years*". Furthermore, it should save our soils to save our planet as stated by Australia's first National Advocate for Soil Health Jeffrey and Achurch (2017). Concerning some climate change issues, McCarl (2017) also stated that it could improve or maintain soil condition and its health including (1) the direct and indirect soil-related effects of extreme events, a warming world and associated changes in precipitation and other climate attributes, (2) soil-related effects of actions to adapt to climate change, and (3) soil-related effects of climate change mitigation efforts in terms of sequestration enhancement and more general greenhouse gas emission control.

The production of crops is extremely susceptible to changes in climate. So, this will cause major shifts in crop distribution due to climate change in the future. Moreover, it has been estimated that climate changes are likely to reduce yields and/or damage crops in the twenty-first century in different parts of the world (Osman 2013). Therefore, the effects of climate changes on agriculture include global warming (Gettelman and Rood 2016; Fieguth 2017; Hope et al. 2017; Salawitch et al. 2017), melting of arctic ice (Alekseev et al. 2015; Hay 2016; Copland et al. 2017; Matishov et al. 2017; Pope et al. 2017), sea level rise (Baumert and Kloos 2017; Cazenave et al. 2017; Jevrejeva et al. 2017; Marzeion et al. 2017; Marcos et al. 2017) and frequent occurrence of extreme events such as flooding and droughts (Aslam et al. 2017; Chen et al. 2017; Debortoli et al. 2017; Jangra et al. 2017; Omambia et al. 2017; Nally et al. 2017).

Concerning the soil biological activity and changing climate, many issues should be addressed including different impacts of global change on soil biota, environmental stress, and its effects on soil enzymes under elevated both temperature and CO_2. Different strategies should be mitigated for soil biological activities under a global change. Therefore, the following facts and perspectives according to Alshaal et al. (2017) could be concluded:

1. Nearly 3.1×10^5 kg of carbon can be found in the soil representing more than the two-thirds from the total C in the terrestrial environments (Davidson et al. 2006).
2. About 10% of the atmospheric carbon dioxide (CO_2) annually passes through the soil.
3. Every year, 5–7.5 Mg of carbon could be emitted or released from the terrestrial surface flux as a carbon dioxide.
4. Every year about 1.2×10^{14} kg of C is taken up by the soil autotrophic microbes, whereas the soil heterotrophic microbes cumulatively emit about 1.19×10^{14} kg of C (Singh et al. 2010).
5. The global CO_2 content in the atmosphere recorded 343 and 406 ppm in 2005 and 2017, respectively (Table 5.2).
6. The amount of nitrous oxide (N_2O) represents more than one-third from the greenhouse gases producing from the nitrification and denitrification from agricultural lands (IPCC 2007).
7. The emission rate from fertilizers is ranged from 1 to 5% as N_2O or deposition rate 10–50 kg N_2O from each 1000 kg fertilizer (Singh et al. 2010)

Fig. 5.10 Average of January and August relative humidity during 1983–1992 (top), 1993–2002 (middle), and 2003–2012 (bottom). NASA POWER

8. The soil biological processes could release some greenhouse gases including CH_4, N_2O, and CO_2 beside the soil represents the pool for C in the terrestrial lands.
9. The annual natural emission rate of the methane (NH_4) is nearly 250 million tons which are produced by rumens, termite, oceans, and paddy rice (Singh et al. 2010).
10. It is predicted that the sea surface temperature or the global warming will be increased from 1.18 to 6.48 °C under low and high CO_2 emission, respectively (Meehl et al. 2007; Dutta and Dutta 2016).
11. The global annual amounts of C produced from the decomposition of soil microbes is 7.5–9 times from the anthropogenic emissions (Crowther et al. 2015).
12. The annual production of CO_2 from decomposition of plant organic matter, microorganisms in soil is 5.5×10^{12} kg and equals 8 times more than that by humans (Zimmer 2010; Dutta and Dutta 2016).
13. Climate changes could cause some modifications and complex changes in the soil microbial communities

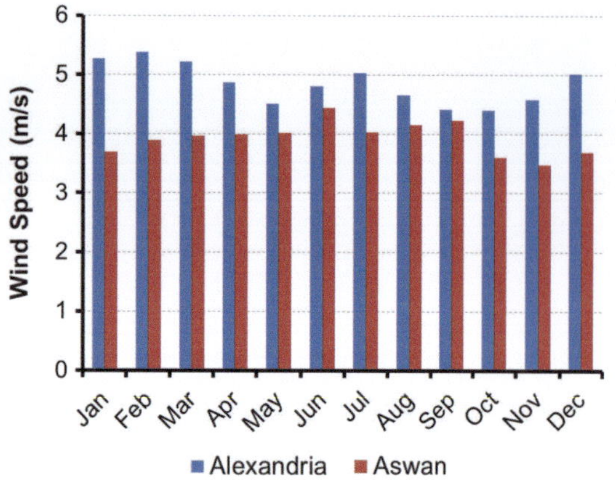

Fig. 5.11 Averages of the monthly wind speed for Alexandria (blue bars) and Aswan governorates (red bars). NASA POWER

(Castro et al. 2010) and their diversity (Shade et al. 2012; Zhang et al. 2016, 2017b).

14. Climate change (i.e., elevated temperature under different levels of soil moisture) could increase the soil respiration, where the soil respiration is very sensitive to changes in climate including soil temperature and moisture as well as the precipitation (Aanderud et al. 2013; Brye et al. 2016; Feng et al. 2017).
15. Different fluctuations in soil moisture and its temperature can change the soil biological activities including enzymatic and microbial activities (Schimel et al. 2007; Rezanezhad et al. 2014; Daou et al. 2016).
16. Climate changes including drought, flooding, freezing stresses as well as rainfall could cause the variability in soil moisture and its respiratory activity (Aanderud et al. 2011, 2013; Classen et al. 2015a, b; Zhang et al. 2017a).
17. The elevated atmospheric CO_2 could increase the plant photosynthesis rate, and hence accelerate the soil biological activities, the carbon availability for the soil microbes enhancing the soil respiration of microbes that release carbon to the atmosphere (Bardgett et al. 2008; Tao et al. 2017).
18. Due to the importance of soil carbon, Framework Convention on Climate Change (COP21) in Paris (2015) has adopted an initiative called "4 per thousand". This initiative already planned to enhance the content of soil organic carbon in world soils at the rate of 0.4% per year to a 40 cm soil depth (Lal 2016; Minasny et al. 2017).
19. The strategy of this initiative depends on promoting the soil organic C through adoption of some recommended agro-management practices including restoration of degraded soils, conservation agriculture, agroforestry, improved grazing, cover cropping, biochar, etc. (Lal 2016; Habtemariam et al. 2017).
20. The main target of the "4 per Thousand" proposal includes the concept of both soils and agriculture which are the main solutions in handling the global issues of changing climate, the insecurity of foods, and the environmental pollution (Lal 2016; Lozano-García et al. 2017).
21. It is reported that the global temperature will increase by 1 °C in the next 25 years (by the end of 2040) and 2–4 °C by the end of 2100 (IPCC 2014; McCarl 2015, 2017).

Concerning some investigations about the climate of Egypt, an increase in the number of publications could be noticed and these studies include different topics, in which climate change has been penetrated directly and indirectly such as:

1. Sustainable agriculture and its relation to climate changes in Egypt (e.g., Eid et al. 2007; Attaher et al. 2009; McCarl 2011; El-Ramady et al. 2013; Froehlich and Al-Saidi 2017; Mahmoud 2017),
2. Climate change impacts and adaptation strategies for crop production in Egypt (e.g., El Afandi 2017),
3. Management of climate-induced drought and water scarcity in Egypt (e.g., Mahmoud 2017),
4. Mitigation of climate change in Egypt (e.g., Hinkel et al. 2012; El Quosy 2013; Ahmed 2017),
5. Impacts of climate change on soil organic carbon in Egypt (e.g., Muñoz-Rojas et al. 2017),
6. Climate changes and food security in Egypt (e.g., Abou Hadid 2013; El-Ramady et al. 2015; Mahmoud 2017),
7. Climate change and extreme events over in Egypt (e.g., Badir and Abd AlRahman 2017),
8. Climate changes and water resources in Egypt (e.g., Elshinnawy 2008; El Quosy 2013; Younos and Grady 2013; Mahmoud 2017) and
9. Using remote sensing following the change in climate in different zones of Egypt (e.g., Hassaan 2013; Elbasiouny et al. 2017; Hassan and Omran 2017).

It could be concluded that climate change is the main reason for the global warming, Arctic Sea Ice would melt away, rising sea level would affect coastal environments, and the extreme events (hurricanes, floods, and droughts) would be more frequent. Several properties and functions of the soil will change in response to the climate changes including (1) increased CO_2 may enhance the production of biomass, (2) these climate changes may lead to enhancement in decomposition of soil organic matter, (3) climate changes

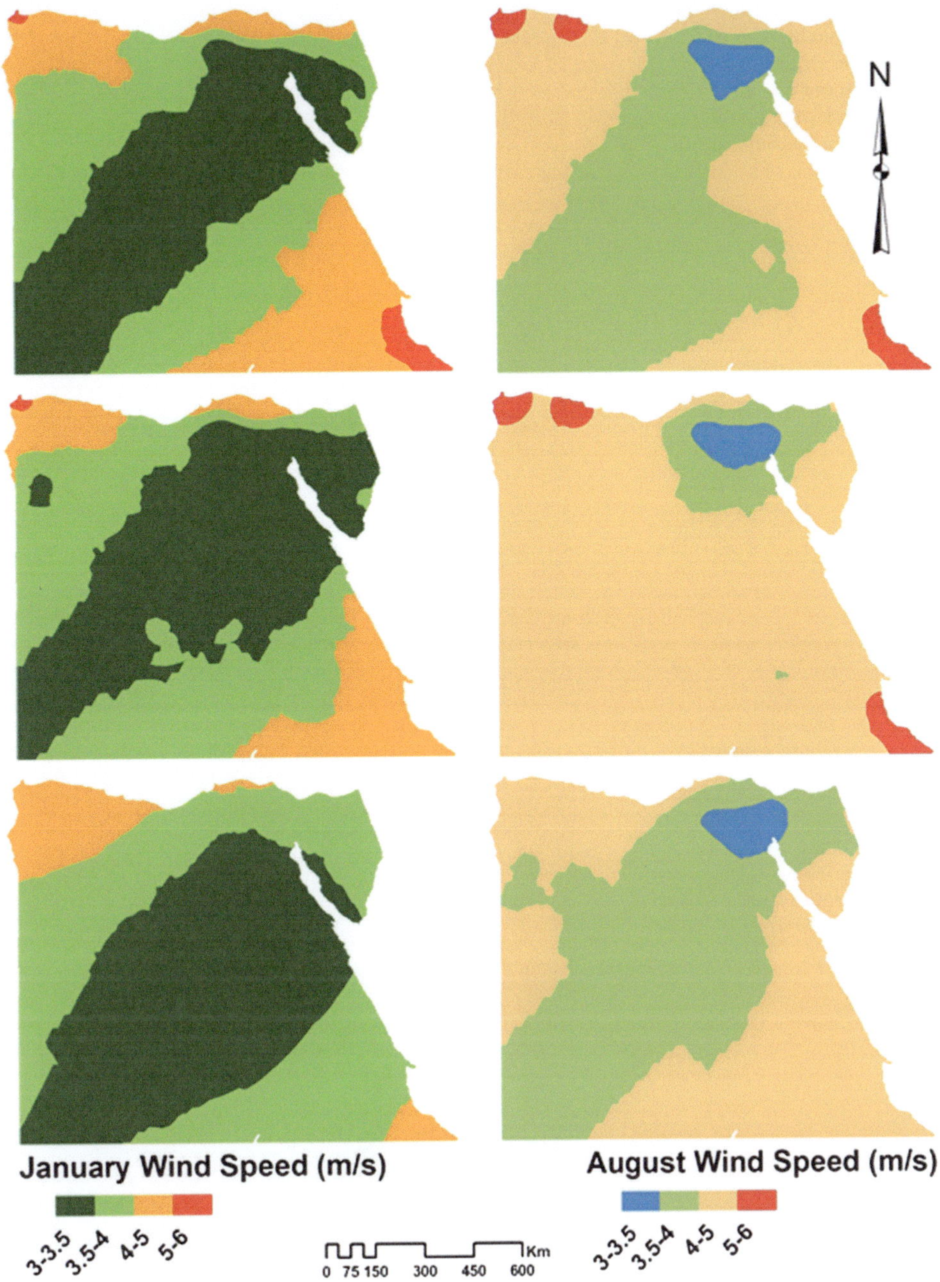

Fig. 5.12 Average of January and August wind speed during 1983–1992 (top), 1993–2002 (middle), and 2003–2012 (bottom). NASA POWER

will increase the evapotranspiration process, (4) climate changes could increase the salinization of many soils, (5) climate changes could alter the composition and the functions of soil microorganisms, and (6) the rising in temperature may lead to permafrost thawing (Osman 2013). It is worth to mention that the Nobel Peace Prize in 2007 was awarded jointly to Intergovernmental Panel on Climate Change (IPCC for Prof. Rattan Lal as a member of the IPCC) and Albert Arnold Gore Jr. "*for their efforts to build up and disseminate greater knowledge about man-made climate change, and to lay the foundations for the measures that are needed to counteract such change*".

5.3.2 Climate Changes and Water Crisis

Egypt is one of the African countries that could be vulnerable to water stress under climate changes. A major challenge is to close the rapidly increasing gap between the limited water availability and the escalating demand for water from various economic sectors. The rate of water utilization has already reached its maximum for Egypt, and climate change will exacerbate this vulnerability. Agriculture consumes about 85% of the annual total water resource and plays a significant role in the Egyptian national economy, contributing about 20% of GDP. Egypt may be

Fig. 5.13 Marsa-Matruh, about 300 km from the west of Alexandria city, is a Mediterranean coastal city may face the risk of sea level rising. Photo by El-Ramady, July 2017

Fig. 5.14 Alexandria is one of the most important coastal Egyptian cities, which faces the risk of sea level rising as well as seawater intrusion and secondary salinization. Photo by El-Ramady 2015

subjected to increasing temperature levels in the upcoming decades. The Egyptian Cabinet's Information and Decision Support Center in 2010 has presented some scenarios including 1.0–1.3 °C by 2025, 1.9–2.6 °C by 2050, and 2.2–4.9 °C by 2100. As a result of the increased temperature levels, the demand for water is expected to increase, especially for crops, while agricultural productivity may decrease. Temperature rise may also cause a decrease in organic matter in soils, reducing soil quality. Also, sudden, unusual temperature drops in the year 2012 damaged fruit trees in Nubaria and increased yields for wheat for the country, meaning that shifting weather patterns caused by climate changes may have negative as well as positive effects on crop production in Egypt in the future. From 2012 to 2017, a distinguished effect has been recorded concerning the low and high temperature in Egypt causing several millions Egyptian pounds losses in the agricultural production.

Fig. 5.15 Some extreme events have been recorded and its impact year to year regarding weather variability on crop yield in Egypt. Photos show heavy rains (in Busily, Rosetta district) and snow rains (the photo of orange) resulted in damage. Photographs by El-Marsafawy

Table 5.2 Atmospheric carbon dioxide (CO_2) concentration (ppm volume) of major greenhouse gases from 1983 to 2017

Greenhouse gas	December 1983	December 1993	December 2003	December 2013	December 2014	December 2015	December 2016	December 2017
CO_2 content	343.05	356.91	375.99	396.84	398.91	401.85	404.48	406.29

Source https://www.co2.earth/annual-co2 accessed on December 1, 2017

The Nile Delta of Egypt is vulnerable due to the impact of climate changes in particular sea level rise. Due to low elevation in the Nile delta region, Egypt is considered as one of the top five countries expected to be mostly impacted with a 1 m sea level rise resulting from global warming. Egypt is ranked as the fifth in the world concerning the impact on the total urban areas, Egypt's GDP would be significantly impacted, and Egypt's natural resources such as coastal

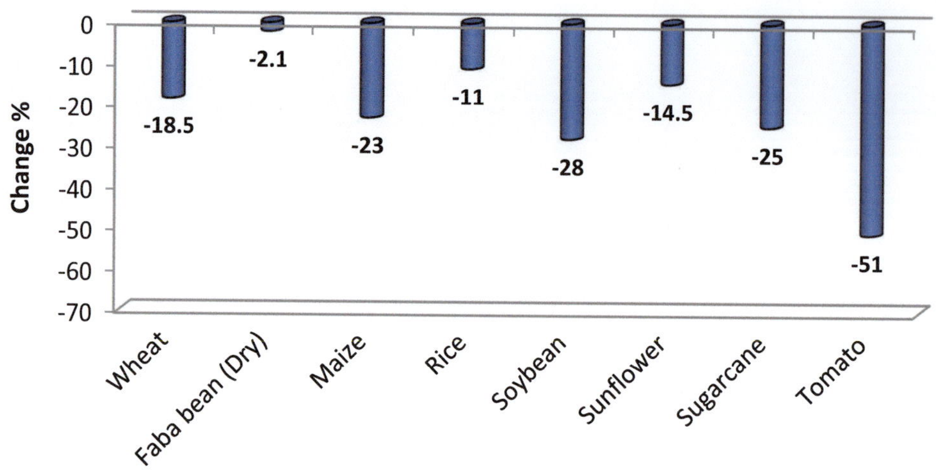

Fig. 5.16 Change percent in crop productivity under climate change conditions as compared with current conditions

zone, water resources, water quality, agricultural land, livestock, and fisheries may be subjected to vulnerability. Also, Egypt may face environmental crises such as shore erosion, saltwater intrusion, and soil salinity. Egypt should realize the threats from climate changes, formulate policies that will minimize the risks, and take action (Batisha 2012). In the same direction, Eldeberky (2011) indicated that global warming and climate change processes are expected to raise sea levels (SLR). The most outstanding impact of SLR on the water resources is that it will increase the occurrence of saline intrusion into coastal aquifers with the contamination of groundwater resources in the upper Nile delta. SLR is expected to cause a landward shift of the salt wedge and increase the rate of saline seepage to the topsoil of the delta. In addition, the salinity in Lake Manzala may increase because of the stronger influence of tidal flows penetrating the lake. El-Raey et al. (1999) mentioned that changes in the salinity conditions of the lake may affect its ecology and fisheries and the accelerated SLR will increase the salinity.

It is indicated that one meter sea level rise will affect an approximate of 6 million people mostly “poor”, living in the Nile Delta basin (Al Olaimy 2013). Thus, this weakening in the Nile Delta which is heavily inhabited and used for agriculture shall impact millions of people. Recent studies have forecasted that Nile River’s flow will decrease by 40–60% which will increase the frequency and intensity of drought, particularly in North African countries causing major socioeconomic and political problem for the region. El-Beltagy and Madkour (2012) showed that nearly 1800 km of land on the north coast of Egypt and the Nile Delta will be under seawater if there is a rise of 50 cm in the sea level, affecting 3.8 million people. Many of the old deltas, where agriculture first started, are among those that are most threatened by a rise in sea levels. Not only will there be a loss of land and its productivity but also people living there will have to migrate to higher (and therefore, safer) lands, adding to social and economical problems there. Tension will increase between communities and wars might even erupt, as has been happening between people who have historically depended on rangelands and those who practice settled agriculture.

Climate change will lead to decrease in crop productivity and increase in irrigation water needs for major crops in Egypt. Crop productivity will decrease from about 2% for Faba bean up to 51% for tomato (Fig. 5.16). In addition, increasing water needs will reach up to 16% (Fig. 5.17).

It is worth mentioning that the climatic changes will lead to a lack of self-sufficiency in major food crops and the shortages will increase if we take into account the population growth in Egypt by the mid of this century.

Table 5.3 shows self-sufficiency in Egypt under current conditions (2015) and climate change, in addition, under conditions of climate change and population growth together. The projected population growth in the year 2050 according to the expectations of the Central Agency for Public Mobilization and Statistics (CAPMAS), which is expected to reach the population growth about 165 million. Under these circumstances, self-sufficiency for most major food crops in Egypt will probably fall to less than 30%.

In this connection, Radhouane (2013) showed that all dimensions of food security (availability, stability, utilization and access) will be affected by climate change. According to many experts because of more frequent drought periods, agriculture performance is projected to drop in the future. The adverse impacts of climate change include reduced crop yield due to drought and reduced water availability. Increasing temperature trend will make crops fail to reach mature due to lack of enough moisture in the soil. On the other side, warmer climate will probably increase crop losses caused by weeds and diseases. A number of countries in North Africa already face semiarid conditions that make agriculture challenging, and climate change will be likely to reduce the length of growing season as well as force large regions of marginal agriculture out of production. Projected

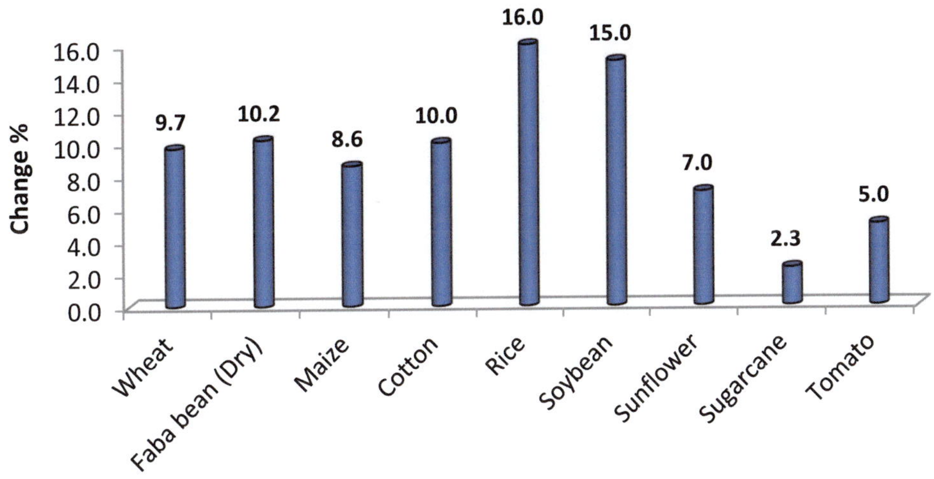

Fig. 5.17 Change percentage in crop water requirements under climate change conditions as compared with current conditions

Table 5.3 Self-sufficient (%) for some major crops in Egypt under current conditions (2015), climate change (CC) conditions (2050) and climate change with population growth

Crop	Self-sufficient (%) in 2015	Self-sufficient (%) under CC	Self-sufficient (%) under CC with pop. Growth
Wheat	52	43	23
Faba bean (dry)	30	29	16
Maize	54	42	22
Rice	104	92	50
Soybean	6	4	2
Sunflower	28	24	13
Sugarcane	69	52	28
Tomato	103	50	27

Results are shown in graphs (Figs. 5.16, 5.17) and Table 3.5 are collected and calculated from previous studies
Current crop productivity (2015) was obtained from Bulletin of the Agricultural Statistics, September 2016, Economic Affairs Sector. Ministry of Agriculture and Land Reclamation
Source of yield and water consumption under climate change: Abou Hadid (2013), Eid et al. (2007), El-Marsafawy (2007) and El-Marsafawy et al. (2007, 2017)

reductions in yield in some countries could be as much as 50% by 2020 and crop net revenues could fall by as much as 90% by 2100, with small-scale farmers being the most affected (Boko et al. 2012). The risks of climate change may include the production capacity of food security, the potential growth in incomes and ability to purchase food of poor people as well as the risk of market disruptions, effects on supply and storage systems, and effects on stability of agricultural and rural incomes as well as nutritional content. The people at greatest risk are those who are dependent on agriculture and natural resources for livelihoods, especially those who are the most vulnerable, and who depend on systems that are the most impacted, and poor people (FAO 2016).

Modeling by IIASA shows that future socioeconomic development and climate change may impact regional and global irrigation requirements and thus agricultural water withdrawals (Fischer et al. 2007). Net irrigation requirements may increase by 45% by 2080. Even with improvements in irrigation efficiency, gross water withdrawals may increase by 20%. Global irrigation requirements with climate change will increase by 20% above the reference base case scenario (without climate change). The simulation shows that the global impacts of climate change on irrigation water requirements could be as large as the projected increase in irrigation due to socioeconomic development.

It is indicated that the population, land use and agriculture and economic activity of Egypt are all constrained along a narrow T-shaped strip of land along the Nile and the coast around its delta (Medany 2016). There is a possibility of significant decline in Nile stream flow under climate change. Nile water availability is likely to be increasingly stressed due to higher water demands and evaporative losses resulting from higher temperatures in the semiarid region, which are projected consistently across various climate models. Coastal zone and water resource impacts also have serious implications for agriculture: sea level rise will adversely impact agriculture: seawater intrusion and increased demand for crops in warmer climate. Extreme weather events are increasingly threatening the farmers as heavy rains hit the delta last October 2015 causing serious damages to more than a million acres, mainly devoted for wheat cultivation.

5.3.3 Adaptation to Climate Change in Egypt

There is a lot we can do to reduce emissions, prevent climate change, and protect our threatened freshwater sources. Using less energy is a great place to start. This can be done by turning off lights, better insulating our homes to conserve heat and air conditioning, and driving more fuel-efficient cars and driving less. Cars and light trucks (like vans and SUVs) are responsible for about 20% of U.S. energy-related carbon emissions. In addition, eating lower on the food chain, even going meatless just one day a week can have a significant impact on environmental resources because industrial meat production has significant greenhouse gas emissions associated with it. Conserving water, food, and other resources is an important step towards reducing overall energy use, because everything that is made, transported, and thrown away requires the use of fuel and water. By carpooling, using public transportation, driving less, and reducing our consumption of food and consumer goods, each individual can make an impact on curbing greenhouse gases (http://www.gracelinks.org). Improving water use and reducing losses is essential for Egypt's future (Research Study Report 2014). This is recognized by the Egyptian National Water Resources Plan (2005), which stresses on better irrigation management through devolving decision-making, adoption of modern pressurized field-level irrigation systems, and introducing more water efficient cropping systems. Egypt's Sustainable Agricultural Development Strategy to 2030 (Abul-Naga 2009) focuses on taking advantage of Egypt's comparative advantages within the global food system to achieve agricultural self-reliance. Comparative advantage in the export crops would ensure sufficient returns to cover the costs of food imports. Many technologies, which facilitate these strategies, are available and are readily adapted to Egyptian conditions. Modern pressurized drip and sprinkler systems have been successfully employed in Egypt's new lands for a generation. Protected agricultural systems showed success in Egypt's fresh produce export sector, although there is room for improving environmental and economic efficiencies. Egyptian scientists recorded successes in breeding improved varieties of major crops that have short duration, or are more drought and heat tolerant. New wheat varieties require 1100 m^3 of water per season instead of the traditional 1560 m^3 on average (personal communication). In addition, in order to maintain the balance in the system, there is a need for new knowledge, alternative policies and institutional changes (El-Beltagy and Madkour 2012). The marginalized people in dry areas are likely to be most seriously hit by the shifts in moisture and temperature regimes as a result of the global climate change. To help them cope with the challenges, there is a need for a new paradigm in agricultural research and technology transfer that makes full use of modern science and technology in conjunction with the traditional knowledge. This necessitates more investment by international agencies and national governments for supporting the relevant integrated research and sustainable development efforts, with full participation of the target communities. Only such an approach can enable the vulnerable communities of the dry land areas to use the natural resources in a sustainable manner and thus help protect the environment for future generations.

5.4 Conclusion

Egypt is nearly a rainless country. The climate of Egypt is generally moderate ranging from semiarid to extremely arid conditions with distinguished features of arid zones. There are three distinguished types of climate in Egypt including the desert climate (in the inland areas), the Mediterranean climate (in the northern coastal areas) and the climate of the coast of the Red Sea. The great challenges faced by Egypt include the limitation of water resources and climate changes. The change in climate is a serious problem in several countries around the world in particular the agricultural and water sectors. Climate changes in Egypt include the risk of sea level rising and the increasing of temperature as well as decrease in the crop productivity. These effects also may include the increase in water requirements of crops, the distribution rate of plant disease and pests, the decrease in precipitation rates or drought, the changing land use, and the secondary salinization due to seawater intrusion. So, there is an urgent need to understand the potential effects of climate changes and to develop adaptation strategies and actions to reduce future climate change risks. Further investigations in Egypt concerning different effects of changing climate on agricultural, energy and water sectors as well as human health should be carried out. These studies are needed for the following issues: a new crop pattern suitable for the available water, selecting new varieties and strains of different crops that tolerate drought, salinity and diseases as well as establish a well-coordinated information system to support decision-makers in water resources management under global climate changes.

References

Aanderud ZT, Jones SE, Schoolmaster DR Jr, Fierer N, Lennon JT (2013) Sensitivity of soil respiration and microbial communities to altered snowfall. Soil Biol Biochem 57:217–227

Aanderud ZT, Schoolmaster DR Jr, Lennon JT (2011) Plants mediate the sensitivity of soil respiration to rainfall variability. Ecosystems 14:156–167

Abdel Meguid M (2017) Key features of the Egypt's water and agricultural resources. In: Negm AM (ed) Conventional water resources and agriculture in Egypt, Handbook of Environmental Chemistry, Springer International Publishing AG. https://doi.org/10.1007/698_2017_41

Abdel-Shafy HI, El-Saharty AA, Regelsberger M, Platzer C (2010) Rainwater in Egypt: quantity, distribution and harvesting. Mediterr Mar Sci 11(2):245–257

Abou Hadid AF (2013) Climate change and food security in the Nile Valley and Red Sea. In: Sivakumar MVK et al (eds) Climate change and food security in West Asia and North Africa. Springer Science + Business Media Dordrecht, pp 193–206. https://doi.org/10.1007/978-94-007-6751-5_11

Abul-Naga AM (2009) Egypt sustainable agricultural development strategy: towards 2030. CIHEAM Analytical Notes, 53, Bari

Ahmed EHM (2017) Indigenous Egyptian Nubians and climate change mitigation. In: Erşahin S et al (eds) Carbon management, technologies, and trends in mediterranean ecosystems, Vol 15. Politik–Economics–Society–Science Series, The Anthropocene

Al Olaimy T (2013) Climate change impacts in North Africa. http://www.3blassociates.com/climate-change-impacts-in-north-africa/

Alekseev GV, Aleksandrov EI, Glok NI, Ivanov NE, Smolyanitsky VM, Kharlanenkova NE, Yulin AV (2015) Arctic sea ice cover in connection with climate change atmospheric and oceanic. Physics 51(9):889–902

Allen DE, Singh BP, Dalal RC (2011) Soil health indicators under climate change: a review of current knowledge. In: Soil health and climate change. Springer, 25 p

Alshaal T, Al-Saeedi AH, Shalaby T, Elsakhawy T, Omara AED, Gad A, Hamad E, El-Ghamry A, Mosa A, Amer M, Abdalla N and El-Ramady H (2017) The rhizosphere and plant nutrition under climate change. In: Naeem M et al (eds) Essential plant nutrients. Springer International Publishing AG. https://doi.org/10.1007/978-3-319-58841-4_11

Aslam Z, Khattak JZK, Ahmed M (2017) Drought tolerance in cereal grain crops under changing climate. In: Ahmed M, Stockle CO (eds) Quantification of climate variability, adaptation and mitigation for agricultural sustainability. Springer International Publishing Switzerland, pp 181–209. https://doi.org/10.1007/978-3-319-32059-5_9

Attaher SM, Medany MA, Abou-Hadid AF (2009) Possible adaptation measures of agriculture sector in the Nile delta to climate change impacts. Adv Sci Res 3:123–126

Badir MK, Abd AlRahman H (2017) Climate change and extreme events over Dabaa Region, Egypt. In: Leal Filho W (ed) Climate change research at universities. Springer International Publishing AG, pp 263–273. https://doi.org/10.1007/978-3-319-58214-6_16

Bardgett RD, Freeman C, Ostle NJ (2008) Microbial contributions to climate change through carbon cycle feedbacks. ISME J 2:805–814

Batisha AF (2012) Adaptation of sea level rise in Nile delta due to climate change. J Earth Sci Clim Change 2012(3):2. https://doi.org/10.4172/2157-7617.1000114

Baumert N, Kloos J (2017). Anticipating Emerging Risks and Vulnerabilities from Sea Level Rise Induced Preventive Resettlement in Greater Alexandria, Egypt. In: Behnassi M, McGlade K (eds) Environmental change and human security in Africa and the Middle East. Springer International Publishing Switzerland, pp 133–157. https://doi.org/10.1007/978-3-319-45648-5_8

Bojko O, Kabala C (2017) Organic carbon pools in mountain soils—sources of variability and predicted changes in relation to climate and land use changes. CATENA 149(Part 1):209–220

Boko M, Niang I, Nyong A, Vogel C, Githeko A, Medany M, Osman-Elasha B, Tabo R, Yanda P (2012) IPCC, 2012: managing the risks of extreme events and disasters to advance climate change adaptation. In: Field CB, Barros V, Stocker TF, Qin D, Dokken DJ, Ebi KL, Mastrandrea MD, Mach KJ, Plattner GK, Allen SK, Tignor M, Midgley PM (eds) A special report of working groups I and II of the intergovernmental panel on climate change. Cambridge University Press, Cambridge, UK, and New York, NY, USA, 582 p

Brye KR, McMullen RL, Silveira ML, Motschenbacher JMD, Smith SF, Gbur EE, Helton ML (2016) Environmental controls on soil respiration across a southern US climate gradient: a meta-analysis. Geoderma Reg 7(2):110–119

Cazenave A, N Champollion, F Paul and J Benveniste (2017) Integrative study of the mean sea level and its components. Space Sciences Series of ISSI, Vol 58. Springer International Publishing AG https://doi.org/10.1007/978-3-319-56490-6

Chen WY, Suzuki T, Lackner M (2017) Handbook of climate change mitigation and adaptation. Springer International Publishing Switzerland. https://doi.org/10.1007/978-3-319-14409-2

Classen AT, Sundqvist MK, Henning JA, Newman GS, Moore JAM, Cregger MA, Moorhead LC, Patterson CM (2015a) Direct and indirect effects of climate change on soil microbial and soil microbial plant interactions: what lies ahead? Ecosphere 6(8):1–21

Classen AT, Sundqvist MK, Henning JA, Newman GS, Moore JAM, Cregger MA, Moorhead LC, Patterson CM (2015b) Direct and indirect effects of climate change on soil microbial and soil microbial-plant interactions: what lies ahead? Ecosphere 6(8):130

Copland L, Mortimer C, White A, McCallum MR, Mueller D (2017) Factors contributing to recent arctic ice shelf losses. In: Copland L, Mueller D (eds) Arctic ice shelves and ice islands. Springer Polar Sciences, Springer Science + Business Media B.V, pp 263–285. https://doi.org/10.1007/978-94-024-1101-0_10

Coyle DR, Nagendra UJ, Taylor MK, Campbell JH, Callaham MA (2017) Soil fauna responses to natural disturbances, invasive species, and global climate change: current state of the science and a call to action. Soil Biol Biochem 110:116–133

Crowther TW, Thomas SM, Maynard DS, Baldrian P, Covey K, Frey SD, van Diepen LTA, Bradford MA (2015) Biotic interactions mediate soil microbial feedbacks to climate change. Proc Natl Acad Sci 112(22):7033–7038

Daou L, Périssol C, Luglia M, Calvert V, Criquet S (2016) Effects of drying-rewetting or freezing–thawing cycles on enzymatic activities of different Mediterranean soils. Soil Biol Biochem 93:142–149

Davidson E, Janssens I, Luo Y (2006) On the variability of respiration in terrestrial ecosystems: moving beyond Q(10). Glob. Change Biol 12:154–164

Debortoli NS, Camarinha PIM, Marengo JA, Rodrigues RR (2017) An index of Brazil's vulnerability to expected increases in natural flash flooding and landslide disasters in the context of climate change. Nat Hazards 86:557–582. https://doi.org/10.1007/s11069-016-2705-2

Durán J, Morse JL, Rodríguez A, Campbell JL, Groffman PM (2017) Differential sensitivity to climate change of C and N cycling processes across soil horizons in a northern hardwood forest. Soil Biol Biochem 107:77–84

Dutta H, Dutta A (2016) The microbial aspect of climate change. Energ Ecol Environ 1(4):209–232. https://doi.org/10.1007/s40974-016-0034-7

Edenhofer O, Pichs-Madruga R, Sokona Y, Farahani E, Kadner S, Seyboth K, Adler A, Baum I, Brunner S, Eickemeier P (2014)

Climate change 2014: mitigation of climate change. In: Contribution of working group III to the fifth assessment report of the intergovernmental panel on climate change, pp 511–597

Eid HM, El-Marsafawy SM, Ouda SA (2007) Assessing the economic impacts of climate change on agriculture in Egypt: a Ricardian approach. World Bank

El-Shaer MH, HM Eid, C Rosenzweig, A Iglesias, D Hillel (1996) Agricultural adaptation to climate change in Egypt. In: Smith JB et al (eds) Adapting to climate change. Springer, New York, Inc., pp 109–127

El Afandi G (2017) Impact of climate change on crop production. In: Chen WY et al (eds) Handbook of climate change mitigation and adaptation. Springer, Science + Business Media New York, pp 723–748. https://doi.org/10.1007/978-3-319-14409-2_64

El Massah S, Omran G (2015) Would climate change affect the imports of cereals? The case of Egypt. In: Leal Filho W (ed) Handbook of climate change adaptation. Springer, Heidelberg. https://doi.org/10.1007/978-3-642-38670-1_61

El Quosy DED (2013) Mitigation and adaptation options of climate change in irrigated agriculture in Arab Countries. In: Sivakumar MVK et al (eds) Climate change and food security in West Asia and North Africa. Springer Science + Business Media Dordrecht, pp 255–267. https://doi.org/10.1007/978-94-007-6751-5_14

Elbasiouny H, Abowaly M, Gad A, Abu_Alkheir A, Elbehiry F (2017) Restoration and sequestration of carbon and nitrogen in the degraded northern coastal area in Nile Delta, Egypt for climate change mitigation. J Coast Conserv 21:105–114. https://doi.org/10.1007/s11852-016-0475-3

El-Beltagy A, Madkour M (2012) Impact of climate change on arid lands agriculture. Agric Food Secur 1:3. http://www.agricultureandfoodsecurity.com/content/1/1/3

Eldeberky Y (2011) Coastal adaptation to sea level rise along the Nile delta, Egypt. In: WIT Transactions on Ecology and the Environment, Vol 149. WIT Press. www.witpress.com, ISSN 1743-3541 (on-line)

El-Marsafawy SM (2007) Impact of climate change on sunflower cropproduction in Egypt. Egypt J Agric Res 85(5)

El-Marsafawy SM, MAM Ibrahim, NG Ainer (2007) Vulnerability and adaptation studies on sugarcane crop in Egypt under climate change conditions and their effects on farm net return. Annals Agric Sci Moshtohor 45(2)

El-Marsafawy SM, Ainer NG, Ibrahim MAM (2017) Vulnerability and adaptation measures of Faba bean under climate change conditions in North Nile Delta and Middle Egypt. N Y Sci J (under publication)

Elnazer AA, Salman SA, Asmoay AS (2017) Flash flood hazard affected Ras Gharib city, Red Sea, Egypt: a proposed flash flood channel. Nat Hazards. https://doi.org/10.1007/s11069-017-3030-0

El-Raey M, Dewidar K, El Hattab M (1999) Adaptation to the impacts of sea level rise in Egypt. Clim Res 12:117–128

El-Ramady H, Abdalla N, Alshaal T, Domokos-Szabolcsy É, Elhawat N, Prokisch J, Sztrik A, Fári M, El-Marsafawy S, Shams MS (2015) Selenium in soils under climate change, implication for human health. Environ Chem Lett 13(1):1–19. https://doi.org/10.1007/s10311-014-0480-4

El-Ramady HR, El-Marsafawy SM, Lewis LN (2013) Sustainable agriculture and climate changes in Egypt. In: Lichtfouse E (ed) Sustainable agriculture reviews, sustainable agriculture reviews Vol 12. Springer Science + Business Media Dordrecht, pp 41–95. https://doi.org/10.1007/978-94-007-5961-9_2

Elshinnawy IA (2008) Coastal vulnerability to climate changes and adaptation assessment for coastal zones of Egypt. Ministry of Water Resources and Irrigation (MWRI). National Water Research Center (NWRC). Coastal Research Institute (CoRI). Cairo, Egypt

FAO (2016) Climate change and food security: risks and responses. Food and Agriculture Organization of The United Nations. ISBN 978-92-5-108998-9, FAO

Feng J, Wang J, Ding L, Yao P, Qiao M, Yao S (2017) Meta-analyses of the effects of major global change drivers on soil respiration across China. Atmos Environ 150:181–186

Fieguth P (2017) Global warming and climate change. In: Fieguth (ed) An introduction to complex systems. Springer International Publishing Switzerland, pp 5–12. https://doi.org/10.1007/978-3-319-44606-6_2

Fischer G, Tubiello FN, van Velthuizen H, Wiberg DA (2007) Climate change impacts on irrigation water requirements: effects of mitigation, 1990–2080. Technol Forecast Soc Chang 74(7):1083–1107

Frank D, Reichstein M, Bahn M, Thonicke K, Frank D, Mahecha MD, Smith P, Velde M, Vicca S, Babst F (2015) Effects of climate extremes on the terrestrial carbon cycle: concepts, processes and potential future impacts. Glob Change Biol 21:2861–2880

Froehlich P, Al-Saidi M (2017) Community-based adaptation to climate change in Egypt: status Quo and future policies. In: Leal Filho W (ed) Climate change research at universities. Springer International Publishing AG, pp 235–250. https://doi.org/10.1007/978-3-319-58214-6_14

Gado TA (2017) Statistical characteristics of extreme rainfall events in Egypt. In: Twentieth international water technology conference, IWTC20. Hurghada, Egypt

Gao J, Wang E, Ren W, Liu X, Yang Y (2017) Effects of simulated climate change on soil microbial biomass and enzyme activities in young Chinese fir (*Cunninghamia lanceolata*) in subtropical China. Acta Ecol Sin 37(4):272–278

Gettelman A, Rood RB (2016) Climate change and global warming. In: Gettelman A and Rood RB (eds) Demystifying climate models, earth systems data and models 2, pp 23–25. https://doi.org/10.1007/978-3-662-48959-8_3

Habtemariam LT, Kassa GA, Gandorfer M (2017) Impact of climate change on farms in smallholder farming systems: yield impacts, economic implications and distributional effects. Agric Syst 152:58–66

Hassaan MA (2013) GIS-based risk assessment for the Nile Delta coastal zone under different sea level rise scenarios case study: Kafr EL Sheikh Governorate, Egypt. J Coast Conserv 17:743–754. https://doi.org/10.1007/s11852-013-0273-0

Hassan MAER, Omran EE (2017) Modelling of land-use changes and their effects by climate change at the southern region of Port Said governorate. Egypt Model Earth Syst Environ 3:13. https://doi.org/10.1007/s40808-017-0276-1

Hay WW (2016) Global and regional aspects of climate change. In: Hay (ed) Experimenting on a small planet. Springer International Publishing Switzerland, pp 743–779. https://doi.org/10.1007/978-3-319-27404-1_32

Hinkel J, Brown S, Exner L, Nicholls RJ, Vafeidis AT, Kebede AS (2012) Sea-level rise impacts on Africa and the effects of mitigation and adaptation: an application of DIVA. Reg Environ Chang 12 (1):207–224

Hope AP, Canty TP, Salawitch RJ, Tribett WR, Bennett BF (2017) Forecasting global warming. In: Salawitch RJ et al (eds) Paris climate agreement: Beacon of Hope. Springer Climate, pp 51–113. https://doi.org/10.1007/978-3-319-46939-3_2

IPCC (2014). Climate change 2014: impacts, adaptation, and vulnerability. Contribution of working group II to the Fifth assessment report of the intergovernmental panel on climate change. Cambridge University Press, Cambridge, UK

IPCC, Intergovernmental Panel on Climate Change (2007) Climate change 2007: the physical science basis. Cambridge University Press, Cambridge

Jangra S, Mishra A, Kamboj D, Yadav NR, Yadav RC (2017) Engineering abiotic stress tolerance traits for mitigating climate change. In: Gahlawat SK et al (eds) Plant biotechnology: recent advancements and developments, pp 59–73. Springer Nature Singapore Pte Ltd. https://doi.org/10.1007/978-981-10-4732-9_3

Jeffrey M, Achurch H (2017) Save our soil to save the planet. In: Field DJ et al (eds) Global soil security, progress in soil science. Springer International Publishing Switzerland, pp 389–395. https://doi.org/10.1007/978-3-319-43394-3_35

Jevrejeva S, Matthews A, Slangen A (2017) The twentieth-century sea level budget: recent progress and challenges. Surv Geophys 38:295–307. https://doi.org/10.1007/s10712-016-9405-z

Kahiluoto H, Kaseva J, Hakala K, Himanen SJ, Jauhiainen L, Rötter RP, Salo T, Trnka M (2014) Cultivating resilience by empirically revealing response diversity. Glob Environ Change 25:186–193

Lal R (2016) Beyond COP 21: potential and challenges of the "4 per thousand" initiative. J Soil Water Conserv 71:20A–25A

Lisson S, Tarbath M, Corkrey R, Pinkard E, Laycock B, Howden S, Acuna TB, Makin A (2016) Ambient climate and soil effects on the headspace under clear mulch film. Agric Syst 142:41–50

Lozano-García B, Muñoz-Rojas M, Parras-Alcántara L (2017) Climate and land use changes effects on soil organic carbon stocks in a Mediterranean semi-natural area. Sci Total Environ 579:1249–1259

Mahmoud MA (2017) Impact of climate change on the agricultural sector in Egypt. In: Negm AM (ed) Conventional water resources and agriculture in Egypt, Handbook Environment of Chemistry. Springer International Publishing AG. https://doi.org/10.1007/698_2017_48

Mäkinen H, Kaseva J, Virkajärvi P, Kahiluoto H (2017) Shifts in soil–climate combination deserve attention. Agric For Meteorol 234–235:236–246

Mäkinen H, Kaseva J, Virkajärvi P, Kahiluoto H (2015) Managing resilience of forage crops to climate change through response diversity. Field Crops Res 183:23–30

Mäkinen H, Kaseva J, Virkajärvi P, Kahiluoto H (2016) Gaps in the capacity of modern forage crops to adapt to the changing climate in northern Europe. Mitig Adapt Strat Glob Change 1–20

Marcos M, Marzeion B, Dangendorf S, Slangen ABA, Palanisamy H, Fenoglio-Marc L (2017) Internal variability versus anthropogenic forcing on sea level and its components. Surv Geophys 38:329–348. https://doi.org/10.1007/s10712-016-9373-3

Mariani L, Parisi SG (2014) Extreme rainfalls in the mediterranean area. In Diodato N, Bellocchi G (eds) Storminess and environmental change, advances in natural and technological hazards research, Vol 39. Springer Science + Business Media Dordrecht, pp 17–37. https://doi.org/10.1007/978-94-007-7948-8_2

Marzeion B, Champollion N, Haeberli W, Langley K, Leclercq P, Paul F (2017) Observation-based estimates of global Glacier mass change and its contribution to sea-level change. Surv Geophys 38:105–130. https://doi.org/10.1007/s10712-016-9394-y

Masud MM, Nurul Azam M, Mohiuddin M, Banna H, Akhtar R, Alam ASAF, Begum H (2017) Adaptation barriers and strategies towards climate change: challenges in the agricultural sector. J Clean Prod 156:698–706

Matishov GG, Dzhenyuk SL, Moiseev DV (2017) Climate and large marine ecosystems of the arctic. Herald Russ Acad Sci 87(1):30–39. https://doi.org/10.1134/S1019331617010087

McCarl BA (2017) Economics, energy, climate change, and soil security. In: Field DJ et al (eds) Global soil security, progress in soil science. Springer International Publishing Switzerland, pp 195–205. https://doi.org/10.1007/978-3-319-43394-3_17

McCarl BA, Musumba M, Smith JB, Kirshen P, Jones R, El-Ganzori A, Ali MA, Kotb M, El-Shinnawy I, El-Agizy M, Bayoumi M, Hynninen R (2015) Climate change vulnerability and adaptation strategies in Egypt's agricultural sector. Mitig Adapt Strateg Glob Change 20:1097–1109. https://doi.org/10.1007/s11027-013-9520-9

McCarl BA (2011) Egyptian agricultural sector climate change analysis. Unpublished, Texas A&M University

McCarl BA (2015). Elaborations on climate adaptation in U.S. Agriculture. Choices 2nd quarter 30

Medany M (2016). Climate change: impacts and responses for sustainable agriculture in Egypt. CIHEAM, Watch Letter no 37—September 2016

Meehl GA, Stocker TF, Collins WD, Friedlingstein P, Gaye AT, Gregory JM, Kitoh A, Knutti R, Murphy JM, Noda A, Raper SCB, Watterson IG, Weaver AJ, Zhao ZC (2007) Global climate projections. In: Solomon et al (eds) Climate change 2007: the physical science basis. Contribution of working group I to the fourth assessment report of the intergovernmental panel on climate change. Cambridge University Press, Cambridge, United Kingdom and New York, NY, USA

Minasny B, Malone BP, McBratney AB, Angers DA, Arrouays D, Chambers A, Chaplot V, Chen Z-S, Cheng K, Das BS, Field DJ, Gimona A, Hedley CB, Hong SY, Mandal B, Marchant BP, Martin M, McConkey BG, Mulder VL, O'Rourke S, Richer-de-Forges AC, Odeh I, Padarian J, Paustian K, Pan G, Poggio L, Savin I, Stolbovoy V, Stockmann U, Sulaeman Y, Tsui C-C, Vågen T-G, van Wesemael B, Winowiecki L (2017) Soil carbon 4 per mille. Geoderma 292:59–86

Muñoz-Rojas M, Abd-Elmabod SK, Zavala LM, De la Rosa D, Jordán A (2017) Climate change impacts on soil organic carbon stocks of Mediterranean agricultural areas: a case study in Northern Egypt. Agr Ecosyst Environ 238:142–152

Najjar S, Bhatia U, Ganguly AR (2017) Climate adaptation: introduction. In: Shekhar S et al (eds) Encyclopedia of GIS. Springer International Publishing AG, pp 174–179. https://doi.org/10.1007/978-3-319-17885-1

Nally RM, Horrocks GFB, Lada H (2017) Anuran responses to pressures from high-amplitude drought–flood–drought sequences under climate change. Clim Change 141:243–257. https://doi.org/10.1007/s10584-016-1890-z

Negm AM, Saavedra O, El-Adawy A (2017) Nile delta biography: challenges and opportunities. In: Negm AM (ed) The Nile delta, handbook environment of chemistry, vol 55. Springer International Publishing Switzerland, pp 3–18. https://doi.org/10.1007/698_2016_62

Olaya-Abril A, Parras-Alcántara L, Lozano-García B, Obregón-Romero R (2017) Soil organic carbon distribution in Mediterranean areas under a climate change scenario via multiple linear regression analysis. Sci Total Environ 592:134–143

Omambia AN, Shemsanga C, Hernandez IAS (2017) Climate change impacts, vulnerability, and adaptation in East Africa (EA) and South America (SA). In: Chen WY et al (eds) Handbook of climate change mitigation and adaptation. Springer International Publishing Switzerland, pp 749–799. https://doi.org/10.1007/978-3-319-14409-2_17

Osman KT (2013) Climate change and soil. In: Osman KT (ed) Soils: principles, properties and management. Springer Science + Business Media Dordrecht, pp 253–261. https://doi.org/10.1007/978-94-007-5663-2_15

Pope S, Copland L, Alt B (2017) Recent changes in sea ice plugs along the Northern Canadian Arctic Archipelago. In: Copland L, Mueller D (eds) Arctic ice shelves and ice islands. Springer Polar Sciences, Springer Science + Business Media B.V, pp 317–342. https://doi.org/10.1007/978-94-024-1101-0_12

Radhouane L (2013) Climate change impacts on North African countries and on some Tunisian economic sectors. J Agric Environ Int Develop JAEID 107(1):101–113

Research Study Report (2014) Climate change and environment in the Arab World Program. In: Impact of population growth and climate change on water scarcity, agricultural output and food security. Issam Fares Institute for Public Policy and International Affairs American University of Beirut

Rezanezhad F, Couture R-M, Kovac R, O'Connell D, Van Cappellen P (2014) Water table fluctuations and soil biogeochemistry: an experimental approach using an automated soil column system. J Hydrol 509:245–256

Rötter RP, Höhn J, Trnka M, Fronzek S, Carter TR, Kahiluoto H (2013) Modelling shifts in agroclimate and crop cultivar response under climate change. Ecol Evol 3:4197–4214

Salawitch RJ, Canty TP, Hope AP, Tribett WR, Bennett BF (2017) Paris climate agreement: Beacon of Hope. Springer Climate Series, Springer International Publishing AG, Switzerland. https://doi.org/10.1007/978-3-319-46939-3

Schimel J, Balser TC, Wallenstein M (2007) Microbial stress response physiology and its implications for ecosystem function. Ecology 88:1386–1394

Shade A, Peter H, Allison SD et al (2012) Fundamentals of microbial community resistance and resilience. Front Aquatic Microbiol 3:417

Singh BK, Bardgett RD, Smith P, Reay DS (2010) Microorganisms and climate change: terrestrial feedbacks and mitigation options. Nat Rev Microbiol 8:779–790

Sisco MR, Bosetti V, Weber EU (2017) When do extreme weather events generate attention to climate change? Clim Change. https://doi.org/10.1007/s10584-017-1984-2

Sluiter R (2009) Interpolation methods for climate data: literature review. KNMI intern rapport; IR 2009-04. De Bilt, Netherlands

Soleimani A, Hosseini SM, Bavani ARM, Jafari M, Francaviglia R (2017) Simulating soil organic carbon stock as affected by land cover change and climate change, Hyrcanian forests (northern Iran). Sci Total Environ 599–600:1646–1657

Streck C, Gay A (2017) The role of soils in international climate change policy. In: Ginzky H et al (eds) International yearbook of soil law and policy 2016. Springer International Publishing AG, pp 105–128. https://doi.org/10.1007/978-3-319-42508-5_10

Tao F, Feng Z, Tang H, Chen Y, Kobayashi K (2017) Effects of climate change, CO_2 and O_3 on wheat productivity in Eastern China, singly and in combination. Atmos Environ 153:182–193

Tataw JT, Baier F, Krottenthaler F, Pachler B, Schwaiger E, Wyhlidal S, Formayer H, Hosch J, Baumgarten A, Zaller JG (2016) Climate change induced rainfall patterns affect wheat productivity and agroecosystem functioning dependent on soil types. Ecol Res 31:203–212. https://doi.org/10.1007/s11284-015-1328-5

Trnka M, Kersebaum KC, Eitzinger J, Hayes M, Hlavinka P, Svoboda M, Dubrovsk´y M, Semerádová D, Wardlow B, Pokorn´y E (2013) Consequences of climate change for the soil climate in Central Europe and the central plains of the United States. Clim Change 120, 405–418

Wesseh PK Jr, Lin B (2017) Climate change and agriculture under CO_2 fertilization effects and farm level adaptation: where do the models meet? Appl Energy 195:556–571

Yigini Y, Panagos P (2016) Assessment of soil organic carbon stocks under future climate and land cover changes in Europe. Sci Total Environ 557–558:838–850

Younos T, Grady CA (2013) Climate change and water resources. In: The handbook of environmental chemistry series, Vol 25, Springer, Heidelberg. https://doi.org/10.1007/978-3-642-37586-6

Zhang P, Zhang J, Chen M (2017a) Economic impacts of climate change on agriculture: the importance of additional climatic variables other than temperature and precipitation. J Environ Econ Manage 83:8–31

Zhang Q, Yang J, Koide RT, Li T, Yang H, Chu J (2017b) A meta-analysis of soil microbial biomass levels from established tree plantations over various land uses, climates and plant communities. CATENA 150:256–260

Zhang Y, Dong S, Gao Q, Liu S, Zhou H, Ganjurjav H, Wang X (2016) Climate change and human activities altered the diversity and composition of soil microbial community in alpine grasslands of the Qinghai-Tibetan Plateau. Sci Total Environ 562:353–363

Zimmer C (2010) The microbe factor and its role in our climate future. http://e360.yale.edu/feature/the_microbe_factor_and_its_role_in_our_climate_future/2279/. Accessed 15 Dec 2015

6 Geology

Heba Elbasiouny and Fathy Elbehiry

Abstract

Egypt consists of four main geological areas, Nile River valley and its delta, Western Desert, Eastern Desert, and Sinai Peninsula. The Nile Valley broadens gradually toward the north of Egypt and it is bounded by several sedimentary basins and desert sands that have been settled upon fluvial soils. The Nile Valley has three geomorphological units: the young alluvial plain, older alluvial plains, and the limestone plateau. The Nile Delta is one of the earliest identified deltaic systems in the world. It was formed by the sedimentary processes between the upper Miocene and present, then it was built up by the alluvium delivered by the old seven active branches of the Nile which flowed through the delta, then humans and nature have closed five branches. The remaining two branches are called Rosetta and Damietta. The Nile Delta is an area of fertile alluvial soils that consists of Nile deposits due to the frequent flooding during geological periods. Several shallow lagoons were developed along the Mediterranean coast and connected to the sea by small openings (Bogaz), in the sand barriers. The Eastern Desert is a part of the Arabian Nubian Shield (Shield is a collage of Neoproterozoic tectonostratigraphic terrains linked to ophiolite-decorated sutures). This desert is generally a huge and rough mountainous terrain composed of Precambrian basement (igneous and metamorphic) rocks. As well, the Eastern Desert is characterized by mountains, plateaus, and vast wadis. The Western Desert as a vast plateau desert involves the most well-known Egyptian oases (i.e., Siwa, Bahariya, Farafra, Kharga, and Dakhla) and some of the coastal basins. The landscape and shape the surface of this desert are modified by fluvial (dominant process shaping the land surface when the water movement is available) and aeolian actions (dominant where water resources are more limited). This desert is characterized by sand dunes and sand seas and many plateaus. Sinai Peninsula is the most attractive region from the geological standpoint in Egypt. This peninsula is located between the Mediterranean Sea on the north and the Red Sea on the south, Suez Gulf on the west, and Aqaba Gulf on the east. Its shape is triangular with apex formed by the connection of two Gulfs: Aqaba and Suez, and base by the Mediterranean coastline. It is characterized by very rough mountains formed by igneous and metamorphic rocks in the south, and limestone plateau in the middle and north.

Keywords

Egypt • Geology • The Nile Valley • The Nile Delta Eastern Desert • Western Desert • Sinai Peninsula

6.1 Introduction

Four main geological areas can be recognized in Egypt that are: Nile River Valley and Delta, Western Desert, Eastern Desert, and Sinai Peninsula (Fig. 6.1). Regarding the altitude, the lowest point is Qattara Depression: −133 m below sea level (b.s.l); while the highest point is Mount Catherine: 2.629 m above sea level (Balderer et al. 2014). The Nile Valley and Delta regions occupy approximately 4% of Egyptian area while other parts depend mainly on groundwater from deep aquifers or harvested rainwater. This area is one of the most population densities in the world. The other area of Egyptian land (i.e., 96%) is arid desert, which means that the need of desert reclamation seems usual considering continues population growth and increasing overcrowding in the old settled lands in the Nile valley and the delta. The Eastern Desert comprises 22% of Egyptian land. It is

H. Elbasiouny (✉)
Environmental and Biological Sciences, Al-Azhar University, Tanta, Egypt
e-mail: Hebayehia79@hotmail.com; Hebaelbasiouny@azhar.edu.eg

F. Elbehiry
Central Laboratory of Environmental Studies, Kafr El-Sheikh University, Al-Geish street, Kafr El-Sheikh, 33516, Egypt
e-mail: Fathyelbehiry@gmail.com

H. El-Ramady et al. (eds.), *The Soils of Egypt*, World Soils Book Series,
https://doi.org/10.1007/978-3-319-95516-2_6

Fig. 6.1 Geological area in Egypt; Nile Delta, Nile Valley, Eastern Desert, Western Desert, and Sinai Peninsula. Modified from: https://visibleearth.nasa.gov/view.php?id=68269, 2018

bounded by the Red Sea and Gulf of Suez in the east and by the Nile Valley in the west. The Eastern Desert region has attracted investments in petroleum exploration, tourism, mining, and agriculture (Abd El-Kawy et al. 2011; Hereher 2014; Mosaad 2017). According to Embabi (2018), the Nile Valley and its Delta represent a unique feature, in Egypt and in the north of Africa. Plateaus, sand dunes, and depressions characterize the Western Desert, while mountains characterize the Eastern Desert from north to south. Triangular shaped region, that are bound by water bodies along long stretches in all directions, characterize Sinai. Moreover, to better understand the geomorphology of Egypt, noting the Mediterranean and the Red Seas should be considered as those four regions.

The Nile delta has received great attention from scholars because of its important role in daily life and development aspects of Egypt (El Banna and Frihy 2009). It is very highly populated and hosts most of the country's population up to

1600 inhabitants/km^2 (Said 1990; El Banna and Frihy 2009; Hereher 2014). It also includes vital industrial and commercial cities, additionally; it is holding main summer tourism centers and recreation areas (El Banna and Frihy 2009). Indeed, these areas were initially recognized as significant agricultural land and ancient place for civilization (El Banna and Frihy 2009). The agricultural land extends for 1000 km long in a north–south direction between the Eastern and Western Deserts. The origin and landforms of the two deserts are different as well as in dominant natural hazards which affect the Nile Valley and Delta (Hereher 2014).

Eastern Desert Geological region of Egypt is one of the regions of the Sahara Desert. It is situated between the Nile Valley from the west and the Red Sea and Gulf of Suez from the east. It expands from Lake Manzala on the north to Sudancsc margin on the south, and further to Eritrea. The main geographical characteristic of this region is the Red Sea mountains that go along the Red Sea coast, with the highest peak, Shaiyb al-Banat (2187 m a.s.l). This region is one of the richest in natural resources such as oil and coal (Balderer et al. 2014). On the other hand, the Egyptian Western Desert belongs to the Lybian Desert, which is one of the 14 major regions of the Sahara (Balderer et al. 2014). The Western Desert of Egypt contains several extensive depressions overlying the Nubian Sandstone Aquifer System which forms isolated oases within a region of intense aridity (El-Saied et al. 2015). This desert covers two-thirds of the total area of Egypt, while the coastal basins, located in the northern half of the Western Desert 75 km to the southwest of Matruh City, Matruh, Shushan, Alamein and Natrun, occupy approximately 3800 Km^2 (Younes 2012). Sinai Peninsula is one of the main geographic regions of Egypt, it covers about 61,000 km^2 and has a triangular shape. It is separated from the Eastern Desert by the Suze Gulf. Sinai is characterized by superficial thermal manifestations characterized by a cluster of hot springs, released out of fractures and dispersed along the eastern shore of the Suze Gulf, with variable temperatures (35–72 °C) (El-Qady et al. 2005).

6.2 The Nile Valley and Its Delta

The Nile Valley is bounded by several sedimentary basins alongside its entire pathway in Egypt, such as Beni Sueif Basin at about 150 km south of Cairo in the northern part of the Nile Valley (Salem and Sehim 2017). Desert sands have been settled on fluvial soils of the Nile Valley (Hereher 2014). The Nile Valley broadens gradually toward north starting with several hundred meters (in the south) to 23 km (at the latitude of Beni Suef city, which located approximately 122 km to the south of Cairo). The Nubian Nile passes through a very narrow valley toward the south of the First Cataract, approximately 60 km south of Aswan, which was encompassed by cliffs from both sides. Currently, this reach is deluged by the Lake Nasser water, which originated after the High Dam construction in the late 1960s of the twentieth century. The whole length of Nasser Lake at 180 m is almost 500 km from Aswan to Dal Cataract in Sudan. Due to previous topography, the width of the lake differs incredibly from one place to other with an average of 12.5 km. The gradient average of the Nubian sector is 1/11,000. In the north of Aswan, the floodplain widens gradually in all locations except for specific sections (such as Silsila gorge; 60 km in the north of Aswan) where the plain disappears. At about 300 km north of Aswan, the Nile makes a large and very sharp bend (i.e., the Qena Bend), which is surrounded by high and steep limestone borders expanding up to 300 m above sea level (a.s.l). The eastern side of the Nile Valley from Aswan toward Cairo is higher and steeper in general than the western one. Most of the important wadis, which were preceding tributaries to the Nile, originate from the eastern side of the Valley. Nevertheless, the Nile Valley sides rise to the surrounding plateaus level (200–400 m a.s.l). Over this distance (1100 km), the Nile takes the eastern side of its floodplain. The width of the Nile canal ranges from 300 m in El-Silsila Gorge to 4–5 km in several parts from the Qena Bend toward the peak of the Delta to the north of Cairo Embabi (2018).

The Nile Valley has three geomorphological units: the young alluvial plain; older alluvial plains; and the limestone plateau. Pleistocene sediments are surrounded on both sides of the modern floodplain and form elevated terraces with elevations extending between 25 and 100 m overhead river level. On the other hand, the surface deposits of the Holocene floodplain of the Nile decline from 43 m above sea level in the south to 39 m in the north of El Minia governorate, the river existing at a typical level of 40 m below these deposits (Zaki 2007). The Nile Delta in Egypt is one of the earliest identified deltaic systems in the world. Herodatus, the Greek historian in 450 BC, was the first to describe its triangular shape as "Delta" because it seems like the reserved Greek letter Δ. It was formed because of sedimentary processes between the upper Miocene and present, then it is built up by the alluvium delivered by the old seven active branches of the Nile. Those distributaries have been consequently silted up and substituted by the present branches (i.e., Damietta and Rosetta). The extinct Sebennetic branch that started during the Holocene between 7500 and 3000 BP crossed over the middle of the delta and shaped its central bulge that reaches its maximum extension directly to the east of Burullus Lake (El Banna and Frihy 2009). According to A.D. reports in the first century, seven branches of the Nile flowed through the delta, then five branches have been closed by humans and nature, however they still exist. The remaining two branches are channels Rosetta, whose mouth is located just east of Alexandria, and Damietta, whose mouth is located at the northeast tip of the Delta (Zeydan 2005). The Nile Delta is a typical wave and the tide is a semi-daily microtidal system with a maximum tidal extension of 50 cm (El Banna and Frihy 2009).

The Nile Delta in Egypt, with its fringes, covers an area of 22,000 km^2. It is the most highly populated governorates in Egypt. About 60% of Egypt's population reside in the Nile Delta region. Agriculture activities are prevalent in the region (around 63% of the total agricultural land) due to the nature of the soil and an irrigation system in this area (Mabrouk et al. 2013), and it is an area of fertile alluvial soils and abundant water (Zeydan 2005). The Nile Delta aquifer is a vast leaky aquifer which is located between Cairo and the Mediterranean Sea. An upper semipermeable layer and lower impermeable rocky layer surround the productive aquifer. This aquifer is recharged by infiltration from surplus irrigation water and the very limited rainfall that infiltrates via the upper clay layer (Mabrouk et al. 2013). The Nile Delta extends about 100 miles/160 km from north to south, and 150 miles/240 km from east to west with its widest in the north (Zeydan, 2005). The length of the Nile delta coast is about 240 km, and it comprises of sandy arcuate beaches. Coastal flats are in the back of the beaches and followed by coastal dunes and three Lakes from east towards to west; Manzala, Burullus, and Idko. Except for the sand dunes, the coastal plain is distinguished by a low-relief surface <3.0 m a.s.l (El Banna and Frihy 2009).

Nile Delta is described as tide-dominated delta where river mouths hit the sea in areas affected by great tidal ranges. The delta shape can be broadly reshaped by the twice a daily flood and ebb tidal streams moving in and out of the river mouth. This often occurs in bays and estuaries where the river mouth is protected from much wave action. The relentless in and out currents of tides can incise the sediment into elongate tidal bars. At the bay heading, there may be a classic delta shape, in this location referred as bay-head delta, but beyond seaward is a zone of lots tidal bars, islands, and inlets produced by tide alteration. A vertical stratigraphic section through this type of deposit will be dominated by a lot of mud and sands that show bidirectional (ebb-directed and flood directed) cross sheet (Zeydan 2005). The Nile Delta is characterized by rich silt soil which considers the most fertile soil in all Africa. The depth topsoil in the Nile Delta varies from 15.24 to 22.86 m, while it is measured in many places in the world in mere inches. The delta area provides plenty of water and rich soils needed for the crops to produce sufficient food which interpret its overcoming the 7-year famine that happened in the past at the time of Jacob's entrance into Egypt (Zeydan 2005).

The Nile Delta forms the vast plain between the two branches of the river. This plain contains the most fertile agricultural and industrial area in Egypt (Said 1990). The geology of the Nile Delta and Nile Valley areas is largely divided into two geologic units; Nile River alluvium and undifferentiated basement rocks. The River Nile alluvium comprises of the River Nile sands and the clay-silt layer. The River Nile sands consist mainly of beds of coarse and fine sands and are mainly covered by the nearby surface clay-silt layer as a cap to the aquifer. The thickness of the alluvium in the Nile Delta varies from approximately 200 to 300 m. The thickness of the clay-silt layer in the Nile Delta is thicker to the north, ranging from about 5 m near Cairo to >50 m close to the Mediterranean Coast. On the other hand, the thickness declines toward the Delta fringes. The clay-silt layer in this area forms the fertile agricultural lands. The soils of alluvial and alluvio-marine deposits contain loam and clay to clay-loam (Moustaf 2000).

According to FAO (1964), most of Nile Delta soils consist of Nile deposits due to the frequent flooding during geological periods. Based on Hamdi and Abdelhafez (1999), the Egyptian soils involve the alluvial soils of the Nile Delta and Valley, the calcareous soils alongside the coastal shoreline of Egypt, the soils of the Eastern, and Western Deserts with the soils of Sinai Peninsula. The major alluvial soils were developed from the suspended solid matter of the Nile, which were deposited every year throughout the flood season. The suspended matter of the Nile is formed from the breakdown of the eruptive and metamorphic rocks of the Ethiopian plateau due to physical, chemical, and biological weathering factors (Hamdi and Abdelhafez 1999). Generally, there are two geomorphic parts in the central part of the Delta; the younger deltaic plain and the Mediterranean coastal plain (Hamdi and Abdelhafez 1999). The main geological features of Egypt are presented in Fig. 6.2.

In the north of Cairo, the Nile Valley opens to the Nile Delta. Because of its shape, Herodotus named the Nile Delta in 450 BC as a first of its kind, after his visit to Egypt and realizing the great similarity between its shape and the shape of the Greek letter D. Before cleaving into two branches, the Nile inflows for about 20 km towards northwest. The two current branches of Delta are eastern branch (i.e., Damietta Branch) and the western branch (i.e., Rosetta branch). Those two branches are the last of the several others that disappeared. The canals of these two branches, like the main Nile canal, are characterized by meandering and the presence of more than 400 silt islands. The delta as a vast depositional plain has gradient about 1/10,000. Several shallow lagoons such as Manzala, Burullus, Idku, and Maryut were developed along the Mediterranean coast and connected to the sea by small openings, called Bogaz, in the sand barriers, which isolate them from the sea (Embabi 2018).

The delta morphology was categorized into six major units based on aerial photographing in 1955 and field observations; the nearshore area up to 7 m water depth, beach and coastal planes, coastal deposit sand ridges,

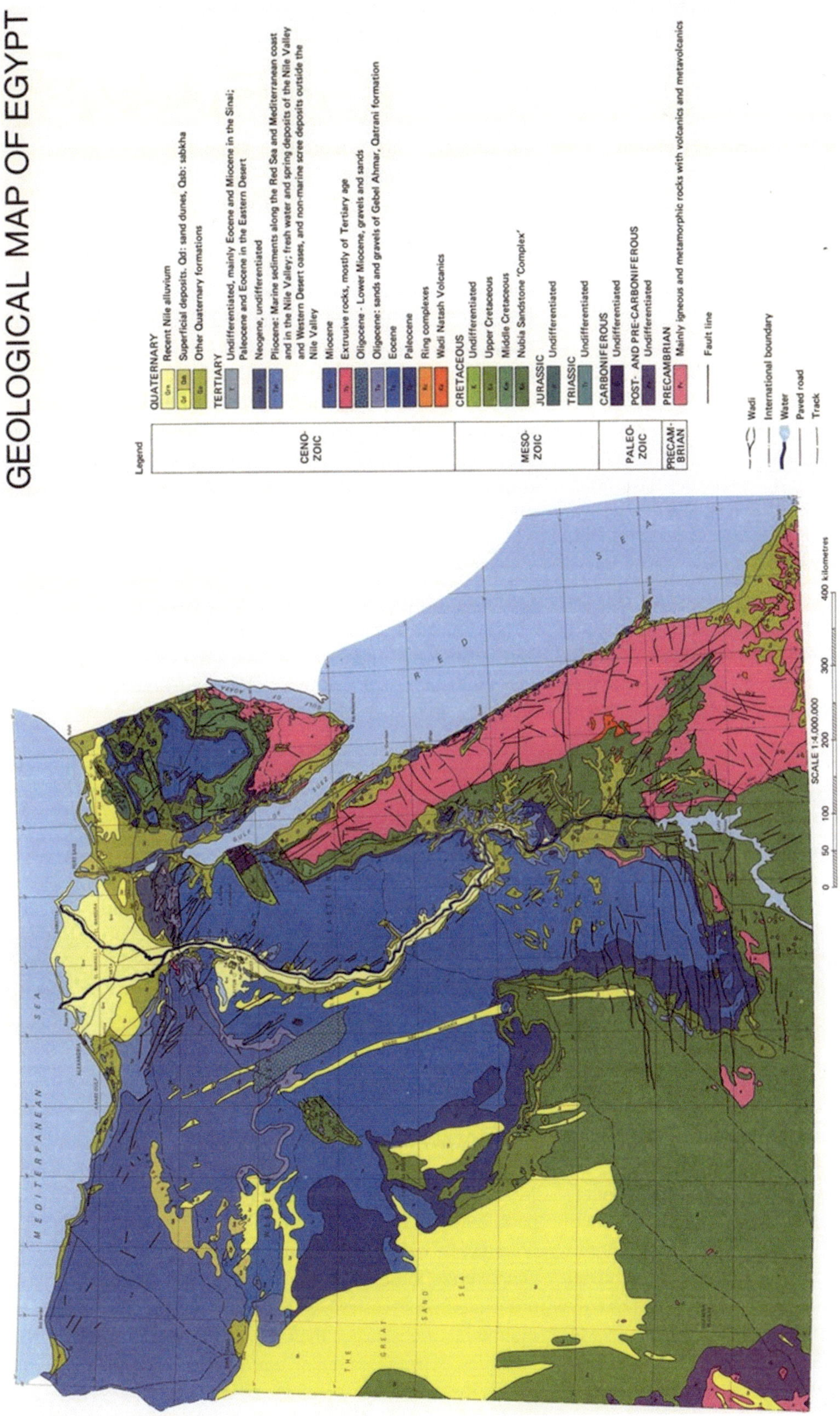

Fig. 6.2 Geology of Egypt Adopted from ESDAC (2017)

carbonate ridges, coastal dunes, and coastal lakes. Dunes in the coastal zone between Baltim and Gamasa have been classified into young active and old stabilized. It was conducted by analyzing the historical maps covering both nineteenth and twentieth centuries, and gradual extension of Rosetta and Damietta delta promontories during the 1800s, into 3–4 km of the Mediterranean Sea. This extension is due to delivering large quantities of sediments to the coast by both Nile branches; Rosetta and Damietta (El Banna and Frihy 2009).

6.3 The Eastern Desert

The Eastern Desert presents 22% of Egypt; it is generally a huge and rough mountainous terrain composed of Precambrian basement (igneous and metamorphic) rocks, which are separated by several dry streams that drain either toward the Nile Valley or at the Red Sea coast. Therefore, flash floods are the most considerable environmental hazards impacting the eastern side of the Nile Valley (Hereher 2014). The Eastern Desert of Egypt is part of the Arabian-Nubian Shield which shaped after Rodinia fragmentation between 900 and 800 Ma, as implied from the oldest ($\sim$870 Ma) juvenile Neoproterozoic rocks in the Shield. Neoproterozoic crustal growth of the Arabian-Nubian Shield was accomplished mostly through the accumulation of island curves to continental boundaries, as the Mozambique Ocean between the West and East Gondwana supercontinents was sealed. Subduction, deposit, and crustal thickening processes took place between about 870 and 570 Ma. The Arabian-Nubian Shield is a collage of Neoproterozoic tectonostratigraphic terrains linked to ophiolite-decorated sutures. Zircon data record ocean crust generation between $\sim$810 and $\sim$730 Ma ago in the Nubian segment of the Arabian-Nubian Shield (El Bahariya 2012). El-Shazly and Khalil (2016) reported that the Eastern Desert of Egypt is part of the Arabian–Nubian Shield, clarified as a series of Neoproterozoic intra-oceanic arcs combined through the climax of the Pan-African Orogeny c. 630 Ma. The Central Eastern Desert is the only terrane in the Eastern Desert with banded iron formations and plentiful ophiolitic rocks, which make it grab the attention of many structural, geochemical, and geochronologic studies. Gad and Kusky (2006) stated that as part of the Pan-African Arabian-Nubian Shield, the Eastern Desert of Egypt is occupied by igneous and metamorphic rocks that were formed in the East African Orogen during the collision between East and West Gondwana and the closure of the Mozambique ocean 600 Ma and are well exposed all over the Red Sea Hills in Egypt, Sudan, Ethiopia, and Saudi Arabia. Eastern Desert lithosphere is of concern to the global geoscience researchers because it is an excellent pattern of the $\sim$4 million km^2 of Neoproterozoic lithosphere in N Africa (Stern 2017).

As the Central Eastern Desert is dominated by Neoproterozoic rocks, two major tectonostratigraphic units characterize it: structural basement of gneisses and related amphibolites and ophiolite island arc associations of low-quality metamorphism. The latter was intensely deformed during oblique collision and accretion of island arcs onto the Saharan forming an ophiolitic mélange (El Bahariya 2012). The Neoproterozoic Egyptian granites are commonly classified into Older, Younger, and Alkali granites. Older granites comprising about 27% of the Precambrian rocks in the Eastern Desert include trondhjemite, tonalite, granodiorite and granite and are described as gray granites, G1 granites and/or subduction-related granite, or island arc plutonic rocks. Younger granites, consisting about 16% of the Eastern Desert basement, are referred to as late to post-orogenic pink and red granites. The younger granites range in constitution from calc-alkaline to alkaline. They are categorized into three phases of I, II, and III or G2 granites. Alkali granites comprise approximately 0.2% of the Precambrian exposure in the Eastern Desert and referred as or G3 granites. They include two-mica granite, alkaline granite, and syenite and represent post-tectonic alkaline to peralkaline rocks in which a primary phase is featured by potassic, subalkaline to peraluminous biotite granite (Asran et al. 2017).

The Eastern Desert is characterized by mountains, plateaus, and vast wadis. The Red Sea Mountains, in this region, and the fringing plateaus are all separated by drainage nets, that drain either to the east in the Red Sea or to the west in the River Nile. Some montane peaks in this region are not continuous but disconnected by basins and deep wadis. The elevation of the peaks may reach more than 2000 m a.s.l, whereas around 30 peaks elevate more than 1000 m a.s.l, (such as Gabal Qattar, Gabal Abu Abid, Gabal Hamata, Gabal Shendib, and Gabal Shayeb Al-Banat where their elevations are 1963, 1900, 1977, 1911, and 2187 m, respectively. The plateaus peaks vary significantly between more than 1000 and some hundreds of meters above sea level. At the south of Suez Gulf, there are few localities which are few meters below sea level. The Red Sea Mountains expand southward from Gabal Um Tenasib in parallel to the Red Sea coastline to and beyond the Egyptian boundaries within Sudan. Igneous and metamorphic rocks characterize these mountains, with a thin sedimentary covering along the eastern wards. The eastern slopes of these mountains are sharper than the western wards because of tectonics, forming a narrow coastal plain along the majority of the Red Sea coast (Embabi 2018).

6.4 The Western Desert

This geological region of Egypt covers approximately 66% of the country's area. It is a vast plateau desert, it extended from the Nile Valley in the east towards the Egypt-Libyan boundary in the west, and from the Mediterranean coast in the north towards the Egypt-Sudan boundary in the south. It is one of the major arid regions in Egypt. However, it involves the most well-known Egyptian oases (i.e., Siwa, Bahariya, Farafra, Kharga, and Dakhla) (Younes 2012; Balderer et al. 2014; Hereher 2014). The coastal basins (Matruh, Shushan, Alamein, and Natrun) are situated in the north of the Western Desert 75 km to the southwest of Matruh City, covering an area of about 3800 Km^2 which forms the major part of the unstable shelf (Younes 2012). Fluvial and aeolian actions are important means to modify landscapes and shape the surface of the Earth, through a range local to regional scales. Fluvial actions are often dominant process shaping the land surface when the water movement is available, while aeolian action is dominant where water resources are more limited. Previous studies that have focused on the interactions between fluvial and aeolian processes tend to investigate the perspective of either fluvial or aeolian research. There is an observation about the effect of flow on the development of aeolian characters. Through the action of fluvial erosion, sand dunes adjacent channels can be altered in both size and location. On the other hand, sand dune can slump into rivers because of sliding of dune edges from river erosion and transporting and redepositing the sand by river which, in turn, determines the location of new dunes obtained by this sediment. Such types of dunes are called source-bordering dunes as they directly border the downwind side of their sediment supply from sand bed streams. In addition, some river systems can interrupt the sediment transport from dune fields to the limit that they block dune movement (Liu and Coulthard 2015). Aeolian sand deposits represented 6% globally and 97% of the deposits in arid lands. Aeolian deposits form sand seas, dune fields and sand sheets. Sand seas represent the largest sand-covered area more than 100 km^2 and dune fields are generally less than 100 km^2. Sand sheets are sand bedforms of significant dune deposits. Sand sea involves not only an area covered by sand dunes but also these dunes should cover at least 50% of the total area. In Egypt—the eastern part of the Sahara—aeolian deposits account for about 16% of total area and they significantly occur in two regions: the Western Desert and the North Sinai. The Western Desert contains a vast sand sea (the Great Sand Sea) and many dune fields (South Qattara, Abu-Muharik, Farafra, Kharga and west Asyut). Sand dunes are a common attribute of many desert regions all over the world and they may occur in many variable types. Barchan dunes are the simplest and most usual dune type as it keeps a crescent shape and exists in areas characterized by steady winds coming from the same direction through the year, if there is no enough sand to cover all regional surface (Hamdan et al. 2016). As well scattered dune accumulations are common in the Nile Delta and Nile Valley (Hereher 2014). Hamdan et al. (2016) indicated that many areas in the Western Desert of Egypt are featured by high accumulations of wind-blown sand. Modern aeolian sand of the Western Desert of Egypt has been in the scientists' interest since the start of the twentieth century.

The Libyan Plateau of Central Egypt in most places is covered by the Eocene Thebes Group limestones. The Libyan Plateau currently a hyperarid environment as well the adjacent Nile Valley cities receive average about less than a millimeter of rainfall per year. This area has most likely been an arid region since the early Holocene, when pluvial conditions were the last present in North Africa. However, a significant humid period occurred across the Sahara ca. 120–140 ka, these wetter conditions may have driven the use of the Libyan Plateau by Middle Paleolithic groups, and would clarify the presence of earlier artifact accumulation in this currently unattractive environment (Adelsberger and Smith 2009). The Libyan Desert occupies a plateau underlied by primarily carbonate rocks of Paleocene and Eocene age (65–35 Ma), which slope very gently north (Brookes 2001). Given that the granite weathering rate is in a similar climate Central Namib Desert above 30–80 cm per 100 ka, surface lowering in the Libyan Plateau where limestone bedrock is bare has most likely averaged above meter per 100 ka. However, plateau surfaces here are practically regularly covered by desert pavement, with very tiny post-Eocene sedimentary deposition other than limited soil formation and local deposition of "ancient" (undifferentiated Oligocene to Pleistocene gravels). The same nature of the pavement surface in this area and the deficiency of lager post-Eocene sedimentation or erosion show widespread soil horizons stability. Surface lowering has probably been limited to exposed (pavement-free) surfaces, whereas sediment deposition occurs only where the surface gravels produce appropriate surface coarseness to act as a dust trap. These processes reduce sediment removal and accumulation because they both require large geologic timescales to effect considerable change on a landscape. Thus, the plateau surface has probably remained rather unchanged since Middle Paleolithic eras (140–40 ka) (Adelsberger and Smith 2009).

The Western Desert is the region of many geomorphological features such as plateaus, depressions, sand dunes, and plains. As well, it is a region of desert oases that rely basically on Nubian aquifers groundwater. Most of the Western Desert does not elevate more than 300–400 m a.s.l, except for the southwestern part of the Gilf Kebir and Uwienat that elevates more than 1000 a.s.l in some places.

However, there are many locations lies below the sea level in this desert. Qattara Depression is the deepest point in this desert −134 m to −145 m, while for Siwa, Fayum, Wadi El-Natrun, and Wadi El-Rayan they are −18.5, −45, −24, and −64 m, respectively. The southern part of Western Desert is a vast plain bounded by Kharga-Dakhla Depression, Gilf Kebir Plateau, Eocene Plateau the Nile Valley in the north, east and west respectively and extended from the south to northern Sudan. This plain is called the Arabin Desert (Forty's Desert) because of the well-known caravan route Darb El-Arbain, which crosses the middle part of this plain. Most of this plain rise from 150 to 200 m a.s.l, however it gets higher in the west until it reaches about 600 m a.s.l at the foot slopes of the Gilf Kebir Plateau. The level of the ground decreases and reaches few meters above sea level in Kharga. Hills in the depressions, such as Gabal El-Ter, Gabal Taref, Gabal El-Qarn and Gabal Um El-Ghanaiem in Kharga and Gabal Edminston in Dakhla may rise to 100–300 m a.s.l (Embabi 2018).

The Western Desert is overlaid by several dune fields and, hence, eroding sand is the obvious environmental threat affecting the western part of the Nile Valley. The predominant wind direction across Egypt's Western Desert is generally northwest, with lesser occurrence of southwest winds in the winter. This pattern of wind regime has a remarkable effect for interpreting the force to sands in the huge Western Desert to move toward southeast (Hereher 2014). Dunes are distributed in the depressions and organized like belts. Barchans move southward by the effect of northern winds representing a permanent hazard for adjacent highways, villages farms, and wells. The Great Sand Sea as an eminent feature in the Western Desert occupies a vast basin covering about 600 km expanding from the northern cliffs of the Gilf Kebir and Abu Ras plateaus (in the south) to the southern borders of Siwa Depression (in the north). The floor of the northern part of the Sand Sea approaches the sea level or located below it. The linear predominant pattern of dune in the Great Sand Sea modifies to parallel and Y-junction from north toward south. This Sand Sea considers the key source of sand, which creates the small dune field extending from the south of the Gilf Kebir Plateau toward Uwienat and northwest Sudan. Several kilometers toward the south of Gilf Kebir, a small unique plateau is located. This plateau composed of igneous and metamorphic rocks and its elevation ranges from 850 to 950 m a.s.l, furthermore it has a single peak elevated to 1114 m a.s.l (Embabi 2018). Sand dune and Sand Sea is presented in Fig. 6.3.

The Western Desert of Egypt consists of many plateaus; the Eocene limestone plateau is the largest one, it occupies about 135 km^2 (i.e., 20% of the Western Desert). Most of this plateau expand between the Nile Valley in the east and the Western Desert depressions (Kharga, Dakhla, Farafra, and Bahariya) in the west. This plateau is presently affected by a hyperarid climate. Therefore, wind action and thermal weathering represent the supreme dynamic geomorphological processes that affect its present surface. However, this plateau has subjected to pluvial periods during the prior geological ages, especially during the Oligocene, Miocene, Pliocene, Pleistocene, and Early Holocene. Several paleokarst features originate in these periods, such as caves, cone karst, karst depressions, shafts, Karren, and red soil. Some of these features have been changed or eroded by wind action and thermal weathering, while some others are still preserved in the host rock. Several paleodrainage systems were shaped in the previous wet periods. All the flows of these systems disappeared and currently are not found in the landscape of the Western Desert; whereas the remaining are many inverted valleys born of topographic inversion in the areas that gravel protected them from erosion (Mostafa 2013).

The south of the Egyptian Western Desert contains significantly a choppy eolian peneplain. It extends to about few hundreds of kilometers, and is intruded around by sand seas, low ridges, cliffs, separated plateaus, and shallow flat-bottomed valleys. Thin, densely packed sand sheets cover the surface, mantled in places by desert lags differs from sand to granule, and cobblestones. The landscape is ancient, while the sand and the lag are mature (El-Baz and Maxwell 1982).

6.4.1 White Desert

White Desert is known as a part of the Egyptian Western Desert, located approximately 45 km northeast of Qasr el-Farafra. The area is famous for its spectacular scenery. The main component of this desert is massive chalk rock formations and has a white, chalk-white, and cream color. The landscape constitutes scattered massive chalk rocks and stones of different, strange shapes. Wind and sand erosion are the main factors for a landscape formation of this area throughout the broken-down chalk plateau, leaving the harder rock shapes settled while the softer parts eroded away. The White Desert is considered as the natural wonder of Egypt, and is now a protection area, known as the White Desert Park (Balderer et al. 2014).

6.4.2 Black Desert

Black Desert is part of the Egyptian Western Desert, situated approximately 50 km to the south of Bawiti, a slight to the north and northeast of the White Desert. The surface layer of the desert is covered by rocks and gravels of brown-orange and black color eroded from volcanic material. Most of these dark rocks are comprised of diabase. The main features of the landscape in this desert are hills and single

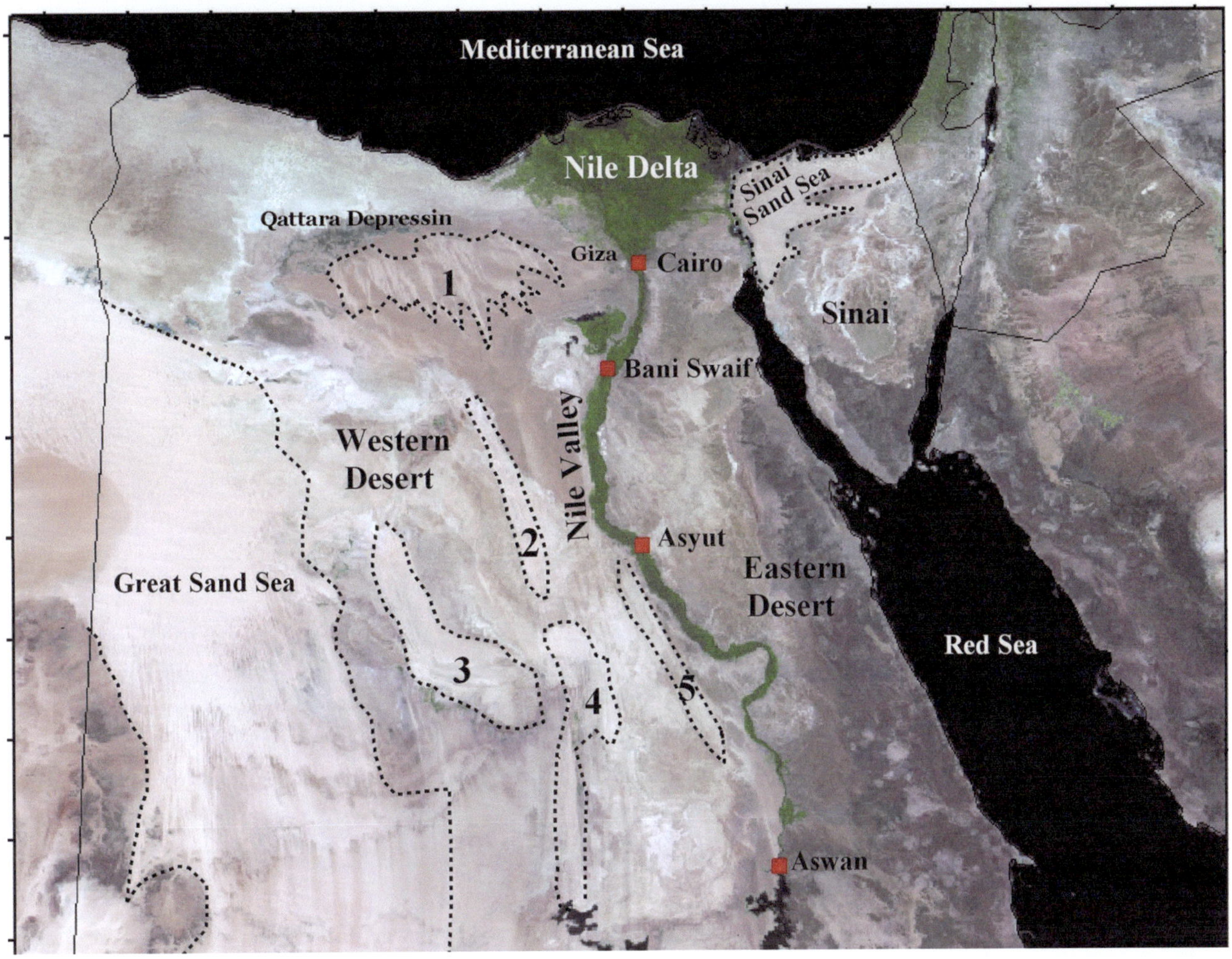

Fig. 6.3 Sand dune accumulations in the Western Desert of Egypt and the north Sinai Sand Sea. Dune fields include: (1) South Qattara, (2) Abu-Muharik, (3) Farafra, (4) Kharga, and (5) west Asyut. Modified from Hereher (2014)

volcano-shaped mountains with a considerable amount of black and brown-colored stones in the upper layers and on the slopes between beige sands. The highest top is the Gebel el-Ingiz (Balderer et al. 2014). Plyusnina et al. (2016) mentioned that the black conical hills, which are spread over the south of the Bahariya Depression, and are known as "the Black Desert". Nemours of these hills are covered by basalt sills which give them a featured black color.

6.4.3 Oases in the Western Egyptian Desert

The Western Desert of Egypt includes numerous extensive depressions (Fig. 6.4) covering the Nubian Sandstone Aquifer System which forms separated oases within intense aridity region. Rashed (2016) defined the Oasis as an isolated fertile area, commonly limited in extent and enclosed by desert. The oases are depressions in the desert encompassing springs, wells, and trees, which reveal the beauty, glamor, and diversity of nature. The oases have an extended history and have supported human populations, trade roads, and agriculture for millennia. Prior to the development of modern rock boring apparatus, agriculture was completely dependent on the utilization of springs and ancient artesian wells, but recently, the oases have been significantly modified through the implementation of ambitious desert reclamation plans. Their social and economic isolation has also been drastically reduced through improved trade, transport, and communication networks as well as tries to bring them under the central administration of the Egyptian government

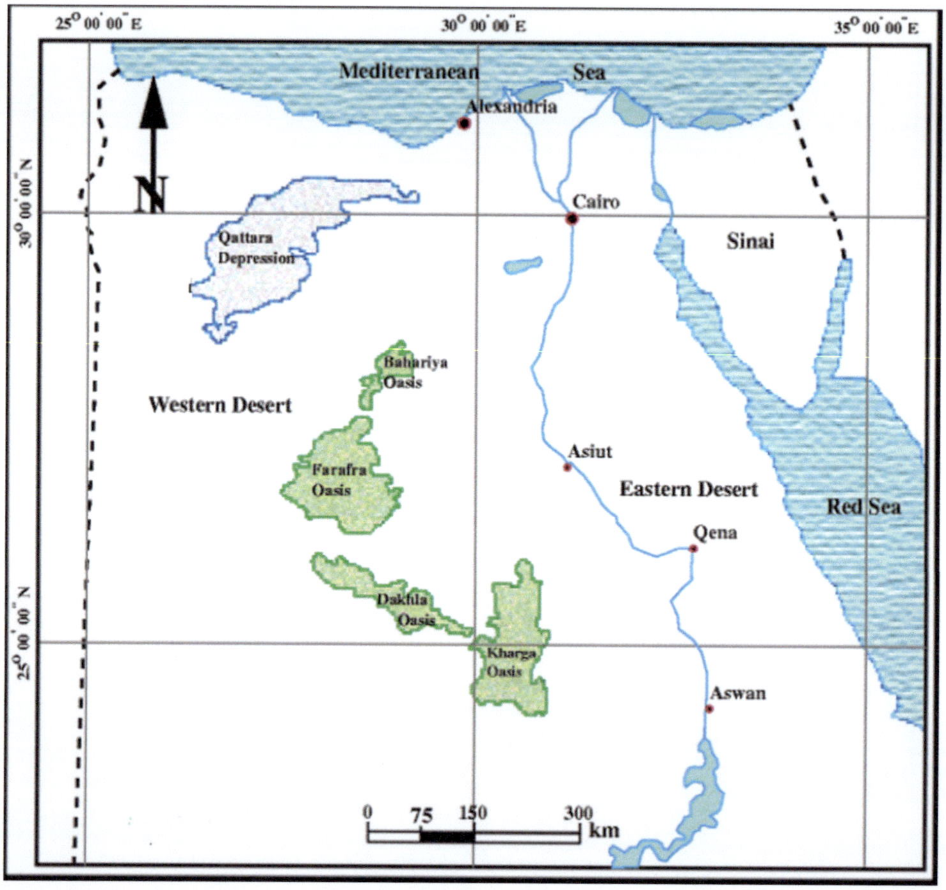

Fig. 6.4 Egyptian oasis in the western desert (green color). Modified from Abdulaziz and Faid (2015)

(El-Saied et al. 2015). The formation of the many oases in the Western Desert is correlated to hard limestones thickness above the clastic beds, as well to underground water table level. Geological structure is very important in the formation of these depressions. In all recognized oases in the Western Desert, the successions include quite thin limestones on a thick succession of clastics (Plyusnina et al. 2016).

Five major depressions dominated the northern section of the Western Desert; Siwa, Qattara, Wadi El-Natrun, Wadi El-Rayan, and Fayum, as well the Miocene plateau. Bahr Youssef bonds the Fayum Depression with the Nile Valley. Therefore, it is the only depression whose soil is similar to the Nile Valley's soil and it is cultivated and irrigated by the Nile water. All these five depressions are located below the sea level. The Qattara Depression area is the largest in terms of the area, 45,000 km^2, and the depth, −145 m not only in Egypt but also in Africa (Embabi 2018).

6.4.3.1 Dakhla Oasis

This oasis is located in the western desert of Egypt, approximately 750 km southwest of Cairo. The study region covers an area of approximately 500 km^2. The bottom of Dakhla depression is bordered from the north and northeast by steep scarps of the Eocene Limestone plateau, but slowly rises to the south where it merges with the plain of Upper Cretaceous Nubian sandstone (Ghadiry et al. 2012). Dakhla Oasis occupies one of a chain of structurally motivated topographic depressions underneath the edge of the Libyan Plateau, centered at 25.5°N, 29°E. The depression is confined within the 140 m contour, containing a larger western basin with least elevation of 92 m and a leaser eastern one down to 121 m. The settled oasis located in the central, better watered parts of those basins (Brookes 1993). It is estimated that 12,140.57 ha of land are being cultivated within the Dakhla Oasis. The population of this oasis are estimated by 70,000 residents who inhibited in the rural areas of the depression, and most of them are farmers who struggle against the threats of mobile sand dunes on their fields. The main dominant wind is North–Northwest; a bidirectional wind system that probably caused the tortuosity of the northern dunes. In the south of the oasis, the variability of wind pattern is reflected by the existence of diverse sand dunes (i.e., Barchan, inter-dune areas). Some of the inter-dune areas have been recently cultivated, which give the dunes the look of being stabilized. However, it is recorded that the dunes are still developing and threatening the adjacent farmlands and urban areas in Dakhla Oases (Ghadiry et al. 2012).

The cuesta landscape of the Dakhla is originated on a slightly north dipping series of sedimentary rocks of Cenomanian to Palaeocene age. Palaeocene Tarawan/Garra/Kurkur Formation limestones underlie the Libyan Plateau, protecting Dakhla Formation shales from erosion alongside the border scarp. Promontories and embayments in the scarp have been established on synclinal and anticlinal swells; slight faults influence some straight scarp fragments in the west. Most of the scarp and the entire oasis depression are covered by Dakhla shales and Quseir. Respectively, formation mudstones, but a smaller scarp intervenes grew on Duwi Formation mudstones, silicified limestones, and phosphorites. The depression to the south is bordered by a low cuesta. Developed on Nubia Group, Taref Formation sandstone, it looks out over an extensive southern plain spotted with hills formed on older Nubia sandstone formations (Brookes 1993).

6.4.3.2 Kharga Oasis

The Kharga Oasis is located in the New Valley Governorate between latitudes 24° and 26° north, approximately 200 km to the west of the Nile Valley. This oasis occurs in the hyperarid climate in the western desert, where life activities rely on deep wells tapping artesian water from the Nubian Sandstone Aquifr. In this Oasis, the upper Palaeocene to lower Eocene sedimentary succession contains: the upper part of the Dakhla Shale, the Tarawan Chalk, the Esna Shale, and the Thebes formations respectively from the base to the top (Mohamaden et al. 2016; Faris et al. 2017). Blackwell et al. (2017) reported that sediments such as lacustrine and palustrine, in addition to artesian spring and tufa units deposited under the Escarpment edge, all reveal that areas in and close to Kharga Oasis had surface water frequently through the Pleistocene. Embabi (2018) stated that both depressions; Kharga and Dakhla could be considered as a very large Depression, because of high escarpments surrounding them in the east and north without any break. Though there is relatively high ground splitting them to the south of the Abu Tartur Plateau, the differentiation between Kharga and Dakhla is indiscernible. Similar forms have developed in both and human occupation history is approximately similar.

6.4.3.3 Siwa Oasis

It is the farest Oasis depression from the Nile Valley towards the west and situated 300 km southern the Mediterranean coast with Latitude from 29°100′ to 29°16′N Longitude from 25°27′ to 25°35′E. As well, it is the most westerly and remotely Egypt's major populated oases (El-Saied et al. 2015). Siwa Oasis occupies an elongated depression in the northwestern part of the western desert, 300 km from Mersa Matruh (Abdulaziz and Faid 2015). The depression extends around 50 km in length, varies from 2 to 20 km in width, and covers about 1000 km^2. Siwa Oasis extends from 0 to 18 m below sea level. It presents a microcosm of current economic, social, and environmental changes (El-Saied et al. 2015). Siwa is the smallest oasis in the Egyptian part of the extensive Libyan Desert and depends exclusively on groundwater resources and drainage water reuse (Abdulaziz and Faid 2015). It displays various landforms including salt lakes, salt marshes (Sabkhas) as well as cultivated lands and orchards new and developed irrigation facilities have occupied widespread strips of desert, however but wasteful groundwater use has resulted in the extension of naturally existing salt lakes and loss of arable land due to waterlogging and land salinization. Many springs were modified and developed since early times and the practice of excavating and lining springs and canals has been shown to eliminate or change the adjacent wetland habitat (El-Saied et al. 2015). The soil of Siwa consists primarily of particles of limestone and sandstone derived from the walls and the floor of the Siwian depression or carried by the winds. It contains small amounts of clay (about 6.9%), larger proportions of sand (59%), and large amounts of soluble matter. The amount of sodium chloride found in Oasis soils ranges from about 0.12 to 59.12% (El-Saied et al. 2015).

The depression is edged by high Miocene escarpments along the northern face and extensive sand dunes along its southern flank (El-Saied et al. 2015). Five local depressions are easily distinguished in Siwa Oasis. These depressions contain the important recognized lakes from west to east as Maraqi, Siwa, Aghourmy, and Zeitoun lakes (Fig. 6.5) (Abdulaziz and Faid 2015). Siwa is also surrounded by several smaller oases which were inhabited in ancient times but are no longer occupied (El-Saied et al. 2015).

6.4.3.4 Baharia Oasis

The Bahariya Oasis is located at the center of the Western Desert of Egypt between 27°48′ to 28°30′N and 28°35′ to 29°10′E Fig. 6.4. It has oval shape. It is stretching northeast to southwest around 94 km and the width of the depression is about 42 km. This Oasis is interesting in many studies aiming at structural geology, stratigraphy, iron ore deposits, sedimentology, paleontology, geoarchaeology, etc. The Bahariya Formation is represented as ferruginous sandstones and shales, and outcrops on the base of the depression. The weathering converted the siliciclastic beds of the Bahariya formation into black conical-like hills, mesas, and buttes. The black conical hills are spread over the south of the Bahariya Depression. The Paleocene is absent in the Campanian, Maastrichtian, and lower Eocene rocks ridge on the edges of the Bahariya Depression. The Bahariya depression superposes the Bahariya anticline that stretches from the north in Gebel Ghorabi, passing throughout the central hills in the depression to the south of the oasis, and extends southward to include the Farafra Oasis (Salama et al. 2012; Plyusnina et al. 2016).

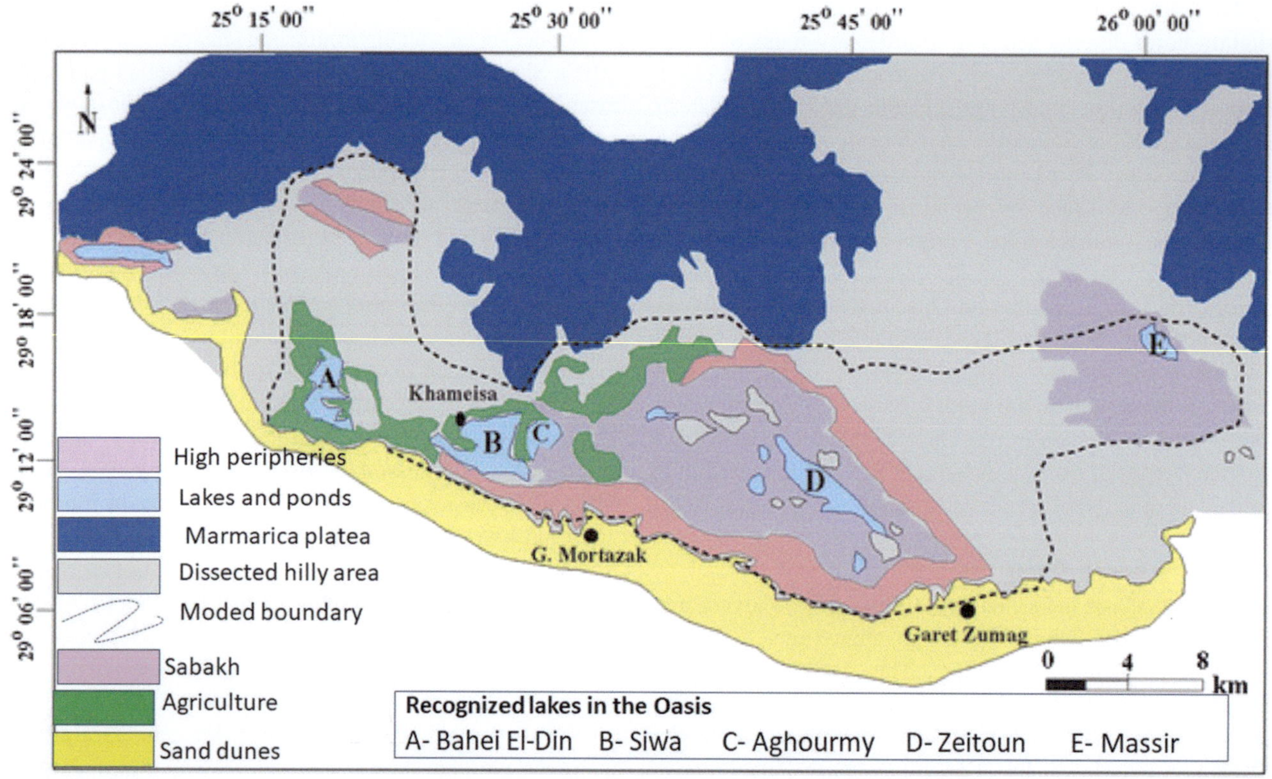

Fig. 6.5 Geomorphological features in Siwa Oasis. Modified from Abdulaziz and Faid (2015)

6.4.3.5 Farafra Oasis

The Farafra Oasis is located about 140 km southwest of the Bahariya Oasis in the central part of the Western Desert between 26°45′–27°40′N and 27°00′–28°50′E (Fig. 6.6). Farafra Osis, such as the Bahariya Oasis, has an oval-shaped depression with an area of approximately 10,000 km^2. The oasis is located about 200 km southeast of Baharia Oasis and about 300 km northeast of Dakhla Oasis. The main settlement in Farafra is Qasr Farafra. On the floor of the Farafra Depression, the Dakhla Shale (Maastrichtian) is outcropped, and it is intertonguing horizontally into the Maastrichtian Khoman Chalk in the center and north of the oasis (Hassan et al. 2001; Plyusnina et al. 2016). It is bounded from three sides; north, east, and west by escarpments of the nearby desert plateaus (270–300 m a.s.l with residual hills as high as 350 m asl). The eastern edge of the Great Sand Sea encroaches on the southwestern border of the greater Farafra depression (Hassan et al. 2001). The Farafra Depression is bounded by high ridges, and its bottom rises slowly to the usual level of the nearby desert toward the south. The scarps of the Farafra Depression consist of the Tarawan Formation (Paleocene) covered by the Esna Shale (Paleocene-lower Eocene) and the Farafra Limestones (lower Eocene). The east of the depression is overlaid by sand sheets, and the depression is surrounded to the west by the Great Sand Sea.

The Farafra Depression forms a cupola structure, which represents the southern expansion of the Syrian Arc System. Its axis stretches in the northeast to southwest direction (Plyusnina et al. 2016).

6.4.4 Toshka Area

The southeastern part of the Egyptian Western Desert is basically exemplified by the Toshka area which covers around 50,000 km^2. It lies between 22° 15′N and 23° 15′N Latitude and 31° 00′E and 32° 00′E Latitude. The bare rock units in Tushka area expand from Precambrian Era to Quaternary Era. However, the resistant Paleocene Esna Shale mostly covers the depression. The Precambrian is characterized by low-lying exposures of the basement rocks which are covered by Palaeozoic and Mesozoic successions. The lower part of these sequences is made of undistinguishable conglomerate, sandstone, and shale (Gilf Formation), while the mid-part is made of coarse gravelly sandstone (Abu Simbel Formation), fine sandstone and shale (Abu Ballas Formation), and the sandstone of El Borg Formation (Bastawesy et al. 2008). The Toshka area is featured by high dune sand accumulation and affected mainly by strong trade winds which use a strong control on the mechanism of dune

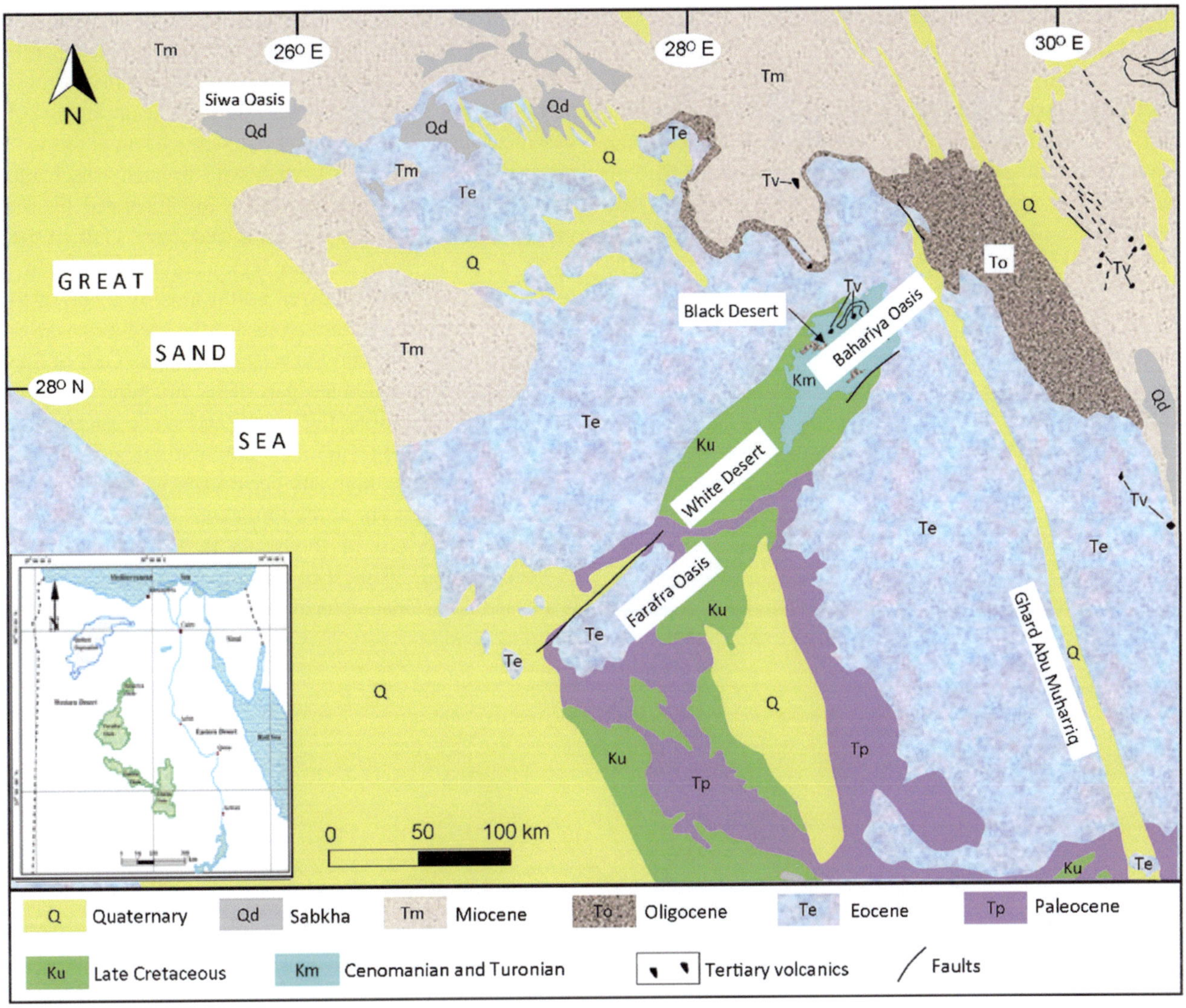

Fig. 6.6 Geological map of the western desert. Modified from Abdulaziz and Faid (2015), Plyusnina et al. (2016)

formation as in the eastern area of the Great Sahara Desert. It also represents the central area of many dune streams and accumulations that transfer from other regions of the Western Desert, Egypt to the northern Toshka area. Hence, the Toshka area is an ideal region for dune extents and can be used as a model for other regions not only in the Western Desert of Egypt but also in North Africa. The area is accessible by many asphaltic, paved roads, and desert tracks. The surface of the area is distinguished by sedimentary rocks varying from the Paleozoic to the Quaternary, with some exposures of igneous and metamorphic rocks attributed to the Late Precambrian, Phanerozoic, Late Cretaceous, and Tertiar (Aggour et al. 2012; Hamdan et al. 2016). Barchan dunes are well developed in many parts of the Egyptian Western Desert, including the Toshka area, as well as in many deserts areas on the surface of the Earth as well on the surface of Mars. Furthermore, many types and forms of barchan occur in these deserts, based on several factors, such as wind intensity and direction, quantity of drifting sand, topography, and vegetation (Hamdan et al. 2016).

The surface of the Toshka depression is approximately flat in its east and is broke up with sand dunes and low hills in west. The depression is bounded by Sinn El Kaddab Eocene limestone plateau at the north and by the Nubian pediplain at the south. The boundary of the depression nearly corresponds to contour line 200 m a.s.l. The Sinn El Kaddab plateau is high (approximately +400 m a.s.l.) and is enclosed by many small structural scarps which form a obvious topographic feature and are parallel to the principle regional E–W faulting system. Numerous rock-cut pediments bound the depression, and from vast flat terraces extending from the Sinn El Kaddab scarp foot and stretching for several kilometers with a slight slope to the depression (Bastawesy et al. 2008).

6.5 The Sinai Peninsula

Sinai is the only part of Egyptian country located in Asia and is separated geographically from the Eastern Desert by Suez Gulf. The peninsula is located between the Mediterranean Sea on the north and the Red Sea on the south, Suez Gulf on the west, and Aqaba Gulf on the east (Fig. 6.7). (El-Qady et al. 2005; Elewa et al. 2012; Balderer et al. 2014; El Tahlawi 2014). Their longitudes and latitudes are 32°20′–34°52′E and 27°45′–31°10′N (Elewa et al. 2012). Sinai Peninsula occupies an area of about 61,000 km^2 and the population is about 0.5 million people, 60% of them is mainly Bedouin and the others are resided in small cities such as El-Arish and Sharm El-Sheikh (Elewa et al. 2012; Abd El-Hameed et al. 2014; Balderer et al. 2014; El Tahlawi 2014). The peninsula has over 900 km of coasts, including approximately 155 km alongside the eastern bank of the Suez Canal (Elewa et al. 2012). The peninsula shape is triangular with apex formed by the connection of two Gulfs; Aqaba and Suez, and base by the Mediterranean coastline. The southern part of the Sinai consists of an intricate set of very rough mountains formed by igneous and metamorphic rocks. The middle and northern parts of the peninsula consist of immensely developed limestone plateau on lapping the basement rocks. The dominant drainage system is formed to the north by the Wadi El-Arish with its many discharges. The eastern and the western edges are separated by deep gorges draining into Aqaba Gulf and Suez Gulf, respectively. In the northern part, the regional dip slope is split up into many large hills followed northwards by a belt of low lands, with high sand dunes alongside the Mediterranean Sea coast. Along the east and the west coast of the Gulf of Suez, the plains or low lands are part of its structural and depositional province. Along the central and southeastern coasts, El-Qaa plain and El-Tur plain are split by an East-West trending subsurface high. Its occurrence is expressed by a regional southwest dip of the formations in El-Tur plain and by a northwest dip of the strata in El-Qaa plain (Abd El-Hameed et al. 2014). The majority of the Sinai Peninsula is a desert area. Geographically, this area is characterized by

Fig. 6.7 Sinai Peninsula (https://visibleearth.nasa.gov/view.php?id=75866, 2018)

highlands at the northern part and mountains at the southern parts, with the highest mountain peak St. Catherine, 2640 m a.s.l. This region is an active seismic area because of the presence of the triple joint of Suez Gulf, Aqaba, and the Red Sea (Balderer et al. 2014). El-Qady et al. (2005) mentioned that Sinai is considered as unstable shelf because of the frequent earthquake activities and its relationship with the geologic setting of the area, which is affected by the tectonic activity of the Red Sea, Suez Gulf, and Aqaba Gulf. As well, Sinai area is characterized by superficial thermal manifestations with a cluster of hot springs with variable temperatures (35–72 °C). These springs are mostly emerging from of fractures and spread along the eastern shore of Suez Gulf (El-Qady et al. 2005).

Sinai Peninsula is the most attractive region from the geological standpoint because it shows a variability of simple and complex structural forms. In general, Sinai exhibits miniature shape of all geologic forms in Egypt (El-Qady et al. 2005). The geology of the Sinai Peninsula varies from the Precambrian basement rocks to Quaternary deposits. The Quaternary deposits extended in the northern part and alongside the Sinai Gulf and the Mediterranean Sea coast. The Mesozoic limestone covers the widespread area from the central part of Sinai Peninsula, while the Precambrian rocks cover a wide area in the southern part (Balderer et al. 2014). It is stated that Sinai Peninsula can be separated into two main parts: The southern part is dominated by Precambrian igneous (magmatic rocks more than 600 million years) and metamorphic rock sand, while the northern part is dominated by plat lying Paleozoic (550 million year) to more recent sediments; Cenozoic sedimentary rocks (Abd El-Hameed et al. 2014; El Tahlawi 2014). This southern portion is an extension of the Arabo-Nubian Desert. A narrow belt of 30 km wide of the soft Nubian sandstone as well comprises majority of the Sinai minerals, such as turquoise and manganese. The Nubian sandstone of lower Cretaceous is one of the most important Mesozoic rock units and represent the main water-bearing unit in the region. Its maximum thickness is about 500 m, whereas at central Sinai, it is about 70–130 m. Oligocene is geothermally the most important in the Cenozoic rooks. The end of the Oligocene period revealed the rifting movements that brought the Gulf of Suez to its current shape (El-Qady et al. 2005).

Seven geomorphological units feature Sinai. Of those units: (1) southern high mountainous area occupies the south of Sinai giving a triangular shape with its apex at Ras Mohamed to the south; (2) central plateau area occupies the center Sinai Peninsula in the shape of two main questas (i.e., El Egma toward the southwest, and El Tieh toward the north); (3) hilly area is located toward the northeast of Sinai Peninsula and is gently sloping to the northeast and is featured by local isolated hills; (4) north and northwest coastal plain area of gently undulated surface characterized by sand ridges of thick sandstones; (5) marshy and sabkhas area occupies the shorelines; (6) alluvial coastal plains area; and (7) lakes (e.g., El Bardaweel, El Temsah and Bitter lakes) (El-Qady et al. 2005; Abd El-Hameed et al. 2014). As well regarding geomorphologic characters of Sinai Peninsula, Elewa et al. (2012) reported that many distinct geographic zones feature the peninsula; in the north, a belt of loose sand and dunes extended internal the coast for 16–32 km and then gives way to a flat, barren plain. This gravel and limestone plain cover around 241 km, rising at its southern fringe to Gebel Al-Tih Plateau. Following this plateau to the southern tip of the peninsula, Sinai is separated by a rough system of mountains and wadies (i.e., channels filled rainstorm water). They also stated that Sinai Peninsula is split into five basic geomorphologic units; southern high mountains, central Sinai plateau, northern Sinai conspicuous unit, mountainous and hills of North Sinai, and Suez Gulf plain. On the other hand, the southern part of Sinai is occupied by the high mountains, such as the Gebel Catherina (2641 m a.s.l), Um Shomar (2586 m a.s.l), and Serbal (2070 m a.s.l). The great plateau of Egma limestone plateau is located toward the north of this mountain mass, with additional than 1000 m slope down toward the Mediterranean Sea. The southern mountainous is greatly separated by watersheds draining either to Suez Gulf or to Aqaba Gulf, whereas most of the drainage basins of the northern plateau are debouching to the north of the Mediterranean Sea (Elewa et al. 2012).

Embabi (2018) added that the southern part of Sinai is the mountainous Sinai, whereas the central part is a tableland region. The northern part is divided into two subdivisions: isolated domal hills and mountains in the southern subdivision, whereas the northern part is mainly covered by sand dunes. Southern Sinai consists of high, complex, and rugged igneous and metamorphic mountains. Most of these mountains rise more than 1000 m a.s.l. However, several peaks rise above 2000 m a.s.l, of which Mount St. Catherine (2641 m a.s.l) is the highest not only in Sinai but also nationwide. Other conspicuous peaks are Gabal Um Shomar (2586 m), Gabal Mousa (2285 m), and Gabal Serbal (2070 m). The mountains are dissected by incised drainage lines which drain either into the Gulf of Aqaba in the east or the Gulf of Suez in the west. Examples of wadis draining into the Gulf of Aqaba are (from north to south) Watir, Dahab, and Kid, whereas wadis Baabaa, Sudr, and Feiran are examples of those draining into the Gulf of Suez. Due to this dissection and the effect of lithological variations, the peaks have the appearance of a jigsaw. However, some of the high areas are beveled and appear as high plains. These are the Haute Surfaces of Awad (1951). They are locally known as farsh, which means flat surface, such as Farsh Zebir, Farsh Itla, and Farsh El-Qasab. Because they are sites for fluvial deposition, intermountain basins lying between peaks have the appearance of small isolated depositional plains. The eastern slopes of the Sinai Mountains directly overlook the waters of the Gulf of Aqaba, except for the areas where small deltas developed along the coastline by the wadis, which drain into the Gulf of Aqaba. In contrast, the western slopes of the mountains are separated from the coastline of the Gulf of Suez by a broad depositional plain (10–30 km) called El-Qaa plain. This plain is featureless, except for the small and faulted sedimentary chains of Gabal Qabilyat, Gabal Abu Dorba, and Gabal Araba, which extend parallel to the coastline. The most common coastal features are coastal inlets (drowned wadis), small deltas, and coral reefs along the Gulf of Aqaba, and tidal flats and coastal lagoons along the Gulf of Suez.

6.6 Conclusion

Egypt is characterized by many geological areas, the main representing areas are Nile River Valley and Delta, Western Desert, Eastern Desert, and Sinai Peninsula. The Nile Valley and Delta occupy about 4% of Egyptian area, however, it has the most fertile soil in Egypt. The Nile Delta is typically fluvial sediments deposited. The remaining 96% of the land is arid desert need for reclamation considering continued population growth rate. The Eastern Desert comprises 22% of Egyptian land, and is bound by the Red Sea and Gulf of Suez in the east and by the Nile Valley in the west. It is one of the regions of the Sahara Desert and it is part of the Arabian-Nubian Shield. It is mainly clarified as a series of Neoproterozoic intra-oceanic arcs. The Eastern Desert of Egypt is occupied by igneous and metamorphic rocks. The western desert of Egypt is featured by many numerous extensive depressions separated by intensive arid oases (i.e., Siwa, Dakhla, Kharga, Farafra, and Baharia). The origin and landforms of the two deserts are varied which affect the Nile Valley and Delta. Sinai Peninsula is the most attractive geological region in Egypt and displays variability of simple and complex structural forms and reflects in miniature form all geologic column of Egypt. Sinai Peninsula can be separated into northern and southern parts. The southern part is dominated by Precambrian igneous and metamorphic rocks, while the northern part is dominated by plat lying Paleozoic and Cenozoic sedimentary rocks. Sinai is characterized by at least five morphological units, as well all other parts of Egypt have its feature morphological characteristics which have effect on the nature of the area and associated with its geology to effect the soil and its properties.

References

Abd El-Kawy OR, Rød JK, Ismail HA, Suliman AS (2011) Land use and land cover change detection in the western Nile delta of Egypt using remote sensing data. Appl Geogr 31(2011):483–494

Abd El-Hameed AG, El-Shayeb HM, Khalid HY (2014) Groundwater exploration by using 1-d resistivity technique at Abu Zenima area, southwestern part of Sinai, Egypt. J Environ Stud Res 1(1):7–22

Adelsberger KA, Smith JR (2009) Desert pavement development and landscape stability on the Eastern Libyan Plateau, Egypt. Geomorphology 107(2009):178–194

Abdulaziz AM, Faid AM (2015) Evaluation of the groundwater resources potential of Siwa Oasis using three-dimensional multilayer groundwater flow model, Mersa Matruh Governorate, Egypt. Arab J Geosci 8(2):659–675

Aggour TA, Korany EA, Mosaad S, Kehew AE (2012) Geological conditions and characteristics of the Nubia Sandstone aquifer system and their hydrogeological impacts, Tushka area, south Western Desert, Egypt. Egypt J Pure Appl Sci 2012:027–037

Asran AM, Emam A, El-Fakharani A (2017) Geology, structure, geochemistry and ASTER-based mapping of Neoproterozoic Gebel El-Delihimmi granites, Central Eastern Desert of Egypt. Lithos 282–283:358–372

Bastawesy MA, Khalaf FI, Arafat SM (2008) The use of remote sensing and GIS for the estimation of water loss from Tushka lakes, southwestern desert, Egypt. J Afr Earth Sci 52:73–80

Balderer W, Porowski A, Idris H, LaMoreaux JW (2014) Thermal and mineral waters—origin, properties and applications. In: Environmental Earth sciences series. Springer, Heidelberg. ISBN: 978-3-642-28823-4. https://doi.org/10.1007/978-3-642-28824-1

Blackwell BAB, Skinner AR, Smith JR, Hill CL, Churcher CS, Kieniewicz JM, Adelsberger KA, Blickstein JIB, Florentin JA, Deely AE, Spillar KV (2017) ESR analyses for herbivore teeth and molluscs from Kharga, Dakhleh, and Bir Tarfawi Oases: constraining water availability and hominin paleolithic activity in the Western Desert, Egypt. J Afr Earth Sci 136:216–238. https://doi.org/10.1016/j.jafrearsci.2017.07.007

Brookes IA (1993) Geomorphology and quaternary geology of the-Dakhla Oasis region, Egypt. Q Sci Rev 12:529–552

Brookes IA (2001) Aeolian erosional lineations in the Libyan Desert, Dakhla Region, Egypt. Geomorphology 39:189–209

El Bahariya GA (2012) Classification and origin of the Neoproterozoic ophiolitic mélanges in the Central Eastern Desert of Egypt. Tectonophysics 568–569:357–370

El Banna MM, Frihy OE (2009) Human-induced changes in the geomorphology of the northeastern coast of the Nile delta, Egypt. Geomorphology 107:72–78

El-Baz F, Maxwell TA (1982) Desert landforms of southwest Egypt: a basis for comparison with Mars. NASA Scientific and Technical Information Branch 1982. National Aeronautics and Space Administration, Washington, DC

Elewa HH, Qaddah AA, El-Feel AA (2012) Determining potential sites for runoff water harvesting using remote sensing and geographic information systems-based modeling in Sinai. Am J Environ Sci 8 (1):42–55

El-Qady G, Salem A, Aboud E, Khalil A, Ushijima K (2005) Geothermal reconnaissance study for Sinai Peninsula, Egypt. In: Proceedings world geothermal congress. Antalya, Turkey, 24–29 April 2005

El-Saied A, El-Ghamry A, Khafagi OA, Powell O, Bedair R (2015) Floristic diversity and vegetation analysis of Siwa Oasis: an ancient agro-ecosystem in Egypt's Western Desert. Ann Agric Sci 60 (2):361–372

El-Shazly AK, Khalil KI (2016) Metamorphic and geochronologic constraints on the tectonic evolution of the Central Eastern Desert of Egypt. Precambr Res 283:144–168

El Tahlawi MR (2014) Sinai Peninsula: an overview of geology and thermal groundwater potentialities. In: Balderer W, Porowski A, Idris H, LaMoreaux JW (eds) Thermal and mineral waters—origin, properties and applications. Environmental Earth sciences series. Springer, Heidelberg. ISBN: 978-3-642-28823-4. https://doi.org/10.1007/978-3-642-28824-1

Embabi NS (2018) Landscapes and landforms of Egypt—landforms and evolution. Series of World Geomorphology. Springer International Publishing AG. https://doi.org/10.1007/978-3-319-65661-8

FAO (1964) High dam soil survey. United Nation Development Program

Faris M, Shabaan M, Shaker F (2017) Calcareous nannofossil bioevents at the Palaeocene/Eocene boundary in the Kharga Oasis, Western Desert of Egypt. J Croatian Geol Surv Croatian Geol Soc 70 (3):179–199

Gad S, Kusky T (2006) Lithological mapping in the Eastern Desert of Egypt, the Barramiya area, using Landsat thematic mapper (TM). J Afr Earth Sci 44(2006):196–202

Ghadiry M, Shalaby A, Koch B (2012) A new GIS-based model for automated extraction of Sand Dune encroachment case study: Dakhla Oases, western desert of Egypt. Egypt J Remote Sens Space Sci 15:53–65

Hamdan MA, Refaat AA, Abdel Wahed M (2016) Morphologic characteristics and migration rate assessment of barchan dunes in the Southeastern Western Desert of Egypt. Geomorphology 257:57–74

Hamdi H, Abdelhafez S (1999) Agriculture and soil survey in Egypt. In: Zdruli P, Steduto P, Lacirignola L, Montanarella L (eds) Soil resources of southern and eastern Mediterranean countries, Options Méditerranéennes. SERIE B: Studies and Research, number 34. Publication of the Centre International de Hautes Etudes Agronomiques Mediterranéennes (CIHEAM), Paris, France. 286 pp. ISBN: 2-853552-227-X

Hassan FA, Barich B, Mahmoud M, Hemdan MA (2001) Holocene Playa deposits of Farafra Oasis, Egypt, and their Palaeoclimatic and Geoarchaeological significance. Geoarchaeol Int J 16(1):29–46

Hereher ME (2014) Assessment of sand drift potential along the Nile Valley and Delta using climatic and satellite data. Appl Geogr 55:39–47

Joint Research Center, European Soil Data Centre (ESDAC) (2017) Geological map of Egypt. https://esdac.jrc.ec.europa.eu

Liu B, Coulthard T (2015) Mapping the interactions between rivers and sand dunes: Implications for fluvial and aeolian geomorphology. Geomorphology 231:246–257

Mabrouk MB, Jonoski A, Solomatine D, Uhlenbrook S (2013) A review of seawater intrusion in the Nile Delta groundwater system —the basis for assessing impacts due to climate changes and water resources development. Hydrol Earth Syst Sci Discuss 10:10873–10911. www.hydrol-earth-syst-sci-discuss.net/10/10873/2013/. https://doi.org/10.5194/hessd-10-10873-2013

Mohamaden MII, Wahaballa A, El-Sayed HM (2016) Application of electrical resistivity prospecting in waste water management: a case study (Kharga Oasis, Egypt). Egypt J Aquat Res 2016(42):33–39

Mosaad S (2017) Geomorphologic and geologic overview for water resources development: Kharit basin, Eastern Desert, Egypt. J Afr Earth Sci 134:56–72

Moustaf MM (2000) A geostatistical approach to optimize the determination of saturated hydraulic conductivity for large-scale subsurface drainage design in Egypt. Agric Water Manag 42:291–312

Mostafa A (2013) Paleokarst shafts in the western desert of Egypt: a unique landscape. ACTA CARSOLOGICA 42(1):49–60

Plyusnina EE, Sallam ES, Ruban DA (2016) Geological heritage of the Bahariya and Farafra oases, the central Western Desert, Egypt. J Afr Earth Sci 116:151–159

Rashed HSA (2016) Evaluation of sustainable land management on some selected soils of Siwa Oasis. Egypt J Soil Sci 56(3):453–470

Said R (1990) The geology of Egypt. Balkema, Rotterdam, The Netherlands, p 734

Salem E, Sehim A (2017) Structural imaging of the East Beni Sueif Basin, north eastern Desert, Egypt. J Afr Earth Sci 136(2017):109–118

Salama W, El Aref M, Gaupp R (2012) Mineralogical and geochemical investigations of the Middle Eocene ironstones, El Bahariya Depression, Western Desert, Egypt. Gondwana Res 22:717–736

Stern RJ (2017) Neoproterozoic formation and evolution of Eastern Desert continental crust—the importance of the infrastructure-superstructure transition. J Afr Earth Sci 1–13 (in press)

Younes MA (2012) Hydrocarbon potentials in the Northern Western Desert of Egypt, crude oil exploration in the world. In: Younes M (ed) InTech. https://doi.org/10.5772/37028. Available from: https://www.intechopen.com/books/crude-oil-exploration-in-the-world/hydrocarbon-potentials-in-the-northern-western-desert-of-egypt

Zaki R (2007) Pleistocene evolution of the Nile Valley in northern Upper Egypt. Quat Sci Rev 26:2883–2896

Zeydan BA (2005) The Nile Delta in a global vision. In: Ninth International Water Technology Conference, IWTC9 2005, Sharm El-Sheikh, Egypt

7 Soil Geography

Abdel-Aziz Belal, Elsayed Mohamed, Ahmed Saleh, and Mohamed Jalhoum

Abstract

Egypt has a total area of about one million km^2, under arid and hyperarid climatic conditions, of which only a small portion (8% of total area) is arable land and the rest of area is barren desert. Almost percent of arable land area are occupied by more than 100 million inhabitants, who are mainly concentrated in the Nile Valley and the delta as well as in the north coastal zone along the Mediterranean Sea. Egypt contains five physiographic zones having specific attributes of resource base, climatic features, soil and geomorphological characteristics, land cover and land use situation, and socioeconomic implications. Therefore, it is found appropriate to formulate programs comprised of subcomponents geared to address the specific attributes in each of the physiographic zones distinguished as follows: Eastern Desert, Western Desert, northern western coast, Sinai Peninsula, and Nile Valley and Delta. The soils types of Egypt are the alluvial soils of the delta and valley, the calcareous soils along the coastal littoral of Egypt, the soils of the Eastern and Western Deserts as well as the soils of Sinai Peninsula. The major alluvial soils were formed from the suspended solid matter of the Nile and dry wadis in the desert, which were deposited during the flood season. Remote sensing (RS), GPS, and geographic information system (GIS) technologies are emerging as indispensable tools in the study and mapping of dynamic phenomenon having spatial and temporal characteristics. The objective of this book chapter is to study the different soil types in Egypt using new information technology such as remote sensing, GPs, and GIS.

Keywords

Soil geomorphology • Soil characteristics
Soil classification • Soil mapping

7.1 Introduction

Soil is our life-support system. It delivers food and freshwater; it recycles wastes; we build on it and with it; and it is not all the same, either globally or locally. Decision-makers need good baseline information about soils and land for development planning, avoidance of natural hazards, investment, and management (Selvaradjou et al. 2005). Egypt has an area of about one million square kilometers or 238 million feddans (one feddan = 0.42 ha). The total agricultural land in Egypt amounts to nearly 8.4 million feddans (3.5 million ha) and accounts for around 3.5% of the total area. One million ha in the irrigated areas suffer from salinization problems, waterlogging, and sodicity. The majority of salt-affected soils are located in the northern-central part of the Nile Delta and on its eastern and western sides. Increased attention is being given to the improvement of salt-affected soils, since they are potentially productive and require less investment, effort, and time for restoring their productivity, than the reclamation of new land (FAO 2005). The identification of the land resources of Egypt for the agricultural development justifies the importance of producing a collective physiographic soil map of Egypt for building up database of land information system. This system is based on the application of reliable remote sensing data as well as using global modern systems for the assessment of land units, features, and soil attributes (Afify et al. 2010).

A.-A. Belal (✉) · E. Mohamed · A. Saleh · M. Jalhoum
Division of Agriculture Applications and Soil Science, National Authority for Remote Sensing and Space Sciences (NARSS), Cairo, Egypt
e-mail: Belalabd@gmail.com

H. El-Ramady et al. (eds.), *The Soils of Egypt*, World Soils Book Series,
https://doi.org/10.1007/978-3-319-95516-2_7

Table 7.1 The major soil groups and land cover in Egypt according to FAO (1998) and their details

Soil groups/land cover	Percent of total (%)	More details
Arenosols	25.80	Sandy-textured soils, significant lack in soil profile development, surface horizon is low in humus, belongs to the Entisol order
Leptosols	24.87	Soils with a very shallow profile depth; contain large amounts of gravel or stone; found in the cold polar regions; being susceptible to erosion or waterlogging
Calcisols	9.12	Soils are developed in mostly alluvial, colluvial, and aeolian deposits of base-rich weathering material; found in arid and semiarid regions
Regosols	8.68	Very weakly developed mineral soil in unconsolidated materials; extensive in eroding lands in arid and semiarid areas and in Entisols
Vertisols	4.85	High content of clay montmorillonite; form deep cracks in drier seasons or years; alternate shrinking and swelling; have an extremely deep A horizon
Fluvisols	0.80	Alluvial deposits from river sediments; occur in lacustrine and marine deposits; they have a good fertility
Solonchaks	0.48	Pale or gray soils found in arid to subhumid regions; poorly drained conditions; saline soils
Calcisols	0.37	Soil with a substantial secondary accumulation of lime; common in calcareous parent materials and widespread in arid and semiarid environments

Where water bodies (15.44%) and soils outside the area surveyed (9.59%)
Adapted from Khalifa and Moussa (2017)

7.2 Egyptian Soils Based on Geological Zones

It could be identified the soil resources in Egypt according to different features of climate, terrain, and landform characteristics. As well known, Egypt has different geological zones including the Nile Delta, the Eastern Desert, the Western Desert, and Sinai. These geological zones mainly differ in their soil fertility due to the difference in their geological deposits such as lacustrine, marine, fluvial, sandy and fluvio-sandy, and calcareous soils. Concerning lacustrine soils, a lot of these soils could be found in north of the Nile Delta like soils from Manzala, Burullus, Edku and Mariut, Bardawil, Nasser, and Qarun lakes. Regarding fluvial soils, they were originated from the Nile Delta, whereas marine soils could occur in the Red and the Mediterranean Sea. About the calcareous soils, these soils could be found in separated areas in Egypt such as north of the Western Desert (Shaheen et al. 2013).

First of all, in terms of cropland base, Egypt could be considered as one of the world's poorest countries because the majority of this country's lands are desert. Therefore, it is estimated the present cultivated area in Egypt to be about 3.66% of the total area (Zaghloul 2013). For more details, the main cultivated lands in Egypt are located close to in (1) the Nile Delta, (2) the banks of the Nile River, (3) and the main branches and canals of the Nile Valley. Regarding the rangelands, they are restricted to a narrow strip representing only a few kilometers wide along the Mediterranean coast and their bearing capacity are quite low. According to the official statistics of the Ministry of Agriculture and Land Reclamation, the agricultural lands increased from 2.38 to 2.78 and about 3.61 million ha (5.71, 6.67 and 8.66 million fed) in 1950, 1982, and 2013, respectively (Khalifa and Moussa 2017). It is reported that the main soil groups in Egypt could be ordered as a percent from the total area as follows in Table 7.1.

Concerning the soils of the Nile Valley and delta zone, most soils are recent Nile alluvium forming from (1) the fluvio-marine and lagoon deposits located adjacent to the northern lakes (Edku, Burullus, and Manzala) and (2) the coastal plain in the extreme north of the delta. These soils are fertile and the basket of food in Egypt. This delta could be divided into the north coastal zone, Eastern Desert zone, Western Desert zone, and inland Sinai. The coastal zone includes northwestern coastal areas and northeastern coastal areas of Sinai. The northwestern coastal areas form a belt about 20-km deep and extend to about 500 km between Alexandria and Salloum near the Libyan borders, covering an area of about 10,000 km^2. This zone includes soils of the old coastal plain (foreshore strip and lagoon depressions or calcic and quartzitic dunes), soils of the alluvial fans and soils of the plateau (Khalifa and Moussa 2017).

Regarding the soils of the Eastern Desert, this area is rich in several wadis (e.g., Wadi Qena, Al-Allaqui, El Assuity, and El Laquita). These wadis are mostly deep and very steep, and their soils are immature (display young stages of development). The soils of these wadis are mainly shallow to deep profile, coarse or moderately fine-textured, with variable content of gravels. Furthermore, the main landforms of these wades include alluvial fans and sand sheets, rubble terraces, plateau, wadi bottoms, outwash plains, and dunes. Regarding the soils of the Western Desert, several natural depressions are scattered in variable areas including the famous oases in the south (Kharga and Dakhla), in the north (Siwa), and in the central area (Bahariya and Farafra). The soils of these oases have distinguished features including particular erosional patterns, special parent materials, different environmental sedimentations, and different eluvial deposit (carbonate, salts, and gypsum). There are many promising reclaimed areas in the Western Desert including Darb El-Arbain, Toshka, East El-Uweinat, and some wadis of High Dam Lake (Khalifa and Moussa 2017).

Concerning the soils of Sinai Peninsula, these soils have borders from the West by the Gulf of Suez and from the East by the Gulf of Aqaba. These soils also have distinguished features including moderately deep-to-deep profile, gravelly, coarse-textured soils (in the southern part of Sinai); deep highly calcareous, gravelly coarse-textured soils; gravelly course over fine-textured soils (in the central part of Sinai); deep, gravelly coarse or moderately textured soils; deep calcareous, coarse to moderately to fine-textured soils (in the alluvial coastal plains); and some soils of the marshy land and alkali soil or sabkhas along the Gulf of Suez as well as some soils sandy deposits as moderately fine-textured with salts and gypsum contents in wide ranges (Khalifa and Moussa 2017).

7.3 Soil of Eastern Desert

The Eastern Desert areas are mostly covered by the Red Sea Mountains consisting of igneous and metamorphic rocks. Surface Quaternary deposits exist in the form of alluvial hills and alluvial terraces consisting of gravels, sands, and cemented by fine clay materials (El-Shamy 1988). The Eastern Desert covers an area of about 223,000 km^2. It is bordered by the Nile Valley on the west and by the Suez Canal, the Gulf of Suez, and the Red Sea on the east. The backbone of this desert is a series of mountain chains (Red Sea mountains), running parallel to the Red Sea and separated from it by a narrow coastal plain (Hegazi et al. 2005). The same authors said that the Eastern Desert is characterized by the hyperarid conditions; with a mild winter and a hot summer. Exceptional being the coastal belt along the Gulf of Suez and highlands of South Sinai, which represent the hyperarid province with a cool winter and hot summer. The Eastern Desert is bordered by the Nile River with high scarps, cut by wadis flowing toward the River Nile and the Red Sea mountains. The wadis of the Eastern Desert are numerous, mostly deep, very steep, and their soils display young stages of development. The most important wadis are, namely, Qena, El Laquita, El Assuity, and Al-Allaqui. The main landforms of these wadis are plateau, rubble terraces, wadi bottoms, outwash plains, alluvial fans and sand sheets, and dunes. Soils of these wadis are mainly shallow to deep coarse or moderately fine-textured with variable content of gravels (Typic Torripsamments, Typic Torriorthents, and Typic Torrifluvents) with six subgroups between Lithic and Typic. Eastern Desert's wadis have severe to very severe soil limitations that may contribute to the desertification processes and maximize the input of the agricultural development and minimize the outputs or the benefits gained. These limitations include: Topography in most cases is rough with moderate to steep slopes. Soil limitations differ widely from one wadi to another and are mostly concerned with parent materials that are either calcareous or siliceous, some cases are mixed with high content of gravels and boulders. Consequently, these soils have low water holding capacity, high infiltration rate, and low fertility status (Hegazi et al. 2005).

7.3.1 Soil of Dry Wadis

Belal et al. (2015) explained the soil of Wadi Al Assiuti in Eastern Desert of Wadi Al Assiuti as one of the most promising developed areas in Egypt, which is located at the east of Nile River, south part of Asyut Governorates. Soil properties and classification: Based on the Soil Survey Staff (1999), two soil orders were identified: Entisols and Aridisols, which were represented by ten great groups (Fig. 7.1). The soil sets of the mapping units were outlined as follows:

Soil of floodplain: The floodplain topographic unit is included: overflow basin, decantation basin, and river terraces. The obtained data indicated that soil depths were ranged between 70 and 150 cm. Soil texture class is clay, sandy loam, and loamy sand. Soil pH value ranged between 7.25 and 8.4, whereas EC value ranged between 0.4 and 1.8 dS m^{-1}. The $CaCO_3$ content ranged between 4.5 and 33.9%. The soil classified as Typic Haplocalcids and Vertic Natrargids.

Soil of recent river terraces: The recent river terraces are included: high and low recent river terraces; these units are originated from Nile sediments before high dam contraction; it is developed from sediments of Ethiopian plateau that transported by Nile River and subsequently deposited in both the valley and delta. Depth of these soils lays around 150 cm. Soil texture class is clay, silt clay, and sand clay loam. Soil pH value ranged between 7.4 and 7.8, whereas EC value ranged between 0.6 and 1.5 dS m^{-1}. The $CaCO_3$ content ranges between 5.5 and 25.7%; the soil is classified as Vertic Haplocalcids. Soil of wadis: These units included wad, wadi bottom, and wadi plain. The soil depths ranged between 100 and 150 cm. Soil texture class is sandy, loamy

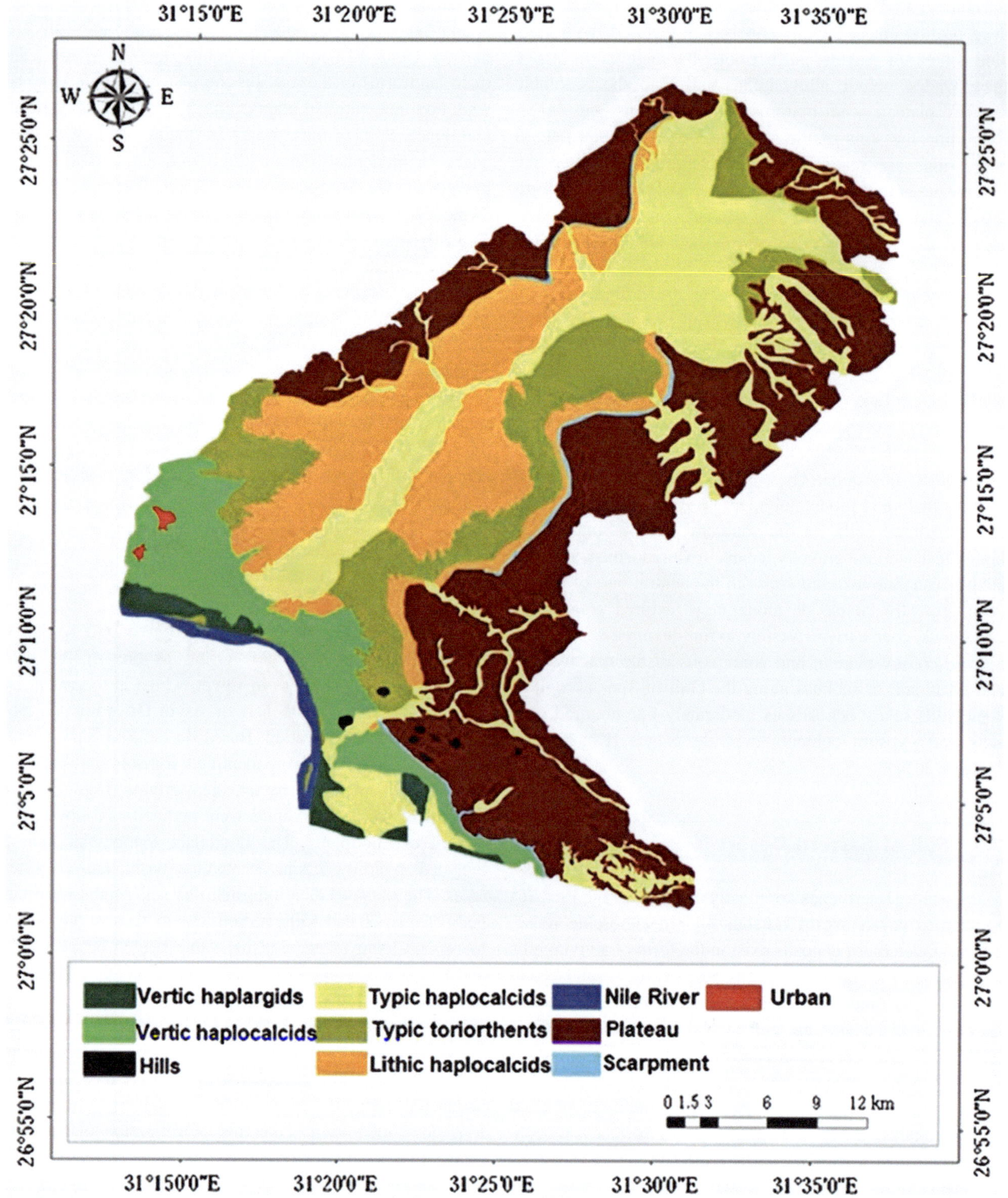

Fig. 7.1 Soil map of Al Assiuti in Eastern Desert. *Source* Belal et al. (2015)

sand, and sandy loam. Soil pH value was neutral to slightly alkaline 7.45 and 8.7. The EC value ranged between 0.35 and 3.5 dS m^{-1}. The $CaCO_3$ content ranges between 7.1 and 41%; the soil is classified as Typic Haplocalcids and Typic Torriorthents. Soil of outwash plain: This unit has a flat surface covered with fine sand and formed by wind action. The soil depth ranged between 50 and 55 cm. Soil texture class is sand, sand clay loam, and loamy sand. Soil pH value ranged between 7.4 and 8.9, whereas the EC value ranged between 0.7 and 2.7 dS m^{-1}. The $CaCO_3$ content ranges between 30.67 and 60%; the soil is classified as Lithic Haplocalcids. Soil of rubble terraces: These areas were slightly saline. Depth of these soils is shallow; it is around 40–50 cm. Soil texture class is sand clay loam, loamy sand. Soil pH value was neutral to slightly alkaline 7.8 and 8.6. The EC value ranged between 0.6 and 6.20 dS m^{-1}. The $CaCO_3$ content ranges between 11.4 and 87.7%; the soil is classified as Lithic Haplocalcid.

Soil of levee: Levee is formed as a result of deposition in relatively low lying area. Depth of these soils lies around 150 cm. Soil texture class is sand clay loam. The pH value ranged between 7 and 8.2 but the EC value was low ranging between 0.6 and 0.8 dS m^{-1}. The $CaCO_3$ content ranges between 7.2 and 14.6%; the soil is classified as Typic Torriorthents.

7.3.2 Soil of Haliab and Shalateen Area

According to NRC (2000), there are two types of soils in Haliab and Shalateen area; the dominant one is the coarse-textured fluvial sediments, while the second one is the salt sabkhas soils along the coastal strip. Soils of wadis were classified under the subgroup Typic Torrifluvents and characterized by a sandy texture with the presence of rock fragments in the high altitude areas. The texture changes to loamy sand in the moderate altitude areas, while at the end of the wadis, there is a thin silty to clayey layer, which covers the soil surface. The soil has a low amount of salts and calcium carbonate. Soils of the deltas and alluvial fans were classified tinder two subgroups: Typic Torripsamments and Typic Torrifluvents. The soils are characterized by a coarse to medium sandy texture. Soils of the sand dunes were classified as Torrifluvents and Typic Quartzipsamments. The soils are sandy in texture and very deep. Soils of the sabkhas were classified under two subgroups: Typic Salorthids and Typic Gypsiorthids. The texture is sandy and contains high amount of soluble salts. The surface is Hat to undulating. The soil depth is shallow to moderate. The level of the water table varies from 30 to 100 cm. The soil profile contains a salic or gypsic horizon. Salt crusts cover the soil surface in most of the area. The sabkhas can be divided into two groups: dry sabkha and wet sabkha.

Also, NARSS (2016) studied the soil in the Haliab and Shalateen area. These soils were classified as follows:

Soil of alluvial fans: These soils are characterized by almost flat to gently undulating. Soil profiles are characterized by deep soils and moderately deep, soil texture is sand, loamy sand, and sandy loam. Soil dry color varies between very pale brown (10 YR 7/4) to yellowish brown (10 YR 5/4), while moist color varies from light yellowish brown (10 YR 6/4) to dark yellowish brown (10 YR 4/6). Chemical analysis of the studied profiles data indicated that calcium carbonate content varies widely from 0.94 to 27%; it is an irregular distribution pattern of total carbonate content with depth, except for some profiles which CaCO3 tends to increase with depth. Gypsum content is considerably low in all soil profiles of this geomorphic unit; it ranges from 0.11 to 1.2%. Considering the depth-wise distribution of gypsum content, data indicate that there is no specific pattern of distribution throughout the entire profiles depths. Organic matter content is very low and does not exceed 0.36%. The pH values range from 7.1 to 7.9. The electric conductivity data of saturation paste extracts (ECe) is nonsaline to saline soils, which ranges between 0.28 and 14 ds m^{-1}. Based on the morphological feature of the representatives profile of this unit and chemical analysis, the soils are classified as Typic Torrifluvents and Typic Torripsamments. Figure 7.2 shows soil profile of alluvial fan.

Soil of alluvial deposits: This physiographic unit is almost flat to gently undulating. Soil profiles are characterized by deep soils; soil texture is loamy sand, sandy loam, and sand clay loam. Soil dry color varies between very pale brown (10 YR 7/4) to yellowish brown (10 YR 5/4), while, moist color varies from light yellowish brown (10 YR 6/4) to dark yellowish brown (10 YR 4/4). Chemical analysis of the studied profiles data indicated that calcium carbonate content varies widely from 0.94 to 25.7%. Gypsum content is considerably very low in all soil profiles of this geomorphic unit; it ranges from 0.11 to 2.2%. Organic matter content is very low and does not exceed 0.36%. The pH values range from 7.15 to 8.1. The electric conductivity data of saturation paste extracts (ECe) is nonsaline to saline soils which it ranges between 0.38 and 13.5 dS m^{-1}. Based on the morphological feature of the representatives profile of this unit and chemical analysis, the soils are classified as Typic Torriorthents.

Soil of wades deposits: This physiographic unit is almost flat to gently undulating. Soil profiles are characterized by deep soils to moderately deep; soil texture is loamy sand, sandy loam, and sand. And it is cover of layer silt or clay on the surface. Soil dry color varies between very pale brown (10 YR 7/4) to yellowish brown (10 YR 5/4), while, moist color varies from light yellowish brown (10 YR 6/4) to dark yellowish brown (10 YR 4/4). Chemical analysis of the studied profiles data indicated that calcium carbonate content varies widely from 0.43 to 3.11%. Gypsum content is considerably very low in all soil profiles of this geomorphic unit; it ranges from 0.11 to 1.25%. Organic matter content is 0.86%. The pH values range from 8.15 to 9.1. The electric conductivity data of saturation paste extracts ranges between 0.12 and 2.25 ds m^{-1}. Based on the morphological feature of the representatives soil profiles of this unit and chemical analysis, the soils are classified as Typic Torrifluvents for all

Fig. 7.2 Typic Torripsamments for soil profile of alluvial fans in Haliab and Shalateen area, Eastern Desert. *Source* NARSS (2016)

profiles except some profiles are classified as Typic Calciorthids and Typic Haplocalcids.

Soil of sand sheet: This physiographic unit is almost flat. Soil profiles are characterized by deep soils, soil texture is loamy sand and sand. Soil dry color varies between very pale brown (10 YR 7/4) to yellowish brown (10 YR 5/4), to dark yellowish brown (10 YR 4/6).

Chemical analysis of the studied profiles data indicated that calcium carbonate content varies widely from 0.77 to 1.52%. Gypsum content ranges from 1.9 to 2.9%. Organic matter content is 0.31%. The pH values range from 7.5 to 7.9. The electric conductivity data of saturation paste extract ranges between 1.02 and 12.6 dS m^{-1}. Based on the morphological feature of the representative's profile of this unit and chemical analysis, the soils are classified as Typic Torripsamments.

Soil of sabkhas: This unit is located at low lying areas and extended with the coastline, exist mainly in depressions, barren or with humid areas natural vegetation is halophytic species, and originated as a result of sea intrusion that rise water table. Water table is rich in soluble salts specially sodium chlorides. This physiographic unit is almost flat. Soil profiles are characterized by moderately deep; soil texture fluctuates between sandy to clay loam. The structure classes are massive and weak subangular blocky. Soil dry color varies between yellowish brown (10 YR 6/4) and light brownish gray (10 YR 7/2). Moist color is yellowish brown (10 YR 5/6) to grayish brown (10 YR 5/2). Chemical analysis of the studied profiles data indicated that calcium carbonate content varies widely from 0.97 to 8.50%. Gypsum contents ranging from 6.9 to 13.9% with there are layers from gypsum in the subsurface for some area. Gypsum content present in the studied layers is in form of small shiny crystals. Organic matter content is 0.31 and 0.51%. The pH values range from 7.15 to 7.7. The electric conductivity data of saturation paste extracts is very high. Where the electric conductivity values range between 30.18 and 178 dS m^{-1}. The relatively high ECe values may be related to the location of the soils with the coastline and the process of soil salinization. Based on the morphological feature of the representatives profile of this unit and chemical analysis, the soils are classified as Typic Aquisalids and Typic Aquisalids. Figure 7.3 shows soil profile of wet sabkha.

Soil of foot slopes: This physiographic unit is representing the foots of the southern triangular mountains, and surfaces of this geomorphic units tend to roll to surrounding pen-plains or sand sheets. This soil is almost flat. Soil profiles are characterized by shallow soil depth, soil texture varies sand in the successive profile layers. Soil dry color varies between very pale brown (10 YR 7/4) to yellowish brown (10 YR 5/4), to dark yellowish brown (10 YR 4/6). Chemical analysis of the studied profiles data indicated that calcium carbonate content varies widely from 1.7 to 9.5%. Gypsum content is 0.36%. Organic matter content is 0.12%. The pH values range from 7.1 to 7.8. The electric conductivity data of saturation paste extracts (ECe) ranges between 0.93 and 1.6 ds m^{-1}. Based on the morphological feature of the representative profile of this unit and chemical analysis, the soils are classified as Lithic Torripsamments.

Soil of deltas: This physiographic unit is almost flat. Soil profiles are characterized by deep soils, soil texture is sandy loam to sand clay loam. Soil dry color varies between very pale brown (10 YR 7/4) to yellowish brown (10 YR 5/4), while, moist color varies from light yellowish brown (10 YR 6/4) to dark yellowish brown (10 YR 4/4). Chemical analysis of the studied profiles data indicated that calcium carbonate content varies widely from 0.40 to 4.40%. Gypsum content is considerably very low in all soil profiles of this geomorphic unit, it ranges from 1.50 to 3.30%. Organic

Fig. 7.3 Typic Aquisalids for soil profile of Sabkha Haliab and Shalateen area, Eastern Desert. *Source* NARSS (2016)

Fig. 7.4 Typic Torrifluvents for soil profile of the Delta in Haliab and Shalateen area, Eastern Desert. *Source* NARSS (2016)

matter content is 0.1 to 1.2%. The pH values range from 7.3 to 8.1. The electric conductivity data of saturation paste extract ranges between 0.12 and 12.25 ds m^{-1} up to 32.5 ds m^{-1} in the subsurface for some profiles. Based on the morphological feature of the representatives profile of this unit and chemical analysis, the soils are classified as Typic Torrifluvents. Figure 7.4 shows soil profile of delta.

Soil of out wish plain: This physiographic unit is gentle sloping surface. Soil profiles are characterized by moderately to deep soils, soil texture is sandy loam to sand. Soil dry color varies between very pale brown (10 YR 7/4) to yellowish brown (10 YR 5/4), to dark yellowish brown (10 YR 4/6). Chemical analysis of the studied profiles data indicated that calcium carbonate content varies widely from 0.77 to 1.52%. Gypsum content ranges from 0.9 to 3.5%. Organic matter content is 0.05 to 1.05%. The pH values range from 7.1 to 7.9. The electric conductivity data of saturation paste extracts ranges between 0.22 and 0.65 dS m^{-1}. Based on the morphological feature of the representatives profile of this unit and chemical analysis, the soils are classified as Typic Torripsamments (Fig. 7.5) and show the soil classification units in Haliab and Shalateen area in Eastern Desert, Egypt based on the American Soil Survey Staff system (2010).

7.4 Soil of Western Desert

The Western Desert extends over a vast area occupying about 681,000 km^2. It is composed of large, rocky surface with the highest portion in the southwestern corner where Gebel

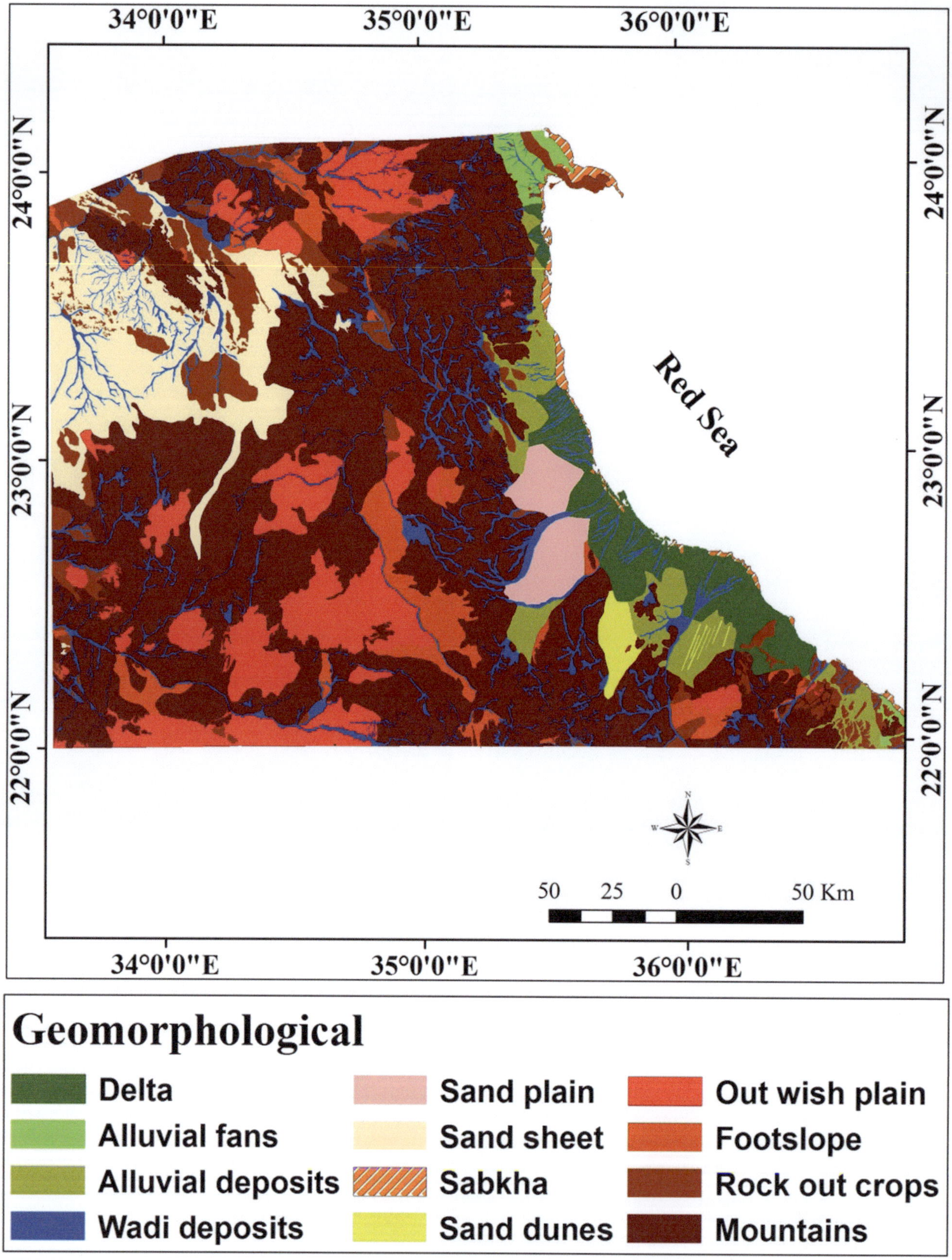

Fig. 7.5 Geomorphological map in Haliab and Shalateen area, Eastern Desert. *Source* NARSS (2016)

(mountain) Uweinat is found. North of Uweinat, the Gilf el-kebir plateau (100 m a.s.l) formed of Nubian Sandstone occurred. This plateau is characterized by scarps which slope sharply toward large depressions in the east and north; Kharga and Dakhla depressions. To the north of this plateau, another plateau with arms extends in several directions. This plateau is composed of limestone and is lower in elevation than the Gilf el-kebir plateau, and constitutes the mainland from feature west of the Nile Valley. Hollowed out in the plateau surface are two great depressions: those of Farafra and Bahariya. The area of the former is more than 3000 km^2, and the latter has an area of about 1800 km^2. The Qattara–Siwa depression is considered to be part of a huge depression in the northern sector of the Western Desert (Hegazi et al. 2005). The same authors said the Western Desert is characterized by hyperarid climatic conditions with rare rainfall and extremely high temperature. The northwestern and the northern winds extend from the Mediterranean over the Western Desert with fallen speed southward. These winds are the major factors of erosion and deposition. The Western Desert has several natural depressions of variable areas which are scattered and include the famous oases: Siwa in the north, Bahariya and Farafra in the central position, and El Kharga and Dakhla in the south. These oases that are distinguished with artesian wells are mainly closed and fragile ecosystems. The Western Desert includes also promising remote areas which are put under reclamation and utilization, i.e., Toshka, Darb El-Arbain, East El-Uweinat, and some wadis of High Dam Lake.

7.4.1 Soil of the Abandoned Oases

The abandoned oases in Western Desert are of four oases, namely, Sitra oasis, El Nuweimisa oasis, El Bahrien oasis, and El Arag oasis. The area of the abandoned oases falls within the southern corner of the northern plateau of the Western Desert of Egypt; the Marmarica Plateau. This area forms the utmost borders of the south of Qattara Depression and situated about 95 km to the southeast of Siwa Oasis and about 200 km to the northwest of El Baharia oasis. The abandoned oases area is bounded to the north by Qattara Depression, while toward the south; it is limited by the field dunes of the Great Sand Sea and extends about 41.8 km to south direction. The area of abandoned oases is bounded by longitudes 26° 15′ 00″ & 27° 18′ 45″ E and Latitudes 28° 37′ 30″ & 29° 00′ 00″ N, Fig. 7.6. The area of the abandoned oases falls within the southern corner of the northern plateau of the Western Desert of Egypt; the Marmarica Plateau. This area forms the utmost borders of the south of Qattara Depression and situated about 95 km to the southeast of Siwa Oasis and about 200 km to the northwest of El Baharia oasis. The abandoned oases area is bounded to the north by Qattara Depression, while toward the south, it is limited by the field dunes of the Great Sand Sea and extends about 41.8 km to south direction. The area of abandoned oases is bounded by longitudes 26° 15′ 00″ & 27° 18′ 45″ E and Latitudes 28° 37′ 30″ & 29° 00′ 00″ N. Saleh (2010) was describe the different landforms of abandoned oases which are, namely, footslopes, pediments, peniplains, sandy plains, sand flats, sandy hummocks, overflow basins, decantation basins, dry sabkhas, and wet sabkhas (Fig. 7.7). The morphological properties of soil profiles of the different landforms that abandoned oases embrace could be considered as relics of the soil forming processes. Soil profiles vary widely with regard to color, texture, structure, consistency, lithology, profile depth, boundaries of underlying horizons, and pedological features through soil profiles. Differences of soil morphology between soil profiles of abandoned oases and also the differences between horizons of the same profile are mainly ascribed to environmental conditions prevailing during soil forming processes. Morphological features of soil profiles of abandoned oases are given here referring to the field studies and observations.

The color of soils (dry) of abandoned oases ranges in general from 2.5Y (yellow) to 10YR (yellow-red). The soils with hue 10YR have values ranging from 5 to 7 and chroma fluctuates between 3 and 8 indicating a color range from yellowish brown to yellow. Soils with hue 2.5Y have a color of yellow. Few of soils of abandoned oases display a white color (N 8/0) while only one soil type in El Bahrien oases shows a color of black (N 2/0). Soil texture differs with location and sometimes with depth (Fig. 7.8). The coarse texture (sandy to sandy loam) represented the most texture in the landforms of abandoned oases except for the footslopes landforms of Sitra oasis whereas the soil texture is loam to silty loam. The soil texture is sand in all soil horizons for footslopes and sand flats of El Nuweimisa oasis; pediments and wet sabkhas of El Bahrien oasis; peniplains, sand flats, and sandy hummocks of El Arag oasis. Soil structure of the profiles of abandoned oases is either single grains or subangular blocky structure. The structure is single grains in the landforms of where the texture is sand and the organic matter is extremely low as organic material contributes in forming soil structure and this may be due to the scanty of natural vegetation in these landforms since the root system would be very little or none existent which contributes to the lack of structure. The dry thermic climatic condition of abandoned oases also contributes to any organic material decline. The other morphological features were observed such as calcareousness, lime nodules and segregations, gypsum crystals, and salt accumulation. Some few mottles of humified organic materials were noticed whereas natural vegetation is common and dense.

On the other hand, the same author described the chemical properties of abandoned oases as follows: Total soluble salts expressed in terms of electrical conductivity (EC) of the

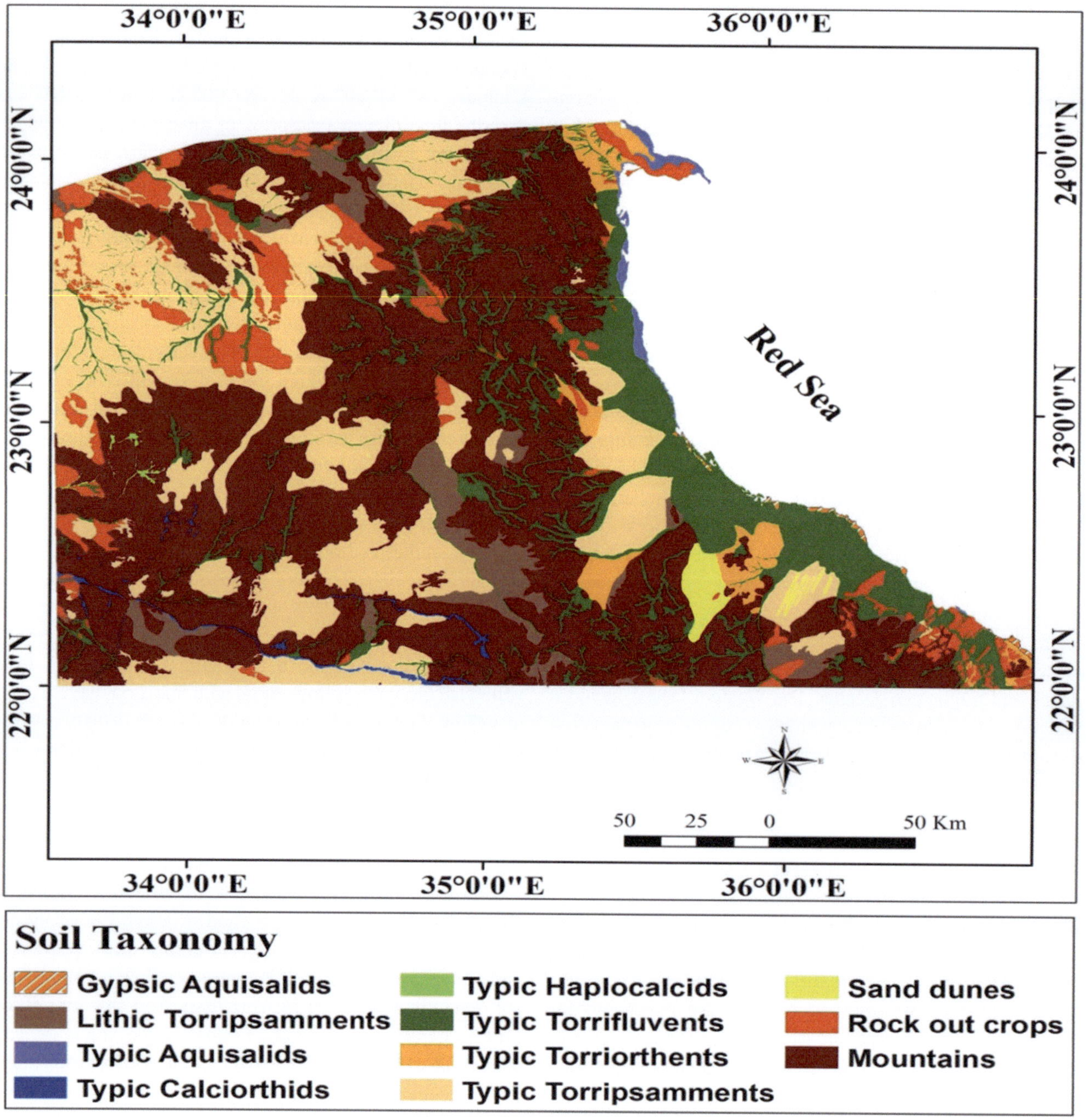

Fig. 7.6 Soil map in Haliab and Shalateen area, Eastern Desert. *Source* NARSS (2016)

extracts of the saturated soil paste vary widely in soils of abandoned oases. EC of the studied soil samples ranges between 1.00 and 155 dS m^{-1} in Sitra oasis, and ranges between 3.00 and 109 dS m^{-1} in Nuweimisa oasis, and ranges between 7.00 and 128 dS m^{-1} in El Bahrien oasis, and ranges between 3 and 117 dS m^{-1} in El Arag oasis. In Sitra oasis, footslopes, pediments, sandy plains, overflow basins, decantation basins, dry sabkhas, and wet sabkhas landforms are strongly saline where the EC is more than 16 dS m^{-1}, while sandy hummocks and sand flats landforms are slightly saline where the EC is between 4 and 8 dS m^{-1}. Peniplains landforms are none saline where the EC is less than 4 dS m^{-1}. In El Nuweimisa oasis, footslopes, pediments, peniplains, sandy plains, sandy hummocks, overflow basins, decantation basins, dry sabkhas, and wet sabkhas landforms are strongly saline while the sand flats landforms are slightly saline.

In El Bahrien oasis, footslopes, peniplains, sand flats, overflow basins, decantation basins, dry sabkhas, and wet sabkhas landforms are strongly saline while the pediments, sandy plains, and sandy hummocks landforms are slightly saline.

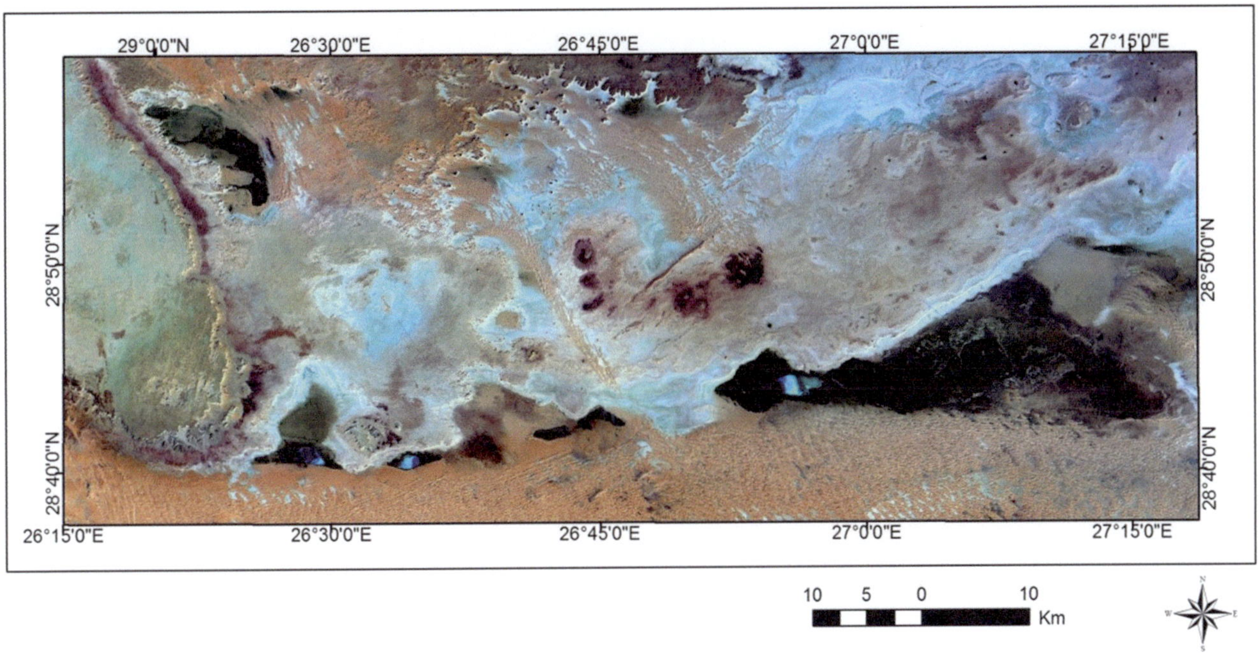

Fig. 7.7 Location of abandoned oases in Western Desert, Egypt. *Source* Saleh (2010)

In El Arag oasis, footslopes, pediments, peniplains, dry sabkhas, and wet sabkhas landforms are strongly saline while overflow and decantation basins landforms are moderately saline. The sand flats landforms are slightly saline while the sandy hummocks landforms are none saline. Sandy plains landforms are non saline.

Organic matter contents in soils of abandoned oases are extremely low as the environmental conditions are dry and hot except for landforms where natural vegetation is dense as in wet sabkhas landforms of Sitra oases; overflow and decantation basins landforms of El Nuweimisa and El Bahrien oases; and decantation basins landforms of El Arag oasis where it exceeds 4.0%. Organic matter content decreases with depth. Gypsum contents in abandoned oases are found in traces in footslopes, pediments, overflow basins, decantation basins, dry sabkhas, and wet sabkhas landforms of Sitra oasis; and pediments, decantation basins, dry sabkhas, and wet sabkhas landforms of El Nuweimisa oasis. Gypsum content ranges between 0.1 and 8.3%. Only the gypsum content in soils of overflow basins landforms of El Nuweimisa oasis qualifies the requirements of gypsic horizon. Calcium carbonate content in soils of abandoned oases varies widely. The Calcium carbonate content ranges between 73 and 445 g kg^{-1} in Sitra oasis, and ranges between 62 and 765 g kg^{-1} in Nuweimisa oasis, and ranges between 27 and 732 g kg^{-1} in El Bahrien oasis, and ranges between 27 and 142 g kg^{-1} in El Arag oasis. Contents of calcium carbonates are mainly dependent on the parent materials. Calcium carbonates content tend to increase with depth in pediments landforms of Sitra oasis; pediments and dry sabkhas landforms of El Nuweimisa oasis; sandy hummocks, overflow basins, and decantation basins of El Bahrien and El Arag oases. The calcium carbonates pattern tendency to decrease with depth is noticed in wet sabkhas landforms of Sitra oasis; peniplains landforms of El Nuweimisa oasis; footslopes and wet sabkhas landforms of El Bahrien oasis; and footslopes, peniplains, dry sabkhas, and wet sabkhas of El Arag oasis. The calcium carbonates pattern of irregular change with depth is found in footslopes, peniplains, sandy plains, sand flats, and sandy hummocks landforms of Sitra oasis; sandy plains, sandy hummocks, decantation basins, and wet sabkhas landforms of El Nuweimisa oasis; sandy plains, sand flats, and dry sabkhas, landforms of El Bahrien oasis; pediments and sand flats of El Arag oasis. There is nearly no change in calcium carbonates content in overflow basins and decantation basins landforms of Sitra oasis; footslopes, sand flats, and overflow basins landforms of El Nuweimisa oasis; and pediments landforms of El Bahrien oasis. The soil sets of the mapping units abandoned oases in Western Desert are as in Table 7.2 after Saleh (2010).

7.4.2 Soils of Bahariya Oasis

Bahariya Oasis is one of the most geologically important areas in the Western Desert. It lies between latitudes 27° 48′–28° 30′ N and longitudes 28° 35′–29° 10′ E. It is located

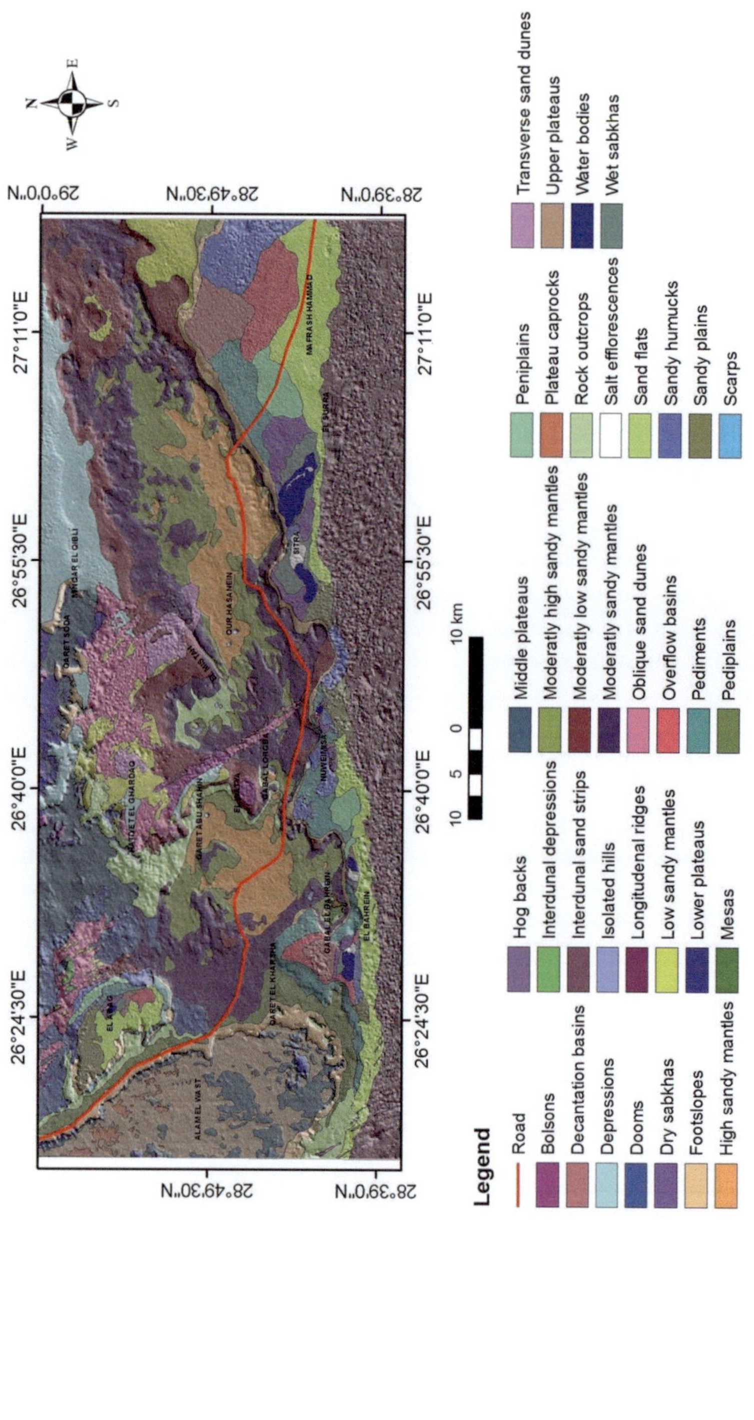

Fig. 7.8 Geomorphological units in abandoned oases. *Source* Saleh (2010)

Table 7.2 Soil taxonomy of abandoned oases in Western Desert, Egypt

Abandoned oases	Geomorphic units	Soil taxonomy
Sitra Oasis	Footslopes	Calcic Haplosalids
	Pediments	Pediments Typic Haplosalids
	Peniplains	Typic Torripsamments
	Sandy plains	Typic Haplosalids
	Sand flats	Typic Haplosalids
	Sandy hummocks	Typic Torripsamments
	Overflow basins	Typic Haplosalids
	Decantation basins	Typic Haplosalids
	Wet sabkhas	Typic Aquisalids
	Dry sabkhas	Typic Aquisalids
El Nuweimisa Oasis	Footslopes	Typic Haplosalids
	Pediments	Lithic Torripsamments
	Peniplains	Typic Torripsamments
	Sandy plains	Sodic Haplocalcids
	Sand flats	Typic Torripsamments
	Sandy hummocks	Sodic Haplocalcids
	Overflow basins	Gypsic Haplosalids
	Decantation basins	Aquic Haplocalcids
	Wet sabkhas	Sodic Psammaquents
	Dry sabkhas	Typic Aquisalids
El Bahrien Oasis	Footslopes	Typic Haplosalids
	Pediments	Lithic Torripsamments
	Peniplains	Lithic Torripsamments, Typic Haplosalids
	Sandy plains	Typic Torripsamments
	Sand flats	Sodic Haplocalcids
	Sandy hummocks	Typic Torripsamments
	Overflow basins	Typic Torripsamments
	Decantation basins	Typic Haplosalids
	Wet sabkhas	Typic Aquisalids
	Dry sabkhas	Typic Torripsamments
El Arag Oasis	Footslopes	Lithic Torripsamments
	Pediments	Typic Haplosalids
	Peniplains	Lithic Torripsamments
	Sandy plains	Typic Torripsamments
	Sand flats	Typic Torripsamments
	Sandy hummocks	Typic Torripsamments
	Overflow basins	Typic Torripsamments
	Decantation basins	Typic Torripsamments
	Wet sabkhas	Typic Aquisalids
	Dry sabkhas	Typic Torripsamments

at about 130 km west of El-Minia governorate in the Nile valley and about 360 km southwest of Cairo and comprising a total area about 2100 km^2. It represents a great depression in the surface of the "Libyan plateau", but it is different from the other depressions of the Western Desert in being surrounded on all sides by high scarps and in having a great number of isolated hills. It is also a promising location for agriculture expansion projects due to the plenty and good quality of groundwater for agricultural activities (Darwish 2004). El-Kafrawy (2013) classified the Bahariya Oasis into

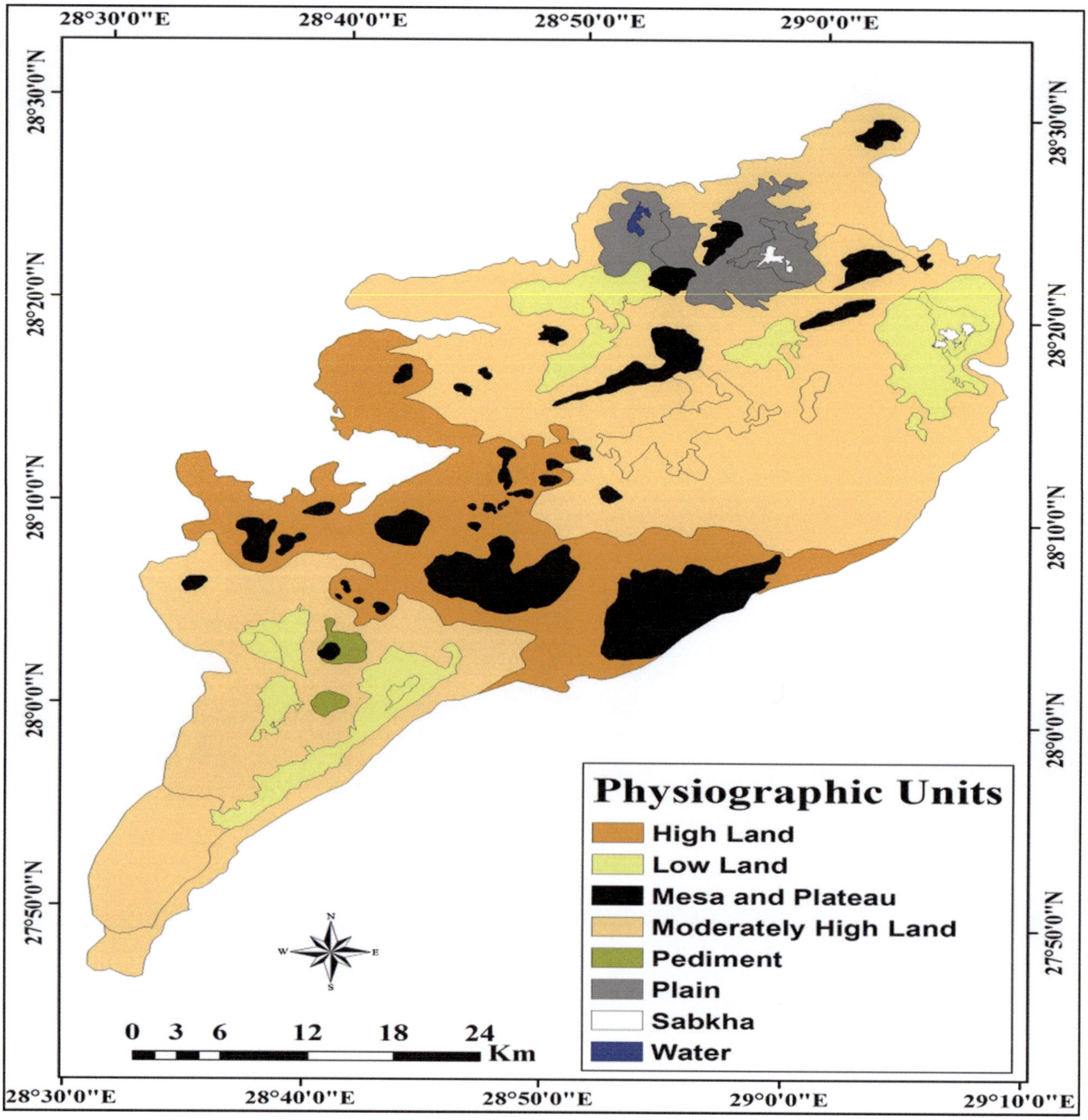

Fig. 7.9 Physiographic units in Bahariya Oasis, Western Desert, Egypt. *Source* El-Kafrawy (2013)

three main physiographic units that were identified in the oasis, which are plains, depression floor with low, moderately high and high lands, and pediment (Fig. 7.9). Plain represents about 4.64% of the studied area. Low land represents about 9.90% of the studied area. Moderately high land represents about 56.31% of the studied area. High land represents about 18.66% of the studied area. Pediment represents about 0.54% of the studied area.

On the other hand, the same author classified the soil as Bahariya Oasis as follows.

Soils in Bahariya Oasis are located under two main soil orders: Aridisols and Entisols. Aridisols represent about 75.05% of the soils in the Oasis, whereas Entisols represent about 15.04% of the soils. Entisols, which are less developed soils, were mainly developed over sand dunes and recently alluvial deposits. Aridisols, which is the major soil order, were developed over relatively old soils in the Oasis (Darwish et al. 2008).

The main diagnostic horizons of Aridisols are calcic, gypsic, salic, and argillic horizons. Calcic horizon is

Table 7.3 Soil types in Bahariya Oasis Western Desert, Egypt

Soil order	Soil suborder	Soil great group	Soil sub-great group
Aridisols	Gypsids	Calcigypsids	Lithic Calcigypsids
	Gypsids	Haplogypsids	Lithic Haplogypsids
	Gypsids	Haplogypsids	Typic Haplogypsids
	Salids	Aquisalids	Typic Aquisalids
	Salids	Haplosalids	Lithic Haplosalids
	Argids	Gypsiargids	Typic Gypsiargids
Entisols	Psamments	Quartzipsamments	Typic Quartzipsamments
	Psamments	Torripsamments	Lithic Torripsamments
	Psamments	Torripsamments	Typic Torripsamments
	Fluvents	Torrifluvents	Typic Torrifluvents
	Orthents	Torriorthents	Typic Torriorthents

Adapted from El-Kafrawy (2013)

common in most of the studied soils. However, it is mostly located in the middle parts of the Oasis. It is usually located at a shallow depth due to the very low rainfall in this very arid region. Gypsic horizon is usually associated with the calcic horizon and it relatively has the same behavior. Salic horizon usually exists in soils close to salt marches and poorly drained area in the Oasis. These areas mostly located in the northern parts of the Oasis close to El-Hara and Mandisha. Soils in these areas are cultivated under poor or no drainage system. Sabkhas and salt marshes are also common in these areas. Salic horizon is also located in the southern parts of the Oasis due to intensive evaporation.

According to Belal and Al-Ashri (2010), the Argillic horizon was observed in one location in the northwest side of the Oasis. This horizon is rich in clay minerals with evidence of clay films over the bed surfaces. It is well known that argillic horizon is usually developed under humid and warm conditions, which currently do not exist in this region. This could indicate that the climatic condition in this region was different in the past than the current situation. In other words, the climate in the Oasis was more humid. Soil suborders under Aridisols include Gypsids, Salids, and Argids. On the other hand, there were three soil suborders under Entisols, which are Fluvents, Orthents, and Psamments.

There were four great groups under both Aridisols and Entisols. Great groups under Aridisols include Aquisalids, Gypsiargids, Haplogypsids, and Haplosalids. Great groups under Entisols include Quartzipsamments, Torrifluvents, Torriorthents, and Torripsamments. Soil sub-great groups under Aridisols include Lithic Calcigypsids, Lithic Haplogypsids, Typic Aquisalids, Lithic Haplosalids, Typic Gypsiargids, and Typic Haplogypsids. Soil sub-great groups under Entisols include Lithic Torripsamments, Typic Quartzipsamments, Typic Torripsamments, Typic Torrifluvents, and Typic Torriorthents. The percent of each soil sub-great groups relative to the total is represented in Table 7.3, and the spatial distribution of these soil sub-great groups is illustrated in Fig. 7.10.

7.4.3 Soil of Wadi EL-Kubbaniya Basin

Recently, both the Egyptian government and private sector have been interested in developing low desert zone areas. The low desert zones are located between the recent Nile floodplain and the limestone plateau, from the east and west sides, and represent an important source of aggregate materials. The development of these areas, however, will reduce the supply of aggregate materials so that new sources must be identified. Therefore, it was decided to study soils and their optimum use as well as surface and underground water that suggested being developed in order to build geodatabase for the horizontal agricultural expansion areas in El Gallaba plain, West of Aswan (Saleh et al. 2015). The climate conditions of Wadi EL-Kubbaniya basins are typically desert-like conditions characterized by an arid climate with long hot rainless summer, mild winter with very low to none amount of rainfall.

According to Ahmed (2017) and Saleh et al. (2015), they classified the soil unites (Fig. 7.11) in Wadi El-Kubbaniya Basin into 12 mapping units that were identified in the study area: piedmont plain covering an area of about 185,996 fed (26.89%), peniplain covering an area of about 68,248 fed (9.78%), sandy plain occupies an area of about 100,286 fed (14.50%), river terraces covering an area of about 63,703 fed (9.21%), foot slopes occupies an area of about 50,591 fed (7.31%), wadis covering an area of about 18,070 fed (2.61%), old lacked occupies an area of 36,951 fed (5.34%), and rock outcrop covering an area of about 117,097 fed (16.93%). These landforms were divided according to its relief to many sub-landforms. In addition to sand sheet, playa, hills, and mesa occupy areas of about 5477, 481, 40,076, and 2913 fed and

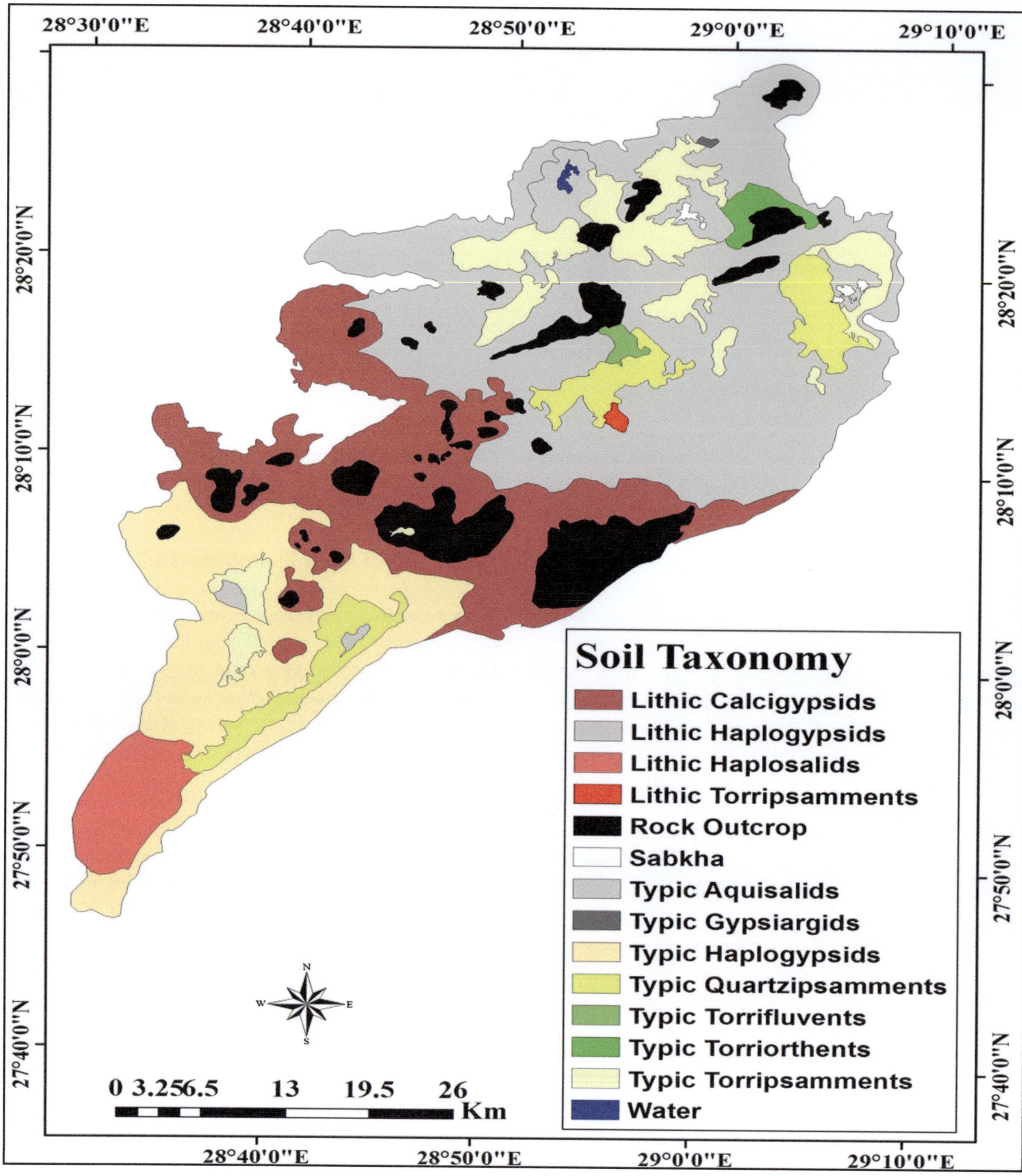

Fig. 7.10 Soil types in Bahariya Oasis Western Desert, Egypt. *Source* El-Kafrawy (2013)

represent 0.79, 0.07, 5.79, and 0.42% of the total area, respectively. The morphological description for soil profiles of the study area showed that the texture class ranged from sand to sand clay loam. Single grain to blocky structure characterized soils of the study area. Also, $CaCo_3$ has been observed differing in varying degrees by location different of segregations, sizes, and shapes. Based on Soil Survey Stuff (2014), two soil orders could be identified: Entisols and Aridisols, which are represented by nine sub-great groups (Table 7.4; Fig. 7.12).

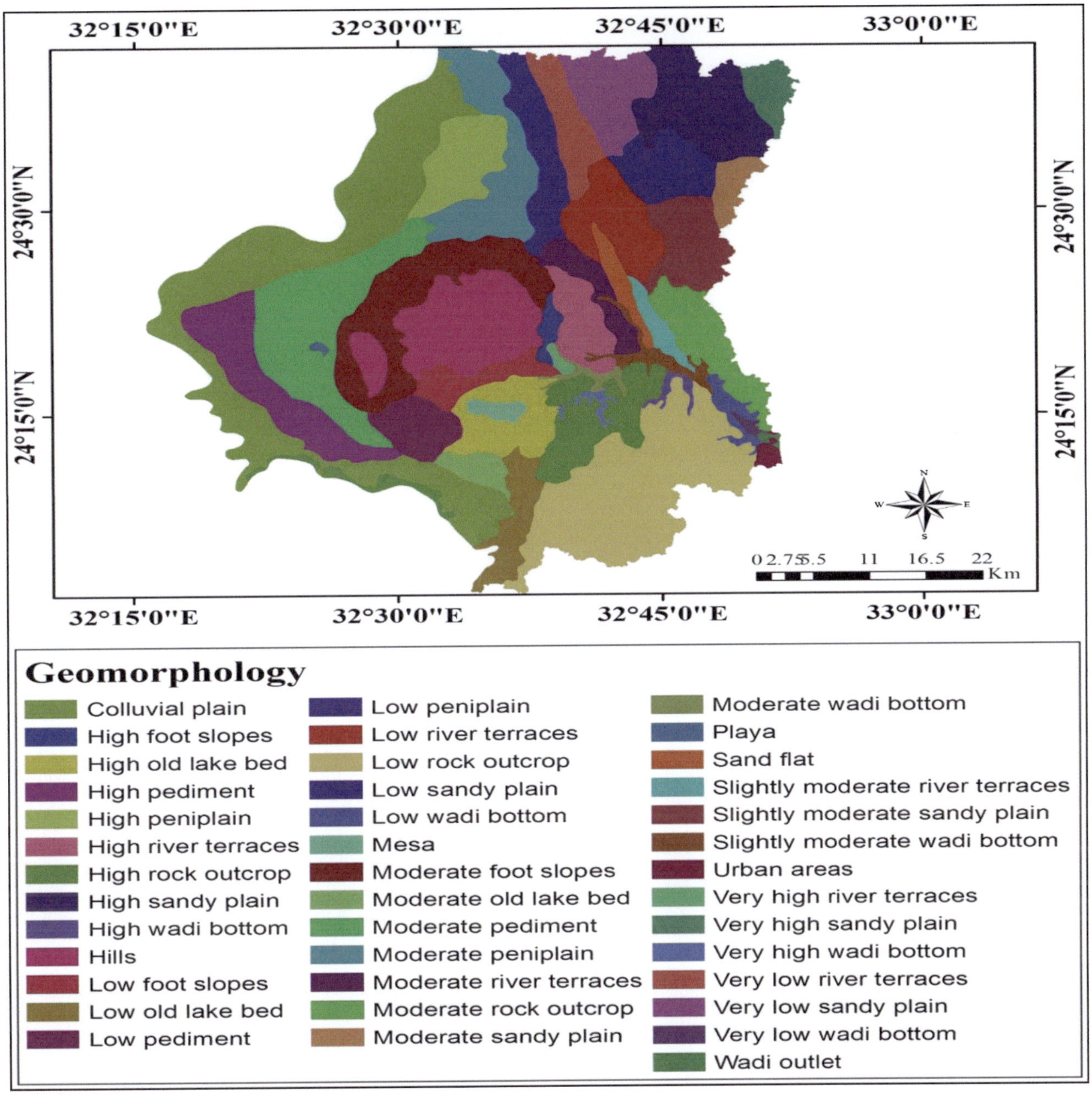

Fig. 7.11 Landform of EL-Kubbaniya basin, Western Desert Egypt. *Source* Ahmed (2017)

Chemical analysis of the studied area indicated that soils are slightly to moderately alkaline with pH values range from 7 to 8.3. Total soluble salts' content differs widely from location to another and has a wide range, as ECe ranges between 0.63 and 55.14 dS m^{-1}. Cation exchange capacity (CEC) differs also from site to another due to the ratio of fine fraction and organic matter content, where it ranges between 0.58 and 4.88 meq/100 g soil. Exchangeable sodium percentage (ESP) ranges between 0.57 and 34.47% in the study area. The calcium carbonates content of the soils has a wide range, as it ranges between 0.36 and 28.4%. The organic matter content O.M % ranged from none organic matter content (0%) to low content (1.38%).

7.5 Soil of the Northwestern Coast

The northwest coastal (NWC) region of Egypt is one of the most promising areas for agricultural development. It extends from Alexandria in the east to Salloum in the west about 500 km and from the Mediterranean coastal plain toward Al-Qattara Depression by about 40 km. According to

Table 7.4 Soil types in Wadi El-Kubbaniya basin, Western Desert Egypt

Landform	Soil sub-great group
Moderate Peniplain	Typic Torriorthents
High Peniplain	Lithic Torriorthents
High pediment, colluvial plain, low peniplain, slightly moderate river terraces, moderate pediment, moderate foot slopes, low wadi bottom, very low wadi bottom, wadi outlet, slightly moderate wadi bottom, sand sheet, high wadi bottom, high wadi bottom, moderate wadi bottom, very high wadi bottom, very high sandy plain, high sandy plain, very low sandy plain, low sandy plain, moderate sandy plain, and slightly moderate sandy plain	Typic Torripsamment
Low foot slopes, low pediment, low river terraces, very low river terraces, high river terraces, very high river terraces, and playa	Typic Haplocalcids
High wadi bottom	Lithic Haplocalcids
Very high sandy plain	Calcic Haplosalids
Low old lake bed	Gypsic Haplosalids
Moderate old lake bed	Typic Haplosalids
High foot slopes	Typic Calciargids

Adapted from Ahmed (2017)

Shalaby and Tateishi (2007), the soils of the northwestern coast are highly calcareous as were formed from calcareous parent materials. The limited translocation of carbonates due to the arid climatic condition results in the of this accumulation formation of. The soil in northwestern coast is classified in two orders as Entisols, which includes Psamments and Orthents suborders, and Aridisols, which include Calciorthids, Paleorthids, and Salorthids suborders. These soils were originated in four parent materials: Alluvial origins: the parent materials are loamy and highly calcareous. They are mostly forming the soils in coastal plain, alluvial fans, outwash plains, and the depressions between eroded and dissected ridges. Marine origin: It is Oolitic sand, which forms the beach ridges and inland consolidated dunes. Aeolian origin: It is covering wide and narrow parts either on the plateau or in the coastal plain. Lagoons origin: It is found in the depressions between the beach ridges.

According to Hegazi et al. (2005), the NWC areas form a belt to about 20 km deep and extend to about 500 km between Alexandria and Salloum near Libya borders, covering an area of about 10.000 km^2 (2.5 million acres). These areas are dominated by different landforms covered with soils that are in equilibrium with the environment. From the pedological point of view, most of these soils are young. Based on the landforms, the soils of NWC areas could be distinguished as follows: soils of the old coastal plain, including foreshore strip and lagoonal depressions (Typic Aqui-Psamments, Typic Haplosalids); soils of the old coastal plains, including calcic and quartzitic dunes as well as the interdunal depressions (Typic Haplosalids, Typic Gypsids, Calcic Haplogypsids, and Typic Torripsamments); and soils of the alluvial fans (Typic Torriorthents) and soils of the plateaux (Lithic Torriorthents).

Jalhoum (2015) and Sayed (2013) are described the main physiographic units in the studied area located between longitude 28° 30′ and 29° 00′ E and Latitudes 30° 45′ and 31° 05′ N and it located in AL-Alamein—SiDI Abderrahman area in NWC, Egypt. This area occupies about 1059.70 km^2 (252,308.79 feddans) and is considered as one of the most promising areas for agricultural expansions in Egypt (Fig. 7.13). Physiographic units were recognized and delineated by analyzing the main landscape that was extracted from visual interpretation of satellite image draped over DTM to get the natural 3D terrain with the aid of the different topographic maps and fieldwork survey. They were identified in the area, which are basin, depressions, terraces (high and low), sand plain, and piedmont plain, in addition to reference terms (shoreline, ridges, lagoons, sabkhas, and tableland). Basin represents about 9.79% of the studied area (about 24,695.24 fed), while depressions represent about 7.43% of the studied area (about 18,752.38 fed). But terraces represent about 29.03%, and it included high terraces (14.70%) and low terraces (14.33%) from the studied area which is about 73,240.00 feddans. On the other hand, sand plain represents about 18.1% of the studied area (about 45,677.54 fed). Finally, the

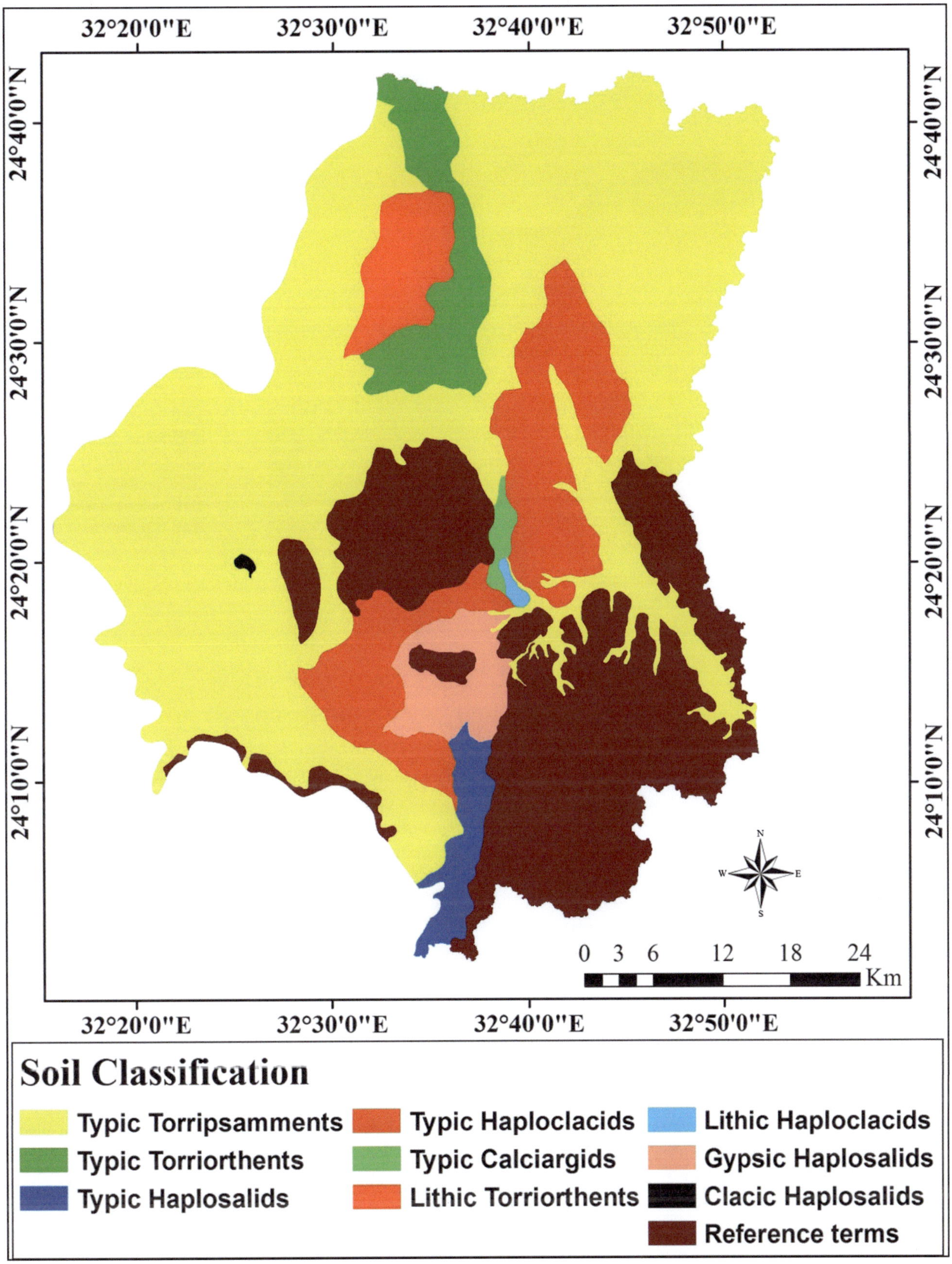

Fig. 7.12 Soil classification map of EL-Kubbaniya basin. *Source* Ahmed (2017)

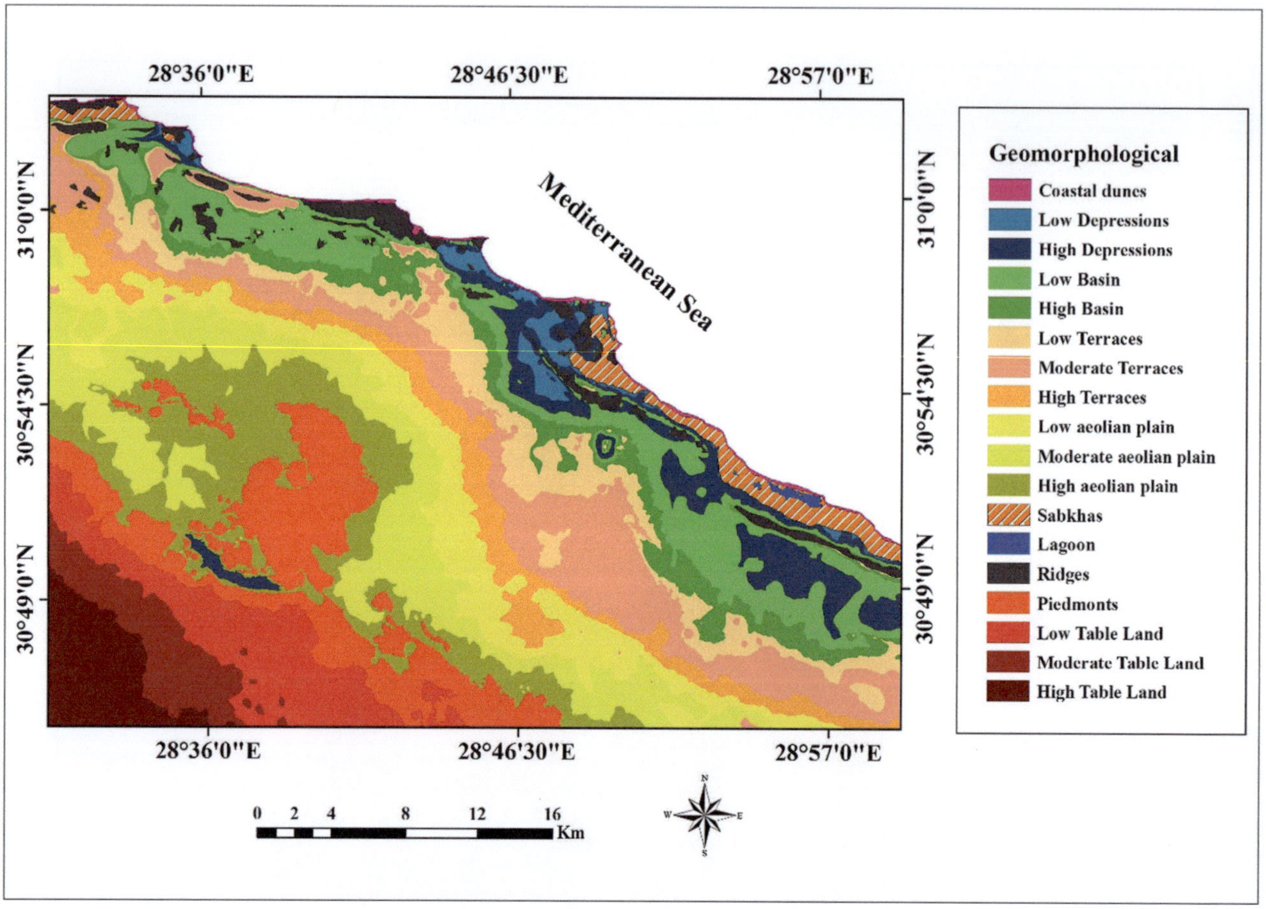

Fig. 7.13 Landforms for AL-Alamein–Sidi Abderrahman area in NWC, Egypt. *Source* Jalhoum (2015)

piedmont plain represents about 17.12% of the studied area (about 43,195.23 fed). The same author classified the soil as follows (Fig. 7.14):

Soils of the basin

The morphological features of soil profiles representing this unit show that the texture class is sandy loam and sand clay loam for the surface layers and sand for some other profiles. Soil profiles are characterized by deep soils, soil structure is subangular blocky and massive in the surface layer. The consistency is soft to slightly hard, hard, and very hard for the subsurface layer when dry, nonsticky to slightly sticky, nonplastic to slightly plastic. Chemical analysis of the studied profiles indicated that calcium carbonate content ranges between 20.5 and 81.8% with a layer of limestone at a depth of 80 cm at profiles 1 and 2. The pH values range between 8.0 and 8.7 in all the studied profiles. The electric conductivity data of saturation paste extracts (ECe) is quite low to moderate in the surface layers. Where the electric conductivity values range between 0.4 and 12.5 dS m^{-1}. Gypsum content ranges between 8.69 and 15.91%. Soils of this unit are classified as Typic Haplocalcids and Typic Torriorthents. Figure 7.15 shows basin of NWC.

Soils of the depressions

The morphological features of soil profiles representing this unit show that the texture class is sandy loam and sand. Soil profiles are characterized by deep soils and moderately deep, soil structure is subangular blocky, massive for the surface layer and single grain structure for subsurface layer. Chemical analysis of the studied profiles indicated that calcium carbonate content ranges between 22.16 and 63.07%. The pH values range between 8.12 and 8.8. The electric conductivity data of saturation paste extracts (ECe) is quite low to moderate in the surface layers. Where the electric conductivity values range between 1 and 13 dS m^{-1}. Soils of this unit are classified as Typic Haplocalcids and Typic Torriorthents.

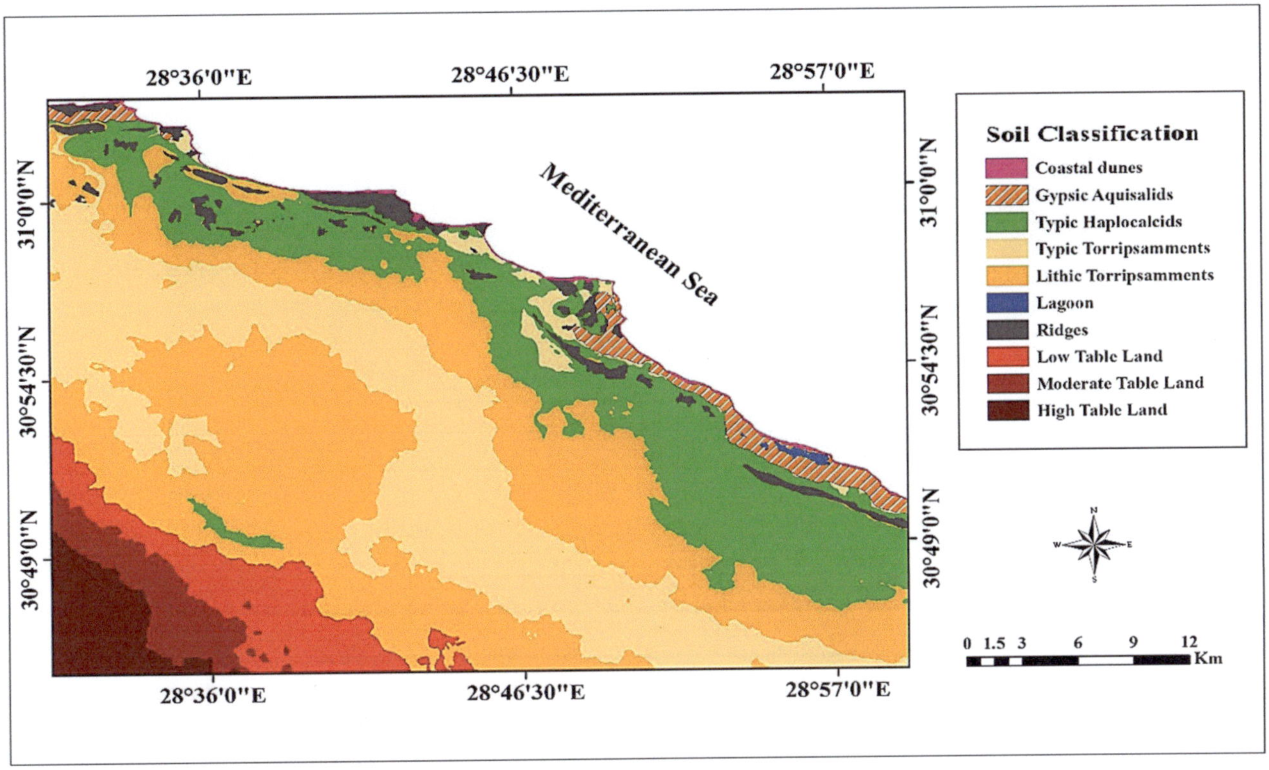

Fig. 7.14 Soil types of AL-Alamein–Sidi Abderrahman area in NWC, Egypt. *Source* Jalhoum (2015)

Fig. 7.15 Basin of AL-Alamein–Sidi Abderrahman area in NWC, Egypt. *Source* Jalhoum (2015)

Soils of the sabkhas

The morphological features of soil profiles representing this unit show that the texture class is sandy loam and sandy clay loam. Soil profiles are characterized by moderately deep soils (50–80 cm). Chemical analysis of the studied profiles indicated that calcium carbonate content ranges between 10 and 39.5%. The pH values range between 8.15 and 8.78. The electric conductivity (ECe) is very high. Where the electric conductivity values range between 30.18 and 178 dS m^{-1}. Gypsum content high ranges between 8.69 and 15.91%. Soils of this unit are classified as Gypsic Aquisalids. Figure 7.16 shows basin of NWC.

Fig. 7.16 Sabkha of AL-Alamein–Sidi Abderrahman area in NWC, Egypt. *Source* Jalhoum (2015)

Soils of the terraces

The soil characteristics of low terraces: The morphological features of soil profiles representing this unit show that the texture class is sandy loam, loamy, and sand. Chemical analysis of the studied profiles indicated that calcium carbonate content ranges between 33 and 75% show up to 95% for the surface layer of some soil profile with a layer of limestone. The pH values range between 8.0 and 8.8. The electric conductivity (ECe) is quite low to moderate in the surface layers. Where the electric conductivity values range between 0.95 and 8.29 dS m^{-1} of all profiles. Exception some soil profile, the electric conductivity (ECe) was high salinity relatively and the ECe values show up to 19.17 dS m^{-1}. The relatively high ECe values may be related to the location of the near the coastline and uncultivated soils. Gypsum content high ranges between 1.3 and 7.05%. Soils of this unit are classified as Typic Calcigypsids, Lithic Haplocalcids, and Typic Haplocalcids.

The soil characteristics of high terraces: The morphological features of soil profiles representing this unit show that the texture classes are sandy loam and sand. Chemical analysis of the studied profile indicated that calcium carbonate content ranges between 30.68 and 68.18%. The pH values range between 8.1 and 8.7. The electric conductivity (ECe) is quite low to moderate in the surface layers. Where the electric conductivity values range between 0.4 and 7.5 dS m^{-1}. Gypsum content high ranges between 1.6 and 7.7%. Soils of this unit are classified as Lithic Torriorthents, Lithic Haplogypsids, and Typic Haplocalcids.

Fig. 7.17 Soil profile of AL-Alamein–Sidi Abderrahman area in NWC, Egypt. *Source* Jalhoum (2015)

Soils of the fluvio-aeolian plain

The morphological features of soil profiles representing this unit show that. The texture classes are sandy loam, sand, loamy, and sandy clay loam. Soil profiles are characterized by deep soil. Soil structure is subangular blocky structure and single grain for the surface layer. Chemical analysis of the studied profile indicated that calcium carbonate content ranges between 30.68 and 57% up to 75.2–85.23% for soil profile 27. The pH values range between 8.2 and 8.3. The electric conductivity (ECe) is quite low to moderate in the surface layers. Where the electric conductivity values range between 0.3 and

7.7 dS m^{-1}. Soils of this unit are classified as Typic Haplocalcids. Figure 7.17 shows the soil profile of NWC.

Soils of the piedmonts

The morphological features of soil profiles representing this unit show that. The texture classes are sandy loam. Soil profiles are characterized by very shallow. Soil structure is subangular blocky structure and single grain for the surface layer. Chemical analysis of the studied profile indicated that calcium carbonate content ranges between 25.5 and 54.5%. The pH values range between 8.1 and 8.6. The electric conductivity (ECe) is quite low in the surface layers. Where the electric conductivity values range between 0.8 and 32.4 dS m^{-1} in the surface layer of profiles 28, 30, and 31. Soils of this unit are classified as Lithic Torriorthents.

7.6 Soil North Nile Delta

According to Aboelmaged (2016), he studied the soil properties in North Nile Delta. The present morphological and pedological study, which is based on interpretation of DEM, satellite images, fieldwork survey, and analytical data, reveals the following:

Soils of alluvial plain

The morphological features of soil profiles shown the color is dark grayish brown (10YR4/2), grayish brown (10YR4/2) and dark brown (10YR4/2) in the subsurface layer. In addition, the moist condition soil color is very dark grayish brown (10YR3/2) to grayish brown (10YR3/2) and very dark gray (10YR3/1); the texture class is clay for most profiles and silty clay for the subsurface layers of profile 1, 2, 8 and 20. Soil profiles are characterized by deep soils, soil structure is moderate to strong structure evolution. The consistency is hard and firm when dry, sticky, plastic, very sticky, and very plastic for the studied profiles. The horizon boundary is clear to diffuse smooth boundary. The bulk density, as an indicator of compaction, ranges between 1.30 and 1.47 g cm^{-3}. It increases with depth in the different layers of the studied profiles. Chemical analysis of the studied profiles data indicates that calcium carbonate content ranges between 1.04 and 5.38% of the fine earth in the different layers of the studied profiles. pH values range between 8.00 and 8.60 in the successive layers of the studied profiles. The electrical conductivity data of saturation paste extracts (ECe) is quite low to moderate for all profiles and range between 0.99 and 7.90 dS m^{-1}. The chemical analysis data of saturated paste extracts indicated that soils are dominated by Na^{+} cation followed by Ca^{++} or Mg^{++} cations alternately, while K^{+} cation is the least soluble component. Anion is dominated by Cl^{-} ion or SO_4^{2-} ion alternately followed by HCO_3^{-}. Thus, sodium chloride or sodium sulfate seems to be dominating in these soils. The organic matter content ranges between 0.34 and 2.78%. The high values of organic matter in the upper layers of the studied profiles may be due to the continuous addition of organic manures for these cultivated soils. The cation exchange capacity ranges between 33.12 and 46.92 meq/100 g soil in the different layers of the studied soil profiles reflecting the effect of clay and organic matter content. The exchangeable calcium forms the main portion of the exchangeable cations in the successive layers of the studied soil profiles. The ESP values range between 3.41 and 14.86% in the different layers of the representative soil profiles. With regard to data of available nutrients, however, soil fertility status depends mainly upon the different levels of organic matter, cation exchange capacity, and the soil pH values. The available nitrogen ranges between 5.60 and 61.6 ppm in the surface layers of the studied profiles and the meanwhile of the available phosphorus ranges between 3.2 and 30.4 layers of the studied profiles. The high value of N and P were located in the surface layers of soil profiles. Available potassium ranges between 233 and 390 ppm in different studied soil profiles. Based on the morphological features of the representative profile of this unit and chemical analysis, the soils are classified as Typic Endoaquerts, Typic Haplotorrerts, and Typic Torrifluvents.

Soils of lacustrine deposits

The morphological features of soil profiles represent this unit are shown the dry soil color is dark grayish brown (10YR4/2), grayish brown (10YR4/2) and dark brown (10YR4/2) in the subsurface layer, while the moist condition soil color is very dark grayish brown (10YR3/2) to grayish brown (10YR3/2) and very dark gray (10YR3/1). The texture class is clay for all profiles. Soil profiles are characterized by moderately deep-to-deep soils. Soil structure is moderate to strong structure evolution and blocky to massive and columnar. The consistency is slightly hard when dry, slightly sticky, and slightly plastic. The horizon boundary is diffuse smooth boundary.

Chemical analysis of the studied profiles data indicates that calcium carbonate content ranges between 0.46 and 3.97% in different layers of all studied soil profiles. The pH values range between 7.80 and 8.50 in all the studied profiles. The electric conductivity data of saturation paste extracts (ECe) is quite low to extremely high. The soil salinity values revealed that the electrical conductivity (ECe) ranges between 1.58 and 79.0 dS m^{-1}. The high value of ECe is found in profile 6, which may be due to the effect of high EC value of water table is under aridity conditions.

Table 7.5 Soil classification of North Nile Delta

Soil order	Soil suborder	Soil great group	Soil sub-great group
Aridisols	Calcids	Haplocalcids	Typic Haplocalcids
Vertisols	Aquernts	Endoaquerts	Sodic Endoaquerts
	Torrerts	Haplotorrerts	Typic Haplotorrerts
			Sodic Haplotorrerts
		Salitorrerts	Typic Salitorrerts
Entisols	Psamments	Torripsamments	Typic Torripsamments
	Fluvents	Torrifluvents	Typic Torrifluvents

Adapted from Aboelmaged (2016)

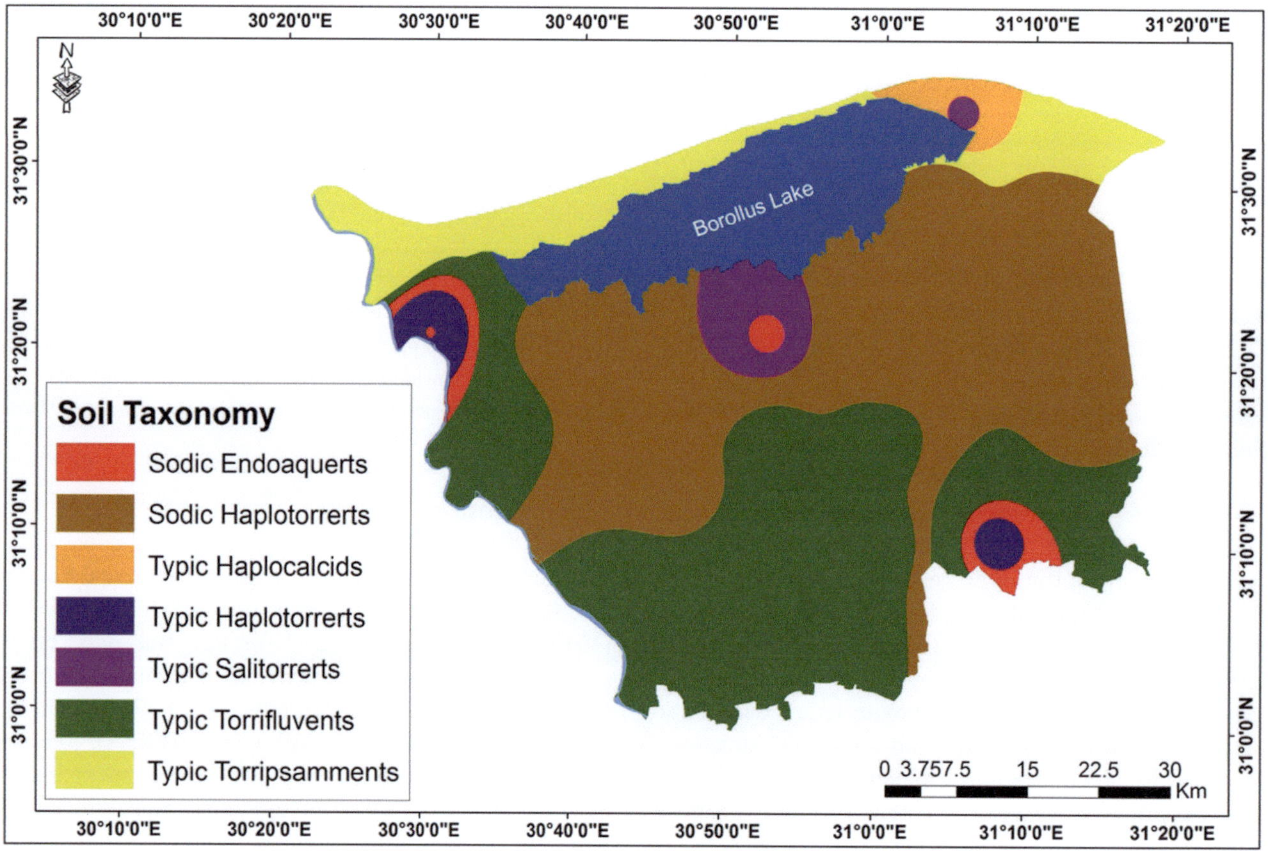

Fig. 7.18 Soil map of sub-great groups of North Nile Delta. *Source* Aboelmaged (2016)

The chemical analysis data of saturated paste extracts indicated that soils are dominated by Na^+ cation followed by Ca^{++} and Mg^{++} cations, while K^+ cation is the least soluble component. Anion is dominated by Cl^- ion followed by SO_4^{2-} ion and HCO_3^-. Otherwise, profile 6 is dominated by Na^+ cation followed by Mg^{++} and Ca^{++}, while K^+ cation is the least soluble component. Anions are dominated by Cl^- ion and SO_4^{2-} ion and HCO_3^-. The organic matter content ranges between 0.33 and 3.03%, and its highest values are in the upper layers of the studied profiles which could be due to the continuous addition of organic manures for these cultivated soils. The cation exchange capacity ranges between 33.12 and 46.92 meq/100 g soil in the successive layers of the studied soil profiles. The high values are correlated with high content of both organic matter and fine-textured soils. The exchangeable calcium forms the main portion of the exchangeable cations in the successive layers of the studied soil profiles. The ESP values range between 6.01 and 43.85% in the different layers of the representative soil profiles. The data of available nutrients in the surface layer revealed that available nitrogen is low to high (ranges between 5.6 and 28.0 ppm). The available phosphorus content is low to medium, and it ranges between 3.2 and 44.4 ppm. Available potassium content is high (more than 250 ppm). Based on the morphological feature of the representative profiles and chemical analysis of this unit the soils are classified as Sodic Haplotorrerts, Sodic Endoaquerts, and Typic Salitorrerts.

Soils of coastal plain

The laboratory analysis indicated that calcium carbonate content ranges between 0.21 and 1.24%. The pH values range between 7.1 and 8.8 in all the studied profiles. The electrical conductivity data of saturation paste extracts (ECe) is quite low to moderate for all profile and it ranges between 1.38 and 3.36 dS m^{-1}. The chemical analysis of saturated paste extracts indicates that soluble cations are dominated by Na^{+}, Mg^{++}, and Ca^{++}, while K^{+} cation is the least soluble component. Anion is dominated by Cl^{-} ion followed by SO_4^{2-} and by HCO_3^{-}. The organic matter content ranges between 0.52 and 2.19%, and it has low values and decreases irregularly with depth. The only anomalous high content of organic matter was found in the surface layer of profile 15, which may be due to application of manure to the soil. The aforementioned evidences of stratification and irregular decrease in organic matter, in addition to the absence of diagnostic horizons, assure that the sand flats occur in Entisols order. The cation exchange capacity varies between 6.9 and 18.36 meq/100 g soil. The low values of CEC referred to lack of clay and organic matter content as well as the dominance of sandy soils. Calcium and magnesium forms represent the main portion of the exchangeable cations in these soils. The exchangeable sodium percentage (ESP) varies between 5.09 and 12.69%.

Macronutrients of the surface layer revealed that the available nitrogen content is low to medium and it ranges between 7.65 and 19.6 ppm, while the higher value may be due to the continuous addition of nitrogenous fertilizers to the cultivated orchards in that site. Available phosphorous and available potassium are also in low values with a content ranged from 3.2 to 16.0 ppm and 71–175 ppm, respectively. Based on the morphological feature of the representative profiles and chemical analysis of this unit, the soils are classified as Typic Torripsamment and Typic Haplocalcids.

Soil classification of the North Nile Delta

The American Soil Survey Staff system (2014) was applied down to the sub-great group level for the studied soil mapping unit. Table 7.5 shows the calculated area for each soil classification unit. The spatial distribution of these soils, sub-great groups, is illustrated in Fig. 7.18.

References

Afify AA, Arafat SM, AboelGhar M, Khadera MH (2010) Physiographic soil map delineation for the Nile alluvium and desert outskirts in middle Egypt using remote sensing data of EgyptSat-1. Egypt J Remote Sens Space Sci 13:129–135

Aboelmaged A (2016) Monitoring rate and type of land degradation in salt affected soils in the north Nile delta using remote sensing and GIS techniques. M.Sc. thesis, Faculty of Agriculture, Kafrelsheikh University, Egypt

Ahmed M (2017) Land resources assessment of wadi ELKubbaniya basin, south Egypt using multi and hyper spectral remote sensing and GIS techniques. M.sc. thesis, Faculty of Agriculture, Tanta University, Tanta, Egypt

Belal ABA, Al-Ashri KMA (2010) Relationship between landforms and soil characteristics in Bahariya Oasis, Egypt. J Soil Sci Agric Eng Mansoura Univ 1(12):1157–11756

Belal AA, Mohamed ES, Abu-hashim MSD (2015) Land evaluation based on GIS-Spatial Multi-Criteria Evaluation (SMCE) for agricultural development in dry Wadi, Eastern Desert, Egypt. Int J Soil Sci 10(3):100–116

Darwish KM (2004) Potential of soil and water resources for agriculture development in Bahariya Oasis, Egypt development of a GIS-based decision support system. Ph.D. thesis, Greifswald University, 335 S

Darwish KM, Rasheed MA, Gad A, Zoelitz-Moeller R (2008) Geographic mapping and analysis using GIS of study areas in Bahariya Oasis, Egypt. Am Eurasian J Agric Environ Sci 4(1):125–130

El-Shamy IZ (1988) Quantitative geomorphometry and surface runoff control for Wadi Qena, Eastern Desert. In: E.G.S. Proceedings of the 6th annual meeting, Mar 1988, pp 13–26

El-Kafrawy MM (2013) Developing soil and soil capability maps of Bahariya Oasis using remote sensing and GIS techniques. Ph.D. thesis, Faculty of Agriculture, Mansoura University, Mansoura, Egypt

FAO (2005) Fertilizer use by crop in Egypt. Rome, Italy

FAO (1998) World reference base for soil resources. World soil resources reports 84. Rome, Italy

Hegazi AM, Afifi MY, EL Shorbagy MA, Elwan AA, El-Demerdashe S (2005) egyptian national action program to combat desertification. UNCCD, Egypt

Khalifa HE, Moussa HA (2017) Soil and agriculture after the Aswan High Dam. In: Satoh M, Aboulroos S (eds) Irrigated agriculture in Egypt. Springer International Publishing, Switzerland, pp 81–124. https://doi.org/10.1007/978-3-319-30216-4_5

Jalhoum M (2015) Land evaluation in Northern Western Coast (NWC) of Egypt using new information technology. M.Sc. thesis, Faculty of Agriculture, Ain Shams University, Cairo, Egypt

NARSS (2016) Planning for the sustainable development of Halaib-Shalatin Area, South Eastern Desert, using RS, GIS and other techniques. Soil report (Part) National Authority for Remote Sensing and Space Sciences (NARSS), Cairo, Egypt, 1120 pp

Saleh A (2010) Studying soil and water resources of some abandoned oases in the Egyptian Western Desert using remote sensing and GIS techniques. Ph.D. thesis, Faculty of Agriculture, Tanta University, Tanta, Egypt

Sayed AS (2013) Evaluation of the land resources for agricultural development—case study: El-Hammam Canal and its extension, Nw Coast of Egypt. Ph.D. thesis, Hamburg University, Germany

Saleh AM, Belal AB, Mohamed ES (2015) Land resources assessment of El-Gallaba basin, South Egypt for the potentiality of agriculture expansion using remote sensing and GIS techniques. Egypt J Remote Sens Space Sci 18(1):S19–S30

Selvaradjou SK, Montanarella L, Spaargaren O, Dent D (2005) European DigitalArchive of Soil Maps (EuDASM)—soil maps of Africa. EUR 21657 EN, 386 pp

Shaheen SM, Abo-Waly ME, Ali RA (2013) Classification, characterization, and management of some agricultural soils in the north of Egypt. In: Shahid SA et al. (eds) Developments in soil classification, land use planning and policy implications: innovative thinking of soil inventory for land use planning and management of land resources. Springer Science + Business Media, Dordrecht, pp 417–448. https://doi.org/10.1007/978-94-007-5332-7_23

Shalaby A, Tateishi R (2007) Remote sensing and GIS for mapping and monitoring LC and land-use changes in the northwestern coastal zone of Egypt. Appl Geogr 27:28–41

Soil Survey Staff (1999) Keys to soil taxonomy. 8th edn., US Department Agriculture and Natural Resources Conservation Service, Washington, DC. Pocahontas Press Inc., Blacksburg, Virgina

Soil Survey Staff (2010) Keys to soil taxonomy. 11th edn., USDA-NRCS, Washington DC

Soil Survey Staff (2014) Keys to soil taxonomy. 12th edn., US Department Agriculture and Natural Resources Conservation Service, Washington, DC

Zaghloul SS (2013) Consideration of the agricultural problems as a base of water resources management in Egypt. In: Seventeenth international water technology conference, IWTC 17 Istanbul, 5–7 Nov 2013, Turkey

Soil Fertility and Its Security

8

Hassan El-Ramady, Tarek Alshaal, Sarwat Yousef, Shaimaa Elmahdy, Salah E.-D. Faizy, Megahed Amer, Hassan Shams El-Din, Ayman M. El-Ghamry, Ahmed A. Mousa, József Prokisch, and Nicola Senesi

Abstract

No doubt that soil and its fertility are considered one of the main factors, which control the agricultural production. So, the soil fertility should be linked with different securities including water, food and soil security to perform the holistic sustainable development. Concerning the fertility of Egyptian soils, this fertility differs with different geological zones of these soils as well as the historical background of soil formation (i.e. factors and processes of soil forming). Generally, the soils of the Nile Delta are the most fertile and mainly represent the basket of food production in Egypt. The fertility of Egyptian soils suffers from the decline after building the Aswan High Dam and stopping the sedimentation of clay and/or silt particles. So, the deterioration of the quality or fertility of Egyptian soils is resulted from the absence of silt from the Nile flooding and crop intensification. Therefore, new approaches should be followed to maintain and sustain the fertility of Egyptian soils such as the precision farming and nanotechnology.

Keywords

Soil fertility • Soil security • Salt-affected soils Depressions • Paddy soils

8.1 Introduction

The fertility of Egyptian soils mainly depended on the Nile River thousand years ago till now. So, most of the cultivated lands are very close to the banks of the Nile River (including all main branches and canals in the Nile Delta). Due to the majority of the country's area is located in desert lands, the fertility of these soils is a little bit low (Noaman 2017; Noaman and El Quosy 2017). It is worth to mention that the well-known name of Egypt '*Kemet*' or the *black land* derived from the alluvial soils deposited during annual floods of the Nile. This annual flooding gave Egypt its fertile lands, helping to expand in cultivating many various crops in the Nile Delta. The Nile Delta is the most fertile area in Egypt due to deposition of silt by the many tributaries of the Nile. After the construction of the Aswan High Dam in 1971, the deposition of silt through the annual floods is confined. Therefore, the fertility of almost Egyptian soils is declined as well as erosion most of the land along the coastal lands. Four salty and shallow lakes have been extended nowadays along the seaward extremity of the Delta (El Gamal and Zaki 2017; Khalifa and Moussa 2017; Noaman 2017).

Due to the presence of the Nile River, God gifts Egypt the fertile soils several thousand years ago. So, this proverb is very common '*Egypt is gift of the Nile*' as mentioned by the Greek historian Herodotus (Zahran 2010c; Negm 2017). This was because the Egyptian ancient civilization depended on the resources of this great river and thrived along the banks of the Nile. Therefore, the fertility of Egyptian soils in the past is absolutely different from the present (Satoh and Aboulroos 2017). It is worth to mention that, the fertility of Egyptian soils differ from its geological zone (i.e. soils of the

H. El-Ramady (✉) · T. Alshaal · S. Yousef · S. Elmahdy
S. E. -D.Faizy
Soil and Water Department, Faculty of Agriculture, Kafrelsheikh University, Kafr El Sheikh, Egypt
e-mail: ramady2000@gmail.com

M. Amer · H. Shams El-Din
Soils, Water and Environment Research Institute (SWERI), Agricultural Research Center (ARC), Giza, Egypt

A. M. El-Ghamry · A. A. Mousa
Soils Department, Faculty of Agriculture, Mansoura University, Mansoura, Egypt

J. Prokisch
Nanofood Laboratory, Faculty of Agricultural and Food Sciences and Environmental Management, Debrecen University, Debrecen, Hungary

N. Senesi
Department of Soil, Plant and Food Sciences, University of Bari, Bari, Italy

H. El-Ramady et al. (eds.), *The Soils of Egypt*, World Soils Book Series,
https://doi.org/10.1007/978-3-319-95516-2_8

Nile Delta, soils of the Eastern Desert, soils of the Western Desert, and soils of Sinai) (Zahran 2010a, b). Egypt has different and distinguished soils including salt-affected soils, natural depressions (oases), soils of waterlogged or submerged, the coastal soils, alluvial soils, lacustrine soils, etc. (Khalifa and Moussa 2017). Thus, a great task for improving soil fertility and its security will face Egypt with different challenges and hopes. These challenges include the decline of soil fertility after the High Dam, high levels in soil pollution, soil erosion and its salinization, desertification, etc. Therefore, this book chapter is an attempt to highlight on the fertility of Egyptian soils and their security from the past to the present, risk and challenges as well as hopes seeking towards the sustainable development in Egypt.

8.2 Fertility of Egyptian Soils: The Past and the Present

The fertility of soil could be defined as the ability of this soil to sustain the growth of cultivated plants and optimize the yield of crops. So, this fertility of soil is the potential nutrient status of a soil to produce crops or the ability or capability of soils to supply cultivated plants with essential and sufficient elements for the growth of plants without a toxic concentration of any element (Fig. 8.1). On the other hand, the productivity of soil is a measure of the ability of soil to produce a particular crop or sequence of crops under a specified management system. The fertility of soils could be enhanced through the application of organic and inorganic

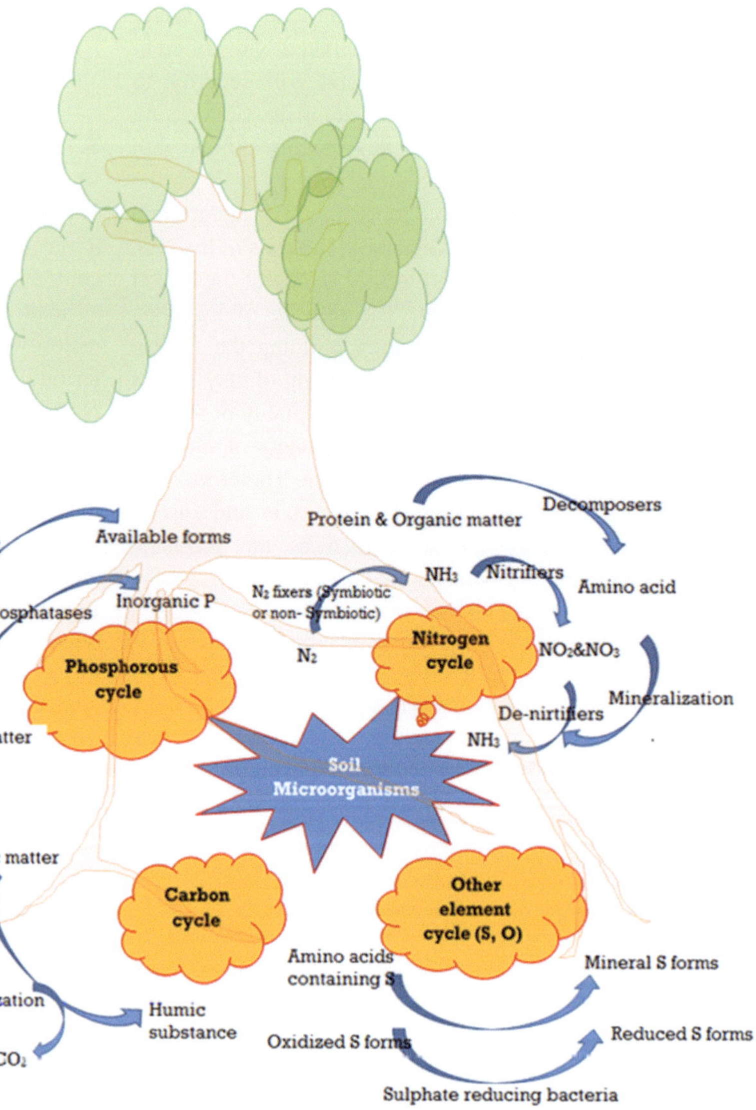

Fig. 8.1 The nutrients in soils and their dynamics definitely depend on several factors including soil microorganisms and cycle of these nutrients in soils. These nutrients are needed for cultivated plants in proper and sufficient amounts for their growth without a toxic concentration of any element. Drawn by Dr. Tamer Elsakhawy, SWERI, Sakha, Kafr El-Sheikh

fertilizers as well as bio-fertilizers to the soil (Huq and Shoaib 2013).

Concerning the global database about the soil fertility, it is found that '*soil fertility*' has been recorded 60,619 in Springer website including 37,674 articles, 22,064 book chapters, 1526 conference papers, 747 reference works and 108 protocols by 15 June 2017. The importance of soil fertility refers to without this fertility we can produce enough and quality food for humans; therefore, hunger and diseases will be spreaded out worldwide. So, we should feed soil to feed humans and also save the soil fertility to save the planet as well known (Jeffrey and Achurch 2017). The potentiality of soil fertility nowadays could be enhanced through new approaches such as precision farming, carbon fertilization, nanofertilizers, etc. Therefore, urgent soil management should be performed to save, sustain and maintain the fertility of soil.

In ancient Egypt, the soil was fertile enough due to the enriching these soils with annual flooding of the Nile with silt particles (suspended solid). So, a great civilization established and flourished on the different branches of the Nile River as painted in the temples in ancient Egypt. After the building of Aswan High Dam, some problems have been taken place for the Egyptian soils including the absence of silt from the Nile as the main reason. Beside this previous reason, there are some practices recently have been deteriorated the agricultural soils because of the crop intensification without proper and enough supply with nutrients, the continuous cultivating of soil top layer making bricks, which increased the loss of fertility in agricultural soils and the rising groundwater level (then soil salinity and waterlogging) because of the failure to apply scientifically recommended crop rotations and the repeated cultivation of particular crops (Khalifa and Moussa 2017). Apart from the bad side for Aswan High Dam, there are many great serves that have been recorded such as saving irrigation water, generating hydropower, controlling the production of annual flood and supplying the water for domestic/industrial uses (Hossain and El-shafie 2014).

Different applications of mineral fertilizers were and still the main approach in offsetting the nutrient depletion in soil worldwide. This supply sometimes suffers from the over-fertilization like in Egypt. Therefore, there is an increase in the consumption of mineral fertilizers year by year in Egypt (Fig. 8.2). According to statistics of FAO, there is over-fertilization in the three main mineral fertilizers (NPK) as reported by FAOSTAT (2017).

Therefore, the fertility of Egyptian soils generally needs more concern, sustainability and promising approaches to improve the nutritional status and the bioavailability of different nutrients in these soils. That means, Egypt has to maintain the current fertility of these soils as well as increasing the fertility of her soils at the same time and seeking for the sustainable land use. Among the new approaches for improving the fertility of Egyptian soils, there is the integration between organic and bio-fertilizers as well as the nanofertilizers, particularly which derived from the biological methods such as nanoselenium, nano-copper, nanosilica, etc.

8.3 Fertility of Egyptian Soils in Different Geological Zones

It could be identified the soil resources in Egypt according to different features of climate, terrain and landform characteristics. As well known, Egypt has different geological zones including the Nile Delta, The Eastern Desert, the Western Desert and Sinai. These geological zones mainly differ in their soil fertility due to the difference in their geological deposits such as lacustrine, marine, fluvial, sandy and fluvio-sandy, and calcareous soils. Concerning lacustrine soils, a lot of these soils could be found in north of the Nile Delta like soils from Manzala, Burullus, Edku and Mariut, Bardawil, Nasser and Qaroun lakes. Regarding fluvial soils, they were originated from the Nile Delta, whereas marine soils could occur in the Red and the Mediterranean Sea. About the calcareous soils, these soils could be found in separated areas in Egypt such as north of the Western Desert (Shaheen et al. 2013).

First of all, in terms of cropland base, Egypt could be considered as one of the world's poorest countries because the majority of this country's lands are desert. Therefore, it is estimated the present cultivated area in Egypt to be about 3.66% of the total area (Zaghloul 2013). For more details, the main cultivated lands in Egypt are located close to in (1) the Nile Delta, (2) the banks of the Nile River, (3) and the main branches and canals of the Nile Valley. Regarding the rangelands, they are restricted to a narrow strip representing only a few kilometres wide along the Mediterranean coast and their bearing capacity are quite low. According to the official statistics of the Ministry of Agriculture and Land Reclamation, the agricultural lands increased from 2.38 to 2.78 and about 3.61 million ha (5.71, 8.67 and 8.66 million fed) in 1950, 1982 and 2013, respectively (Khalifa and Moussa 2017). It is reported that the main soil groups in Egypt could be ordered as a percent from the total area as follows in Table 7.1.

Concerning the soils of the Nile Valley and Delta Zone, most soils are recent Nile alluvium forming from (1) the fluvio-marine and lagoon deposits located adjacent to the northern lakes (Edku, Burullus and Manzala) and (2) the coastal plain in the extreme north of the Delta. These soils are fertile and the basket of food in Egypt. This Delta could be divided into the north coastal zone, Eastern Desert zone, Western Desert zone and inland Sinai. The coastal zone

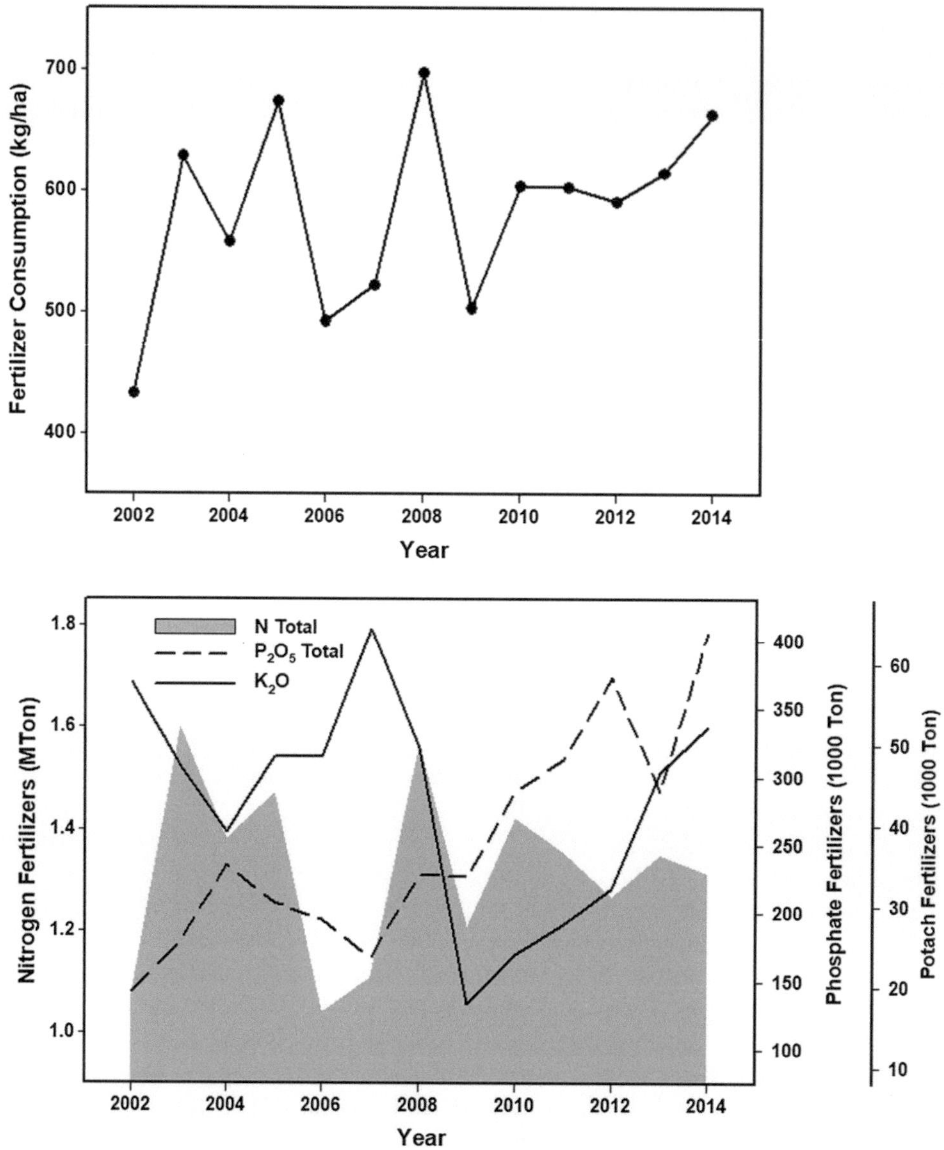

Fig. 8.2 The change in the total consumption of mineral fertilizers (kg ha^{-1}) in Egypt from 2002 to 2014 according to the statistics of FAO (the top figure), whereas the bottom one belongs the average of main fertilizers (NPK) uses per ha. Both figures have been drawn by Dr. Noura Bakr; created based on FAO data that downloaded from FAOStat; http://www.fao.org/faostat/en/#country/59

includes north-western coastal areas and north-eastern coastal areas of Sinai. The north-western coastal areas form a belt about 20-km deep and extend to about 500 km between Alexandria and Sallum near the Libyan borders, covering an area of about 10,000 km^2. This zone includes soils of the old coastal plain (foreshore strip and lagoon depressions or calcic and quartzitic dunes), soils of the alluvial fans and soils of the plateau (Khalifa and Moussa 2017).

Regarding the soils of the Eastern Desert, this area is rich in several wadis (e.g. wadi Qena, Al-Allaqui, El Assuity and El Laquita). These wadis are mostly deep and very steep and their soils immature (display young stages of development). The soils of these wadis are mainly shallow to deep profile, coarse or moderately fine-textured, with variable content of gravels. Furthermore, the main landforms of these wadis include alluvial fans and sand sheets, rubble terraces, plateau, wadi bottoms, outwash plains and dunes. Regarding the soils of the Western Desert, several natural depressions are scattered in variable areas including the famous oases in the south (Kharga and Dakhla), in the north (Siwa), and in the central area (Bahariya and Farafra). The soils of these oases have distinguished features including particular erosional patterns, special parent materials, different environmental sedimentations and different eluvial deposits (carbonate, salts and gypsum). There are many promising reclaimed areas in the Western Desert including Darb El-Arbain, Toshka, East El-Uweinat and some wadis of High Dam Lake (Khalifa and Moussa 2017).

Concerning the soils of Sinai Peninsula, these soils have borders from the West by the Gulf of Suez and from the East by the Gulf of Aqaba. These soils also have distinguished features including moderately deep to deep profile, gravelly, coarse-textured soils (in the southern part of Sinai), deep highly calcareous, gravelly coarse-textured soils and gravelly course over fine-textured soils (in the central part of Sinai), and deep, gravelly coarse or moderately -textured soils, deep calcareous, coarse to moderately to fine-textured soils (in the alluvial coastal plains), and some soils of the marshy land and alkali soil or sabkhas along the Gulf of Suez as well as some soils sandy deposits as moderately fine-textured with salts and gypsum contents in wide ranges (Khalifa and Moussa 2017).

8.4 Fertility of Some Common Egyptian Soils

Egypt has different types of soils and each kind of these soils has particular characterizations such as salt-affected soils, waterlogged or paddy soils, coastal soils, soils of the Nile Delta, soils of oases or depressions, calcareous, sandy, heavy clayey, alluvial, floodplain, lacustrine, marine, fluvial and fluvio-sandy, etc. These soils could be considered one of the most important features of soils in Egypt, which are may develop and influence by the type of parent materials and conditions of depositional environments. These soil parent materials may form different sources like mineral mixtures, rocks and sediments resulting from glacial and aeolian deposits as well as floods. Concerning fluvial deposits, they include all sediments by which deposited the running waters and are generally poorly sorted. Regarding the alluvial soils, they are developed through the alluvium deposition brought by streams to the land. About the lacustrine and marine deposits, lacustrine are settled out of stagnant lakes, but marine deposits are originated from stream action and deposited in the sea (Shaheen et al. 2013).

8.4.1 Salt-Affected Soils and Their Fertility

It is well known that salt-affected soils are common soils could be found under arid and semi-arid conditions. These soils have adverse effects on the growth of most cultivated plants due to the presence as excess of soluble salts and/ or exchangeable sodium, or both. These soils also include mainly saline, sodic and saline-sodic soils and non-saline–sodic soils as presented in Table 8.1 (Mohamed 2017a). Globally, it is estimated that about 400 million hectares (over 6% of the world land area) represent salt-affected soils (Arora 2017). Great problems could cause and result from salt-affected soil conditions such as excess sodium on the soil exchange complex and/or soluble salts in the soil reduces (1) the productivity of these soils, (2) soil physical conditions (i.e. soil structure, poses problem of water and nutrient availability) and (3) then these soils show micronutrient deficiency (Figs. 8.3 and 8.4; Arora 2017). Several investigations regarding salt-affected soils and their management have been published (e.g. Arora 2017; Arora et al. 2017a; Chen et al. 2017; Mohamed 2017a; Rezapour et al. 2017; Sharma and Singh 2017; Yadav et al. 2017) including phytoremediation of salt-affected soils (Arora and Rao 2017; Purakayastha et al. 2017), bioremediation of salt-affected soils (Arora and Vanza 2017; Arora et al. 2017b), using of soil and water amendments (Choudhary 2017), enhancing the fertility status of salt-affected soils (Singh 2017a, b), etc.

The common and dominant types of salt-affected soils in the Nile Delta region include sodic, saline and saline-sodic soils. These soils spread in the middle and north of the Nile Delta, which suffer from its location close to the Mediterranean coast and saline waterlogging. It is reported that salt-affected soils in the Nile Delta land exceed 30% of its total area, whereas 46% from the total area of the Nile Delta represents saline soils and other salt-affected soils in the north delta due to the intrusion of the Mediterranean seawater (Mohamed 2017a). It is worth to mention that these saline-sodic soils could be found in the coast of north lakes (Maryut and Idku) in the western delta near Alexandria governorate, in the coast of Burullus Lake in Kafr El-Sheikh governorate in the north middle delta and finally in east north delta in the coast of Manzala Lake especially in south Port Said and Damietta governorates (Mohamed 2017a).

Concerning the fertility of the salt-affected soils in Egypt, as well documented by 1971 (the construction of Aswan High Dam), the flooding of the Nile River has been stopped and its frequent annual leaching of the soil delta from the salts and pollution wastes also has been ended. So, great problems have been established for these soils as well as the

Table 8.1 Properties of different kinds of salt-affected soils

Properties of soils	Non-saline, nonalkaline	Saline soils	Saline–sodic soils	Alkaline soils
Soil salinity (as EC in dS m^{-1})	<4.0	>4.0	>4.0	<4.0
Soil reaction (pH)	<8.5	<8.5	>8.5	>8.5
Exchangeable sodium percent (ESP)	<15	<15	>15	>15
Sodium adsorption ratio (SAR)	<13	<13	>13	>13

Sources Daliakopoulos et al. (2016) and Arora (2017)

Fig. 8.3 Some distinguished features of salt-affected soils in Kafrelsheikh University at experimentation field including from top to bottom the effects of sodium in dispersion of aggregates, accumulation of salts on surface and effects of soil salinity and alkalinity on plant growth. Photos by El-Ramady 2014

Fig. 8.4 Some cultivated crops (tomato, lettuce and maize from to top to bottom) in salt-affected soils in Kafrelsheikh University at experimentation field, where the salts cover surface of soils even under surface or drip irrigation for heavy clay soils. Photos by El-Ramady 2015

salinity and/or sodicity problems. One of the most important solutions for this problem is the cultivation of rice in the north of delta. This cultivation of rice could replace the Nile flooding role in the leaching of salt accumulation in the Nile Delta soils especially in the north and coastal delta lands. The benefits of rice cultivation in this area include (1) could leach the delta soils helping in the reclamation of saline-sodic soils, (2) improving the groundwater table quality, (3) stopping the intrusion of seawater during the rice growing season and (4) supporting both the Egyptian demand and supply of one million tonnes of rice for export (Mohamed 2017a). It is worth to mention that, the common constraints of salt-affected soil include (1) the presence of excess sodium on the soil exchange complex and/or soluble salts in the soil reduces the productivity of these soils, (2) the soil physical conditions pose problem of water and nutrient availability for plants, particularly soil structure and these soils show micronutrient deficiency in soil fertility point of view (Arora 2017).

The main salt-affected soils in Egypt could be found in the northern-central part of the Nile Delta and in its Eastern and Western sides. There are some practices in Egypt that enforce the soils for the salinization including the deficient drainage system, continued irrigation with low-quality groundwater, the recycling of agriculture drainage water and inappropriate infield management of applied water. These practices could help in salt loading of the soil system, which has significant negative effects on its productivity (Ouda et al. 2016). It is reported that a great threat is facing the soils of the Nile Delta of Egypt and their fertility representing in land degradation due to salinization and waterlogging (Shalaby et al. 2012). Therefore, a comprehensive programme should be performed in order to reclaim the salt-affected soils in Egypt and more control of salinity and sodicity including different approaches chemical methods (like gypsum), hydro-technical methods or removing salts by mechanical ways (like flushing and leaching), drainage in surface, subsurface and mole drainage methods, biological management (application of organic fertilizer, compost and bio-fertilizers), physical methods (e.g. land levelling, tillage, deep ploughing and subsoiling) and using of nanomaterial (Ouda et al. 2016; El-Ramady et al. 2017). It could be also used crop rotation in improving soil structure and reduces soil degradation in salt-affected soils. It was also reported that it could be improved soil fertility by the inclusion of legume crops in a rotation implemented in salt-affected soil more than the application of farmyard manure due to improvement in organic matter and physical conditions of the soils (Ouda et al. 2016).

Concerning the integrated management of salt-affected soils, this management includes physical or mechanical, hydraulic, chemical and biological approaches. Regarding the biological methods for management of salt-affected soils, these methods include the application of organic matters, mulching, green manuring, tree plantation, applications of blue-green algae, and the bio-saline agriculture. The previous methods could help in increasing the uptake of nutrients and the biological activity in these soils. Thus, these methods also could be considered methods of removal of excess salts from the soil surface and the root zone in these saline soils (Shankar and Shikha 2017).

8.4.2 The Fertility of Oases and Depression Soils

It is well known that Egypt has several oases and depressions in the Western Desert (i.e. the Libyan portion of the Sahara) including the oases of Siwa, Bahariya, Kharga and Dakhla, which are be found in great depressions and their groundwater supplies can rise to the surface, but the vast intervening areas of high plateau are waterless (Figs. 8.5 and 8.6). Regarding the depressions, also many depressions could be found in the Western Desert (e.g. the Qattara Depression) including saline (or salt marshes) and non-saline depressions. The saline depressions are a common habitat of the Western Mediterranean coastal belt, which could be divided into two groups (series): the first one is formed from depressions directly adjacent to the dune strips. Salinity of this group has been resulted from the evaporation of seepage water, where the water table is exposed or near the surface and where there is poor drainage. The soil is mostly calcareous-sandy due to the encroachment of sand from the neighbouring dunes. These soils mainly cultivated with olive, vine and date palm (Zahran and Willis 2009).

Concerning the non-saline depressions, called the barley fields, they are the most fertile areas of the Western Mediterranean coastal belt of Egypt. The non-saline depressions provide favourable conditions for cultivation; extensive areas are occupied by barley, figs and olives. These depressions are mainly limited to the plains south of the second ridge in the eastern section of the coast, but are widespread in the valley and plains of the western section. The soils of these depressions (e.g. Abu Sir Depression) are variable and highly calcareous soils are derived from drifted oolitic grains of the coastal ridge in some parts; in other parts, alluvial, less calcareous, loamy soils are derived from the Abu Sir ridge (Zahran and Willis 2009).

Therefore, the soils of oases and depressions could be used in cultivating barley, olive, vine, etc. (in case of non-saline soils) or in salt production if these soils are saline. Several studies involved these oases and depressions in the Western Desert including the hydro-chemical characteristics of groundwater in oases (e.g. Abdulaziz and Faid 2015; Aly 2015), the fate of the Nubian sandstone aquifer springs in the oasis depressions (Powell and Fensham 2016), the adaptation of oases to changing climatic conditions (Hawkins 2012), using GIS and remote sensing in different studies like land degradation (e.g. Gad 2015; Elbeih 2017; Kotb et al. 2017; Salem 2017; Youssef et al. 2017).

Therefore, the fertility of Egyptian soils in oases and depressions should be increased the concern about them because simply these soils could be considered treasure for the Egyptian agriculture. That means, a very clear case could be mentioned for Siwa oasis, where this oasis is located in the Western Desert but has enough and suitable water for agriculture and a fertile soil, which needs only the proper drainage. Siwa is one of several oases with a little bit of concern could be not only the basket of food for entire of Egypt but also many industries could be established in parallel. Generally, depression soils are filled with alluvia of various natures including sandy, silty, loamy and clayey. Many depression soils have saline or alkaline properties, with a shallow saline water table and are covered with halophytic vegetation, mainly of Chenopodiaceae (Zahran 2010b).

It is reported that the Western Desert of Egypt could be characterized by the large depressions and oases occupying about 36% of its area. The northern oases include Siwa, Moghra, Wadi EL-Natrun and Bahariya, while the southern oases include Kharga, Dakhla and Dungul. It is reported also that the soils of the halophytic communities in these oases are characterized by a high content of $CaCO_3$ and low content of organic carbon (El-Khouly and Zahran 2002). In these oases and depressions, four main types of vegetation could be characterized including (1) salt marsh vegetation,

Fig. 8.5 A comparison between some soils with distinguished features of cultivated areas in Siwa Oasis (Egypt), where sandy soils contain a little bit organic matter resulting from the continuous cultivation (right photos) and some soil profiles with different climatic conditions with high rainfall rate and high fertility of soils (left photos) in Göttingen, Germany. Photos by El-Ramady 2017

Fig. 8.6 Three different locations for Siwa oasis represent different features of soils (left photos), where this oasis located at 20 m down the sea level and the water table is nearly close to the soil surface and only a good drainage is the urgent practice for maximizing the agricultural production of these soils. The right photos represent some natural and cultivated plants like hibiscus in Matrouh and Siwa. Photos by Alshaal, April 2017

(2) iced swamp vegetation, (3) sand dune vegetation, and (4) desert plain vegetation (El-Khouly and Zahran 2002; El Shaer 2009).

8.4.3 Waterlogged Soils and Their Fertility

Waterlogging process refers to the saturation of soil with water. The waterlogged soil could be formed when the water table of the groundwater is very high for a long time and soil pores in the root zone will be saturated, resulting in restriction of the normal circulation of the air, decline in the level of oxygen and increase in the level of carbon dioxide (Arora 2017). This process helps in prevailing anaerobic conditions and embedding any anticipated activities like the agricultural activity. In some extreme cases of waterlogging, the plant roots will suffer from deficiency of oxygen and then occur different chemical reduction processes including denitrification and methanogenesis as well as the reduction of iron and manganese oxides. Waterlogging of the soil stops air getting into this soil and reaching the water table near to the soil surface will prevent plant roots from the growth. It should also keep in mind that there is a wide difference in the crop tolerance for waterlogging conditions, which may vary between seasons of the year such as the growing of crop rice depending on upon the crop types, the soil types and the quality of underground water (Fig. 8.7). In general, the waterlogging process is often dominant and accompanied by the salinity of soils as waterlogged soils prevent leaching of the salts imported by the irrigation water. Concerning the water table, it is considered harmful, which may vary over a wide range from zero for rice crop, 1.5 m for other arable crops and more than 2 m for horticultural and forest plantations (Table 8.2).

Fig. 8.7 The waterlogging process could be found in case of paddy rice cultivation in Egypt (upper photo), very high of groundwater with high pollution level (middle photo) and soil profile characterized by high water table (70 cm) as presented in lower photo in Kafr El-Sheikh. Photos by El-Ramady 2013

Therefore, the development of waterlogging and soil salinization upon introduction of irrigation in arid and semi-arid regions is a global phenomenon. Due to waterlogging and soil salinization, it is estimated that about 10–33% of irrigated lands have adversely been affected in various countries (Arora 2017). There are many areas in Egypt is affected by the problem of waterlogging and soil salinization too. With regard to soil fertility, the waterlogged soils are usually suffered from the availability of nutrients, whereas suitable for cultivating submerged crops (rice) and tolerant crops like date palm. The major problem of these soils is the reduction conditions. In Egypt, these soils are already used in rice and date palm production.

A number of studies have been conducted on the waterlogging of soils and many practices and management for these soils such as effects of liming on this waterlogging (Haddad et al. 2017), the bio-management of waterlogged soils (Qureshi 2016), waterlogging and solubility of nutrients in paddy soils (Matin and Jalali 2017), waterlogging, salinization and drainage (Gupta 2016; Gebrehiwot 2017), nitrogen fertility under waterlogged soils (Khan et al. 2017), etc. It is estimated that about 2.4 million feddans from the irrigated agricultural areas in Egypt suffer from problems of waterlogging, salinization and sodicity. The majority of salt-affected soils are located in the northern central part of the Nile Delta. It is also recorded that only 5.4% of the land

Table 8.2 Classification of waterlogged soils and their potentiality according to depth of water table

Depth to water table	Nomenclature of different waterlogged soils
<1.5 m	Waterlogged soils
1.6–3 m	Potentially waterlogged soils
>3 m	Safe soils

Adapted from Arora (2017)

resources at present in Egypt are excellent in their quality, while about 40% are of poor quality mainly due to the development of sodicity and salinity problems. Furthermore, the productive lands in Egypt are finite and irreplaceable and should be protected against degradation such as soil fertility depletion, soil crusting, compaction and pollution (Negm et al. 2017a).

8.4.4 Coastal Soils and Their Fertility

Globally, the coastal soils have a great deal of diversity in terms of physical characteristics, climate and physiography as well as in terms of rich stock of fauna and flora (Rao et al. 2013). These soils comprise different types of features including lacustrine fringes, lagoons, deltas, coastal marshes and narrow coastal plains or terraces along the creeks (Arora 2017). These coastal soils generally may be either saline or acid sulphate in their nature. Concerning the saline soils, they are dominant with NaCl and Na_2SO_4 with abundance of soluble cations in the order of Na > Mg > Ca > K and chloride as the predominant anion. The main problems encountered in these areas include (1) shallow water table enriched with salt contributes to increase in soil salinity, (2) these lands are subjected to the influence of tidal waves and seawater intrusion, (3) heavy rainfall may result in excess water, (4) poor surface and subsurface drainage conditions, (5) lack of good quality irrigation water and acute salinity, and (6) poor socio-economic conditions of the farming community limiting introduction of high investment technologies (Arora 2017). With regard to soil fertility of these lands, the coastal soils are usually poor in their content of soil organic matter as well as the essential nutrients in proper and sufficient amounts.

It is well known that Egypt has a total area estimate of 1,001,449 km^2 (about 1 million km^2). Egypt also is well known with long coastlines of both Mediterranean and the Red Sea. The longest straight-line distance in Egypt from north to south is 1024 km, while that from east to west measures 1240 km. The coastal lands in Egypt represent a very important source for the Egyptian economic and agriculture as well. These coastal lands definitely have particular differences representing distinguished features, which are mainly located in the Mediterranean and Red Sea coasts. The Red Sea coasts are located in some governorates, i.e. Red Sea and Suez governorates, whereas the coasts of governorates of the Mediterranean Sea include North Sinai, Port Said, Damietta, Dakahlia, Kafr El-Sheikh, Beheira, Alexandria and Matruh.

Concerning Red Sea coast, it occurs along about 1100 km including the western coast of the Gulf of Suez, 340 km (Zahran 2010a). This land of Red Sea coast extends from Suez (Lat. 30° N) to Mersa Halaib (Lat. 22° N) at the Sudanese borders. These lands adjacent to the sea are generally mountainous, flanked on the western side by the range of coastal mountains. Along the Egyptian Red Sea coast, the mean annual rainfall ranges from 25 mm in Suez, 4 mm in Hurghada (Lat. 27° N) to 3.4 mm in Qusseir (Lat. 26° N). The main bulk of rain occurs in winter as Mediterranean affinity and is rainless in summer generally (El-Khouly and Zahran 2002). The soil features of the littoral vegetation are significantly affected by the climatic aridity of the Egyptian Red Sea coasts. There is insufficient leaching and salts accumulate as crusts on soil surface due to low precipitation. The soil features are apparently one of the main factors influencing the plant cover, the plant growth, the zonation pattern as well as the geographical distribution of the littoral halophytes. Three main littoral halophytic types within the Red Sea coastal belts are recognized including mangrove and reed swamps as well as salt marshes (El Shaer 2009; Zahran 2010a). Regarding the land adjacent to the Red Sea is generally mountainous, flanked on the western side by the range of coastal mountains (1705–2187 m above sea level). In the deep, trough between the shoreline and the highlands extends a gently sloping plain. This coastal plain, which varies in width, is covered with sand, over which the drainage systems (wadis) meander with their shallow course. These wadis run eastward to drain their water into the Gulf of Suez and the Red Sea (Zahran 2010a).

It is well documented that Mediterranean means 'in the midst of the land'. The Mediterranean Basin includes 21 countries with total surface areas under the influence of the Mediterranean climate is 7,357,000 km^2 distributed 9% in Europe (690,000 km^2), 46% in Africa (3,347,000 km^2) and 45% in Asia (3,320,000 km^2). It has unique original region due to its geography landscape that surrounds it as well as a significant role in the flourishing of ancient civilizations. This Mediterranean Basin also is one of the most complex regions on Earth in terms of history, geographic, geological, morphology and natural history due to its fantastic variety of plants, animals, people and landscapes. It is characterized by various habitats including coastal wetlands, coastal dunes, high mountains, seashores, desert wadis, small islets, dry shrublands, grasslands, etc. The East Mediterranean Basin is called *Fertile Crescent*, where established and flourished a great civilization in Egypt and Mesopotamia in this region since millennia. So, an endless number of investigations have been published about the Mediterranean Sea and the lands around it (e.g. Zahran 2010b).

In three sections the Mediterranean coast of Egypt extends for 970 km from Sallum eastward to Rafah including the western coast (550 km), the middle (deltaic) coast (180 km) and the eastern (Sinai) coast (240 km). The Mediterranean coastal lands of Egypt could be geographically divided into three sectors western or the Mareotis District (related to Mariut Lake), middle or Deltaic and eastern or Sinai sectors. The first two sectors (western and middle) belong to the North African Mediterranean coast, while Sinai sector belongs to the South West Asian Mediterranean coast. The main land uses of western areas include grazing and rain-fed farming through irrigation by underground and run-off water. The main annual crop is barley crop, whereas figs have been successfully grown on calcareous coastal dunes as well as olives, almonds and pistachio in inland alluvial depressions. Irrigated agriculture of grain crops, pasture and fruit trees (mainly vine) are spreading after the extension of irrigation canals from Alexandria to Sallum (Zahran 2010b).

Five main types of ecosystems could be recognized in Mediterranean coast of Egypt including sand dunes (coastal calcareous and inland siliceous), rocky ridges and plateau with skeletal shallow soils, saline and non-saline depressions and wadis. Concerning the soils of Mediterranean coast of Egypt, they are usually alkaline with pH from 7.5 to 9.0, calcium carbonate content is high and sulphates are common especially gypsum and sodium chloride is almost always present. In general, phosphorus and potassium are adequate for plant growth but nitrogen is usually deficient, whereas the deficiencies of trace elements may be occurred. The main soil factor should be considered in these Arid Zones is the moisture regime, which is closely related to ecosystem structure and productivity. In this context, the most significant soil characteristics include (1) soil depth or its storage capacity, (2) the topographic or geomorphic position (run-off or run-on), (3) soil texture or permeability of the upper layer, (4) infiltration rate, and (5) field capacity, etc. (Zahran 2010b).

A chain of intensely white calcareous granular sand dunes could be noticed along the Western Mediterranean coasts, which formed of calcium carbonate. The dunes are more humid and exposed to the immediate effect of the northerly winds owing to proximity to the sea (Figs. 8.8 and 8.9). A short distance from the beach, freshwater is frequently obtained by digging carefully in the sand to a depth 3–4 m. This freshwater is undoubtedly rainwater, which, having a lower specific gravity than saline water below, can form a layer above it; there may be a hard pan of limestone rock underlying the sand which prevents percolation of rainwater, the sand acting as a reservoir of freshwater (Zahran 2010b). These sand dunes need more concern and they could be used in agriculture (to be the basket of food in Egypt), industry (sand and silicon as a great treasure), nanotechnology, etc.

Fig. 8.8 The Western Mediterranean coastal belt of Egypt is mainly sandy soils, low in organic matter content and suitable for date palm and olive cultivation. These are the photos of some of the coastal soils in Baltim (Kafr El-Sheikh) where some vegetables could be cultivated on stored rainfall water. Photos by El-Ramady 2015

Fig. 8.9 Three features of coastal soils near Matrouh distract as a coastal soil, where the effect of rain is clear in the top and middle photos but in the bottom sandy dunes suffer from the deficit of water. Photos by Alshaal, April 2017

Concerning the investigations about the Egyptian coastal soils, several studies have been published including the study of restoration of the degraded coastal soils under climate changes and different assessments of risks for soils using GIS (e.g. El-Shaer and El-Morsy 2008; Elshinnawy 2008; El-Nahry and Doluschitz 2010; Zahran 2010a, b; Abdel-Kader 2013; Hassaan 2013; Darwish et al. 2015; Yousif et al. 2016; Arnous et al. 2017; Elbasiouny et al. 2017a; Koraim and Negm 2017).

8.4.5 Soils of the Nile Delta and Their Fertility

The Nile Delta of Egypt is one of the oldest Deltas worldwide. It has been formed by sedimentary processes through the alluvium brought by the old former seven active branches of the Nile (El Banna and Frihy 2009). At present, these branches have been silted up and replaced by Damietta and Rosetta branches. The Nile Delta hosts more than 41% of the country's population, covers only about 2% of Egypt's area but and comprises nearly 63% of its agricultural land (Hereher 2010; Mabrouk et al. 2013a, b, c). About 40% of all Egyptian industries are located in the Nile Delta as well as the most investments and economic activities go towards the Delta region. Therefore, it is among the most densely populated agricultural areas in the world, with 1360 inhabitants per km^2 (Negm et al. 2017a). This Nile Delta is characterized by many distinguished features such as:

(1) The total area of the Nile Delta is about 22,000 km^2 presenting about two-thirds of Egypt's agriculture and the most fertile areas in Egypt (Kashef 1983).
(2) The Delta aquifer is a unique and considered one of the largest groundwater reservoirs in the world with a huge volume and storage capacity with a wide (245 km) and a deep (more than 900 m) exposure to the Mediterranean Sea.
(3) This aquifer may be is the only coastal aquifer, in which the seawater has migrated to a distance of more than 100 km from the shore boundary (Sherif et al. 2012).
(4) The length of this Delta is approximately 160 km from north to south and it covers about 240 km of coastline at the Mediterranean Sea from west to east.
(5) This Delta is included three coastal lakes, i.e. Lake Burullus, Lake Idku and Lake Maryut at the northwest of Egypt.
(6) The quality of soils of the floodplains has declined and large amounts of fertilizers are now used because of stopping receive the annual supply of nutrients and sediments from upstream due to the construction of the Aswan High Dam (Negm et al. 2017a).
(7) Due to deposition during thousands of years of Nile flooding, soils in the Nile River and Delta are silt–clay mixtures. Most cultivated soils in Egypt are clayey to loamy in texture and only about one million feddans (420,000 ha) are sandy and calcareous (FAO 2005).

The main sources for the soil degradation in the Egyptian Nile Delta region include (1) the coastal erosion due to the natural processes of wave-induced longshore currents and sediment transport, (2) the land use changes due to the urban sprawl, (3) potential impacts of sea level rise due to climate changes, (4) seawater intrusion and water quality deterioration due to contamination of surface water canals, and (5) soil compaction and nutrient depletion (Mohamed 2017b; Negm et al. 2017a). Therefore, many studies have been focused on the fertility of the Nile Delta soils such as the challenges for the Nile Delta (Negm 2017; Negm et al. 2017a), water and salt movement in soils of the Nile Delta (Kubota et al. 2017), management of salt-affected soils in the Nile Delta (Elbasiouny et al. 2017b; Mohamed 2017a), the degradation of soils in the Nile Delta (Mohamed 2017b), the land-use and the distribution of soil organic carbon (Eid et al. 2017), transport of fertilizes in soils (Negm and Eltarabily 2017), the methods of irrigation, management and water requirements in the Nile Delta (El-Kilani and Sugita 2017; Khadr et al. 2017), the An integrative approaches for reclaiming desert land of the Nile Delta (Attwa and El-Shinawi 2017), the groundwater and its remediation in the Nile Delta (El-Agha et al. 2017; Elbeih 2017; Negm and Armanuos 2017; Sharaky et al. 2017; Zeidan 2017), different GIS studies on the Nile Delta (Elbeih 2017; Gemail et al. 2017; Masria et al. 2017; Negm and Armanuos 2017; Negm et al. 2017b), climate changes (El-Ramady et al. 2013; Eid et al. 2017; Khir-Eldien and Zahran 2017; Koraim and Negm 2017), etc.

8.5 Soil Fertility and Its Security: Challenges and Hopes

The fertility of soils was and still one of the most important issues in soil sciences. That means, it is impossible to achieve any security in agriculture without starting from the soils and their fertility (or security). These soils are the main source in providing nutrients, water and air for the growth of plants (Fig. 8.10). Therefore, the capacity of soils to sustain the growth of plants as well as the converting of solar energy into biomass energy is called soil fertility (Table 8.3). Furthermore, this term of soil fertility is an old as humankind. It is the main source of almost satisfaction of human needs including human security, human health and human creation. The fertility of soils is linked also to both plant and animal needs through many indicators for the quantity and quality of plant and animal growth as well as the potential of soil fertility on the global biodiversity and its conservation (Blum 2014).

The fertility of soils has been broadly discussed in different ancient civilizations, ancient literatures, world beliefs and religions all over the world (e.g. Warkentin 2006; Winiwarter and Blum 2006; Blum 2014). Therefore, soil fertility should not only manage and conserve but also sustain as well. The common conversation of soil fertility includes the application of different soil amendments, biostimulators and fertilizers such as organic, inorganic and bio-fertilizers, safe waste cycling, conserving different soil management practices like tillage or ploughing. Furthermore, all adverse effects on soil fertility should be avoided like soil erosion, pollution, degradation, etc. Globally, there are several threats could be observed concerning the soil fertility including soil sealing, pollution, urban sprawl, soil erosion, salinization, climate changes, the loss of organic matter, loss of biodiversity, floods and landslides. Therefore, it could be stated that, the fertility of soils is an environmental resource and a predictor of quality of life at the same time, defining to population growth, a large extent the quality of life, the life quality index and public health (Blum et al. 2010; Blum and Nortcliff 2011; Blum 2014).

As well known, the soil security has different dimensions in a close link to soil fertility and it could be stated that the security of soil may be started from its fertility. Regarding these dimensions, they include (1) *soil capability* or the ability of soils to produce enough and high-quality food and to produce clean and safe water (Bouma et al. 2017), (2) *soil condition* or the soil condition includes different soil properties physical, chemical and biological, which impact on the nutritional quality of agricultural products (McBratney et al. 2017a), (3) *soil capital* or the soil capital has a value and soil promotes human health through healthy food and water (McBratney et al. 2017b), (4) *soil connectivity* or the interactions between soils and surrounding ecosystem (Carré et al. 2017; Kim et al. 2017) and (5) *soil codification* or through conservation and improvement water and soil quality (Amundson 2017; Koch 2017).

Several studies have been published regarding the soil fertility and its security worldwide with stressing on the root microbe interactions under different soil conditions (Somenahally 2017), different links between soil security and its sustainable management practices (Murphy 2017), different links between the security of soil and its influence of soil on human health (Brevik et al. 2017), different assessments of applicability of soil evaluation systems to soil security including security of food, water and energy, human health, climate change abatement and biodiversity protection (Liniger et al. 2011; McBratney et al. 2014), the role of soil security in more agricultural productivity (Koch 2017), etc. Therefore, it could be concluded that, the security of soils

Fig. 8.10 The soils are the main source in providing nutrients, water and air for the growth of plants or trees. The role of soils in nutrition of plants is very clear as shown in the photos presenting some cultivated crops under Ismailia Governorate and Nubaria area in Beheira Governorate in Egypt as photographed by El-Ramady in 2016 and 2012, respectively. Photos by El-Ramady

Table 8.3 Different categories of soil test values including low, medium and high levels as an important criterion for evaluating of soil fertility

Element and its form	Different soil nutrient content and its categories (mg kg^{-1})		
	Low	Medium	High
NO_3–N	<10	10–20	>20
Olsen P	<10	10–20	>20
Extractable K	<150	150–250	>250
Extractable Ca	<1000	1000–2000	>2000
DTPA extractable Fe	<60	60–300	>300
DTPA extractable Zn	<2.5	2.5–4.5	>4.5
DTPA extractable Cu	<0.3	0.3–1.0	>1.0
DTPA extractable Mn	<1.0	1.0–2.5	>2.5

From Osman (2013)

requires first maintenance and then improvement of different soil resources to produce our needs (food, feed, fibre, fuel and freshwater) seeking for the sustainable production of food and energy as well as the adaptation to climate changes and the maintenance of the biodiversity, human health and function in ecosystems. Concerning the challenges of soil fertility and its security, many goals should be designed to achieve including different environmental, social and economic dimensions. Furthermore, these goals should be identified to secure soil and its fertility so that it can contribute in solving other global issues. It was agreed that we should work towards making soil security a recognized sustainable development goal in its own right (Morgan et al. 2017).

In Egypt, the soil security and its fertility should be considered an important issue to feed around 90 million people. So, the food security in Egypt definitely starts from the soils and their fertility, where there is no food security without soil security and concerning its fertility. The several problems are facing the fertility and security of Egyptian soils include soil pollution, erosion, depletion of soil fertility, urban sprawl, salinization, climate changes, etc. These problems should be solved through a holistic approach taking into account all environmental, social, economical and political issues. One of the great challenges facing Egypt nowadays is the using of precision farming. Some studies have been published regarding the application of precision farming in the agriculture in Egypt and its importance for improving this sector (e.g. El Nahry et al. 2011; Abd El-kader and El-Basioni 2013; Mohamed et al. 2014; Omran 2017).

Recently, many studies have been published focusing on the spatial distributions for many nutrients in Egypt (studies published by NARSS, Egypt). So, the spatial distribution of N, P and K nutrients and others in Egyptian soils is an emerging issue. Generally, nutrients content in desert lands is very low compared to the valley and delta. Soils in the Nile Delta are characterized by high content of nutrients; nitrogen, for example, reaches to 60 mg kg^{-1} in the eastern Delta whereas maybe even higher in middle Delta. Furthermore, phosphorus concentration reaches 25 mg kg^{-1}, while potassium concentration reaches 600 mg kg^{-1} in the Delta region (Fig. 8.11). On the other hand, the spatial distribution of N, P and K does not exceed 12, 11 and 226 mg kg^{-1}, respectively, in the desert soils of Kharga oases, southern west Egypt. The heterogeneous concentrations of nutrients between the Delta and desert lands are due to several factors including soil texture and human activities. Concerning soil texture, the soil of the valley and Delta was formed through the mud sediments and are characterized by higher adsorption capacity for nutrients and higher water retention, thus providing a suitable environment for plant and microorganisms growth. Human activities have had a great impact in increasing the development of soil profile and concentrating of nutrients in the soil of the valley and Delta, where organic manure and mineral fertilizers were used for crop production.

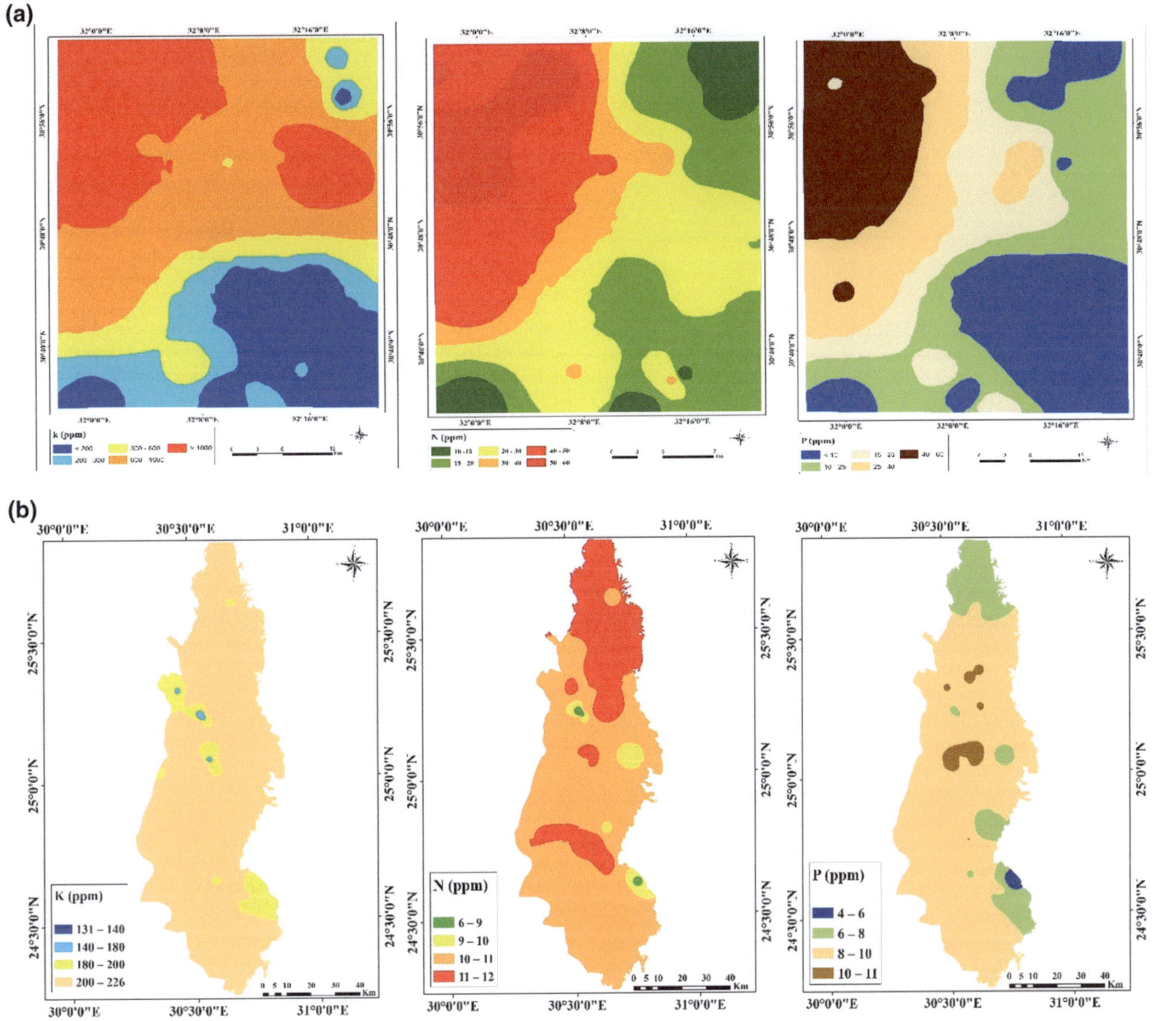

Fig. 8.11 The spatial distribution of N, P and K in soils of both **a** the eastern Delta and **b** Kharga oases, New Valley in southern Egypt. Created by both Dr. Abdel-Aziz Belal and Dr. El-Sayed S. Mohamed from NARSS, Egypt

8.6 Conclusion

The well-known call '*feed the soil to feed the human*' is common nowadays. This means that the world is requested to start from soil fertility to save the safe and sufficient food (i.e. food security) for all people. The great concern with soils will guarantee us enough bioenergy (security of energy), safe and enough food (security of food), the sustainable use of water (security of water) and then the security of soils. Therefore, great challenges faced and still facing Egypt concerning the soil fertility and its security including the depletion of Egyptian soil fertility, soil pollution, urban sprawl and swallow the fertile soils, fragmentation of agricultural tenure, etc. Hence, great tasks face Egypt nowadays to protect the fertility of her soils some of these tasks related to the new technologies (e.g. the nanotechnology and precision farming) and others related to the good management of both soil and water resources. The awareness concerning the soil fertility and its security in Egypt must be activated on all levels in education, in farm, in environment, in society, etc. It could be also concluded that about 90% of the Nile sediments already accumulate behind the High Dam after its construction by 1970. These sediments are originating from the Ethiopian High Plateau and very rich in both phosphorus and potassium. So, this decline in Egyptian soil fertility is expected several years ago and should overcome this problem by supplement and provide cultivated soils with essential nutrients.

Acknowledgements El-Ramady and Alshaal thank the outstanding contribution of STDF research teams (Science and Technology Development Fund, Egypt) and MBMF/DLR (the Federal Ministry of Education and Research of the Federal Republic of Germany) (Project ID 5310) for their help. Great support from this German-Egyptian Research Fund (GERF) is gratefully acknowledged. Authors acknowledge Dr. Abdel-Aziz Belal and Dr. El-Sayed S. Mohamed for their creation and dedication Fig. 8.11.

References

Abd El-kader SM, El-Basioni BMM (2013) Precision farming solution in Egypt using the wireless sensor network technology. Egypt Inform J 14:221–233

Abdel-Kader FH (2013). Digital soil mapping using spectral and terrain parameters and statistical modelling integrated into GIS-northwestern coastal region of Egypt. In: Shahid SA et al (eds), Developments in soil classification, land use planning and policy implications: innovative thinking of soil inventory for land use planning and management of land resources. Springer Science + Business Media, Dordrecht, pp 353–371. https://doi.org/10.1007/978-94-007-5332-7_19

Abdulaziz AM, Faid AM (2015) Evaluation of the groundwater resources potential of Siwa Oasis using three-dimensional multi-layer groundwater flow model, Mersa Matruh Governorate, Egypt. Arab J Geosci 8:659–675. https://doi.org/10.1007/s12517-013-1199-4

Aly A (2015) Hydrochemical characteristics of Egypt western desert oases groundwater. Arab J Geosci 8:7551–7564. https://doi.org/10.1007/s12517-014-1680-8

Amundson R (2017) Whose security is important? Communicating environmental risk about soil to a diverse audience. In: Field DJ et al (eds) Global soil security, progress in soil science. Springer International Publishing, Switzerland, pp 383–387. https://doi.org/10.1007/978-3-319-43394-3_34

Arnous MO, El-Rayes AE, Helmy AM (2017) Land-use/land-cover change: a key to understanding land degradation and relating environmental impacts in Northwestern Sinai, Egypt. Environ Earth Sci 76:263. https://doi.org/10.1007/s12665-017-6571-3

Arora S (2017) diagnostic properties and constraints of salt-affected soils. In: Arora S et al (eds) Bioremediation of salt affected soils: an Indian perspective. Springer International Publishing AG, Switzerland, pp 41–52. https://doi.org/10.1007/978-3-319-48257-6_2

Arora S, Singh AK, Singh YP (2017a) Bioremediation of salt affected soils: an Indian perspective. Springer International Publishing AG, Switzerland. https://doi.org/10.1007/978-3-319-48257-6

Arora S, Singh AK, Sahni D (2017b) Bioremediation of salt-affected soils: challenges and opportunities. In: Arora S et al (eds) bioremediation of salt affected soils: an Indian perspective. Springer International Publishing AG, Switzerland, pp 275–301. https://doi.org/10.1007/978-3-319-48257-6_14

Arora S, Rao G (2017) Bio-amelioration of salt-affected soils through halophyte plant species. In: Arora S et al (eds) bioremediation of salt affected soils: an Indian perspective. Springer International Publishing AG, Switzerland, pp 71–85. https://doi.org/10.1007/978-3-319-48257-6_4

Arora S, Vanza M (2017) Microbial approach for bioremediation of saline and sodic soils. In: Arora S et al (eds) bioremediation of salt affected soils: an Indian perspective. Springer International Publishing AG, Switzerland, pp 87–100. https://doi.org/10.1007/978-3-319-48257-6_5

Attwa M, El-Shinawi A (2017) An integrative approach for preliminary environmental engineering investigations amidst reclaiming desert-land: a case study at East Nile Delta, Egypt. Environ Earth Sci 76:304. https://doi.org/10.1007/s12665-017-6627-4

Blum W (2014) Soil fertility. In: Michalos AC (ed) Encyclopedia of quality of life and well-being research. Springer Science + Business Media, Dordrecht, pp 6212–6214. https://doi.org/10.1007/978-94-007-0753-5

Blum WEH, Nortcliff S (2011) Soils for the production of food and/or renewable energy—the future perspectives. Ann Agrarian Sci 9 (4):9–15

Blum WEH, Gerzabek MH, Hacklander K, Horn R, Rampazzo-Todorovic G, Reimoser F et al (2010) Ecological consequences of biofuels. In: Lal R, Stewart BA (eds) Soil quality and biofuel production, Advances in soil science, CRC Press, Boca Raton, pp 63–92

Bouma J, van Ittersum MK, Stoorvogel JJ, Batjes NH, Droogers P, Pulleman MM (2017) Soil capability: exploring the functional potentials of soils. In: Field DJ et al (eds) Global soil security, Progress in soil science. Springer International Publishing, Switzerland, pp 27–44. https://doi.org/10.1007/978-3-319-43394-3_3

Brevik EC, Steffan JJ, Burgess LC, Cerdà A (2017) Links between soil security and the influence of soil on human health. In: Field DJ et al (eds) Global soil security, Progress in soil science series. Springer International Publishing, Switzerland, pp 261–274. https://doi.org/10.1007/978-3-319-43394-3_24

Carré F, Caudeville J, Bonnard R, Bert V, Boucard P, Ramel M (2017) Soil contamination and human health: a major challenge for global soil security. In: Field DJ et al (eds) Global soil security, Progress in soil science. Springer International Publishing, Switzerland, pp 275–295. https://doi.org/10.1007/978-3-319-43394-3_25

Chen J, Gu W, Tao J, Xu Y, Wang Y, Gu J, Du S (2017) The effects of organic residue quality on growth and reproduction of *Aporrectodea trapezoides* under different moisture conditions in a salt-affected agricultural soil. Biol Fertil Soils 53:103–113. https://doi.org/10.1007/s00374-016-1158-9

Choudhary OP (2017) Use of amendments in ameliorating soil and water sodicity. In: Arora S et al (eds) Bioremediation of salt affected soils: an Indian perspective. Springer International Publishing AG, Switzerland, pp 195–210. https://doi.org/10.1007/978-3-319-48257-6_10

Daliakopoulos IN, Tsanis IK, Koutroulis A, Kourgialas NN, Varouchakis AE, Karatzas GP, Ritsema CJ (2016) The threat of soil salinity: a European scale review. Sci Total Environ 573:727–739

Darwish KhM, Rashad M, Mohamed SZ, Gad A (2015) Spatial distribution analysis of soil variables for agronomic development in El-Omayed Area, North-Coastal of Egypt. Environ Earth Sci 74:889–901. https://doi.org/10.1007/s12665-015-4095-2

Eid EM, Moghanm FS, Shaltout KH (2017) Effect of the different types of land-use on the distribution of soil organic carbon in north Nile Delta, Egypt. Rend Fis Acc Lincei, Egypt Rend. https://doi.org/10.1007/s12210-017-0621-3

El Banna MM, Frihy OEO (2009) Human-induced changes in the geomorphology of the northeastern coast of the Nile delta, Egypt. Geomorphology 107(1–2):72–78. https://doi.org/10.1016/j.geomorph.2007.06.025

El Gamal T, Zaki N (2017) Egyptian irrigation after the Aswan High Dam. In: Satoh M, Aboulroos S (eds) Irrigated agriculture in Egypt. Springer International Publishing, Switzerland. https://doi.org/10.1007/978-3-319-30216-4_4

El Nahry AH, Ali RR, El Baroudy AA (2011) An approach for precision farming under pivot irrigation system using remote sensing and GIS techniques. Agric Water Manage 98(4):517–531

El Shaer HM (2009) Potential role of Sabkhas in Egypt. an overview. In: Ashraf M et al (eds) Salinity and water stress. Springer Science + Business Media B.V., Berlin, pp 221–228

El-Agha DE, Closas A, Molle F (2017) Below the radar: the boom of groundwater use in the central part of the Nile Delta in Egypt. Hydrogeol J. https://doi.org/10.1007/s10040-017-1570-8

Elbasiouny H, Abowaly M, Gad A, Abu-Alkheir A, Elbehiry F (2017a) Restoration and sequestration of carbon and nitrogen in the degraded northern coastal area in Nile Delta, Egypt for climate change mitigation. J Coast Conserv 21:105–114. https://doi.org/10.1007/s11852-016-0475-3

Elbasiouny H, Elbehiry F, Abowaly M (2017b) Soil quality indices—special focus on salt-affected soil: review and case study in Northern Egypt. Environ Biodivers Soil Secur 1:85–100. https://doi.org/10.21608/jenvbs.2017.1085.1005

Elbeih SF (2017) Mapping of groundwater in Egypt using RS/GIS: case studies, areas surrounding Nile Valley and its Delta. In: Negm AM (ed) The Nile Delta, The handbook of environmental chemistry. Springer International Publishing, Switzerland. https://doi.org/10.1007/698_2016_52

El-Khouly AA, Zahran MA (2002) On the ecology of the halophytic vegetation of the oases in Egypt. In: Proceedings of international symposium on "the optimum resources utilization in salt-affected ecosystems in arid and semi arid regions", Cairo, Egypt, pp 277–286, 8–10 Apr 2002

El-Kilani RMM, Sugita M (2017) Irrigation methods and water requirements in the Nile Delta. In: Satoh M, Aboulroos S (eds) Irrigated agriculture in Egypt. Springer International Publishing, Switzerland, pp 125–151. https://doi.org/10.1007/978-3-319-30216-4_6

El-Nahry AH, Doluschitz R (2010) Climate change and its impacts on the coastal zone of the Nile Delta,Egypt. Environ Earth Sci 59:1497–1506. https://doi.org/10.1007/s12665-009-0135-0

El-Ramady HR, El-Marsafawy SM, Lewis LN (2013) Sustainable agriculture and climate changes in Egypt. In: Lichtfouse E (ed) Sustainable agriculture reviews, vol 12. Springer Science + Business Media, Dordrecht, pp 41–95. https://doi.org/10.1007/978-94-007-5961-9_2

El-Ramady H, Alshaal T, Abowaly M, Abdalla N, Taha HS, Al-Saeedi AH, Shalaby T, Amer M, Fári M, Domokos-Szabolcsy É, Sztrik A, Prokisch J, Selmar D, Pilon Smits EAH, Pilon M (2017) Nanoremediation: towards sustainable crop production. In: Ranjan S et al (eds) Nanoscience in food and agriculture, vol 5, Sustainable agriculture reviews, vol 26, Springer International Publishing AG, Switzerland. https://doi.org/10.1007/978-3-319-58496-6_12

El-Shaer HM, El-Morsy MH (2008) Potentiality of salt marshes in Mediterranean coastal zone of Egypt. In: Abdelly C, Öztürk M, Ashraf M, Grignon C (eds) Biosaline agriculture and high salinity tolerance. Birkhäuser Verlag, Switzerland, pp 207–219

Elshinnawy IA (2008) Coastal vulnerability to climate changes and adaptation assessment for coastal zones of Egypt. Ministry of Water Resources and Irrigation (MWRI), National Water Research Center (NWRC), Coastal Research Institute (CoRI), Cairo, Egypt

FAO (1998) World reference base for soil resources. World soil resources report 84. Rome, Italy

FAO (2005) Fertilizer use by crop in Egypt. Food and Agriculture Organization of the United Nations Rome, Land and Plant Nutrition Management Service, Land and Water Development Division, Rome

FAOSTAT (2017) Egypt: inputs. This website access on the following link: http://www.fao.org/faostat/en/#country/59

Gad A (2015) Land capability classification of some western desert Oases, Egypt, using remote sensing and GIS. Egypt J Remote Sens Space Sci 18(Supplement 1):S9–S18

Gebrehiwot KA (2017) A review on waterlogging, salinization and drainage in Ethiopian irrigated agriculture. Sustain Water Resour Manage, Sustain. https://doi.org/10.1007/s40899-017-0121-8

Gemail KS, El Alfy M, Ghoneim MF, Shishtawy AM, Abd El-Bary M (2017) Comparison of DRASTIC and DC resistivity modeling for assessing aquifer vulnerability in the central Nile Delta, Egypt. Environ Earth Sci 76:350. https://doi.org/10.1007/s12665-017-6688-4

Gupta SK (2016) integrated drainage solutions for waterlogged saline lands. In: Dagar JC et al (eds) Innovative saline agriculture. Springer, India, pp 183–200. https://doi.org/10.1007/978-81-322-2770-0_9

Haddad SA, Tabatabai MA, Loynachan TE (2017) Effects of liming and selected heavy metals on ammonium release in waterlogged agricultural soils. Biol Fertil Soils 53:153–158. https://doi.org/10.1007/s00374-016-1163-z

Hassaan MA (2013) GIS-based risk assessment for the Nile Delta coastal zone under different sea level rise scenarios case study: Kafr EL Sheikh Governorate, Egypt. J Coast Conserv 17:743–754. https://doi.org/10.1007/s11852-013-0273-0

Hawkins AL (2012) The aterian of the oases of the western desert of Egypt: adaptation to changing climatic conditions? In: Hublin JJ, McPherron SP (eds) Modern origins: a north African perspective, vertebrate paleobiology and paleoanthropology. Springer Science + Business Media B.V., Berlin, pp 157–175. https://doi.org/10.1007/978-94-007-2929-2_11

Hereher ME (2010) Vulnerability of the Nile Delta to sea level rise: an assessment using remote sensing. Geomat Nat Haz Risk 1(4):315–321. https://doi.org/10.1080/19475705.2010.5169125

Hossain MS, El-shafie A (2014) Performance analysis of artificial bee colony (ABC) algorithm in optimizing release policy of Aswan High Dam. Neural Comput Appl 24:1199–1206. https://doi.org/10.1007/s00521-012-1309-3

Huq SMI, Shoaib JUM (2013) Soil fertility. In: Huq SMI, Shoaib JUM (authors) The soils of Bangladesh, World soils book series, vol 1. Springer Science + Business Media, Dordrecht, pp 99–111. https://doi.org/10.1007/978-94-007-1128-0_10

Jeffrey M, Achurch H (2017) Save our soil to save the planet. In: Field DJ et al (eds) Global soil security, Progress in soil science. Springer International Publishing, Switzerland, pp 389–395. https://doi.org/10.1007/978-3-319-43394-3_35

Kashef AI (1983) Salt water intrusion in the Nile Delta. Ground Water 21:160–167. https://doi.org/10.1111/j.1745-6584.1983.tb00713.x

Khadr M, Zeidan BA, Elnmer A (2017) On-farm water management in the Nile Delta. In: Negm AM (ed) The Nile Delta, The handbook of environmental chemistry, vol 55. Springer International Publishing, Switzerland, pp 325–344. https://doi.org/10.1007/698_2016_56

Khalifa HE, Moussa HA (2017) Soil and agriculture after the Aswan High Dam. In: Satoh M, Aboulroos S (eds) Irrigated agriculture in Egypt. Springer International Publishing, Switzerland, pp 81–124. https://doi.org/10.1007/978-3-319-30216-4_5

Khan A, Tan DKY, Afridi MZ, Luo H, Tung SA, Ajab M, Fahad S (2017) Nitrogen fertility and abiotic stresses management in cotton crop: a review. Environ Sci Pollut Res 24:14551–14566. https://doi.org/10.1007/s11356-017-8920-x

Khir-Eldien K, Zahran SA (2017) Nile Basin climate changes impacts and variabilities. In: Negm AM (ed) The Nile river, The handbook of environmental chemistry, vol 56. Springer International Publishing AG 2016, Switzerland, pp 533 566. https://doi.org/10.1007/698_2016_116. Published online: 1 March 2017

Kim SC, Lim KJ, Yang JE (2017) The measurement of soil security in terms of human health: examples and ideas. In: Field DJ et al (eds) Global soil security, Progress in soil science. Springer International Publishing, Switzerland, pp 297–303. https://doi.org/10.1007/978-3-319-43394-3_26

Koch A (2017) Soil security for agricultural productivity: the policy disconnect and a promising future. In: Field DJ et al (eds) Global soil security, Progress in soil science. Springer International Publishing, Switzerland, pp 425–435. https://doi.org/10.1007/978-3-319-43394-3_39

Koraim AS, Negm A (2017) Protection methods against sea-level rise caused by climatic change: case study of the Nile Delta Coastal zones. In: Negm AM (ed) The Nile Delta, The handbook of environmental chemistry, vol 55. Springer International Publishing AG, Switzerland, pp 397–424. https://doi.org/10.1007/698_2016_120

Kotb MM, Ali RR, El Semary MA (2017) Use of remote sensing and GIS for land degradation assessment of Qarun Lake Coastal Area, El-Fayoum, Egypt. In: Ouessar M et al (eds) Water and land security in drylands. Springer International Publishing AG, Switzerland, pp 251–267. https://doi.org/10.1007/978-3-319-54021-4_22

Kubota A, Zayed B, Fujimaki H, Higashi T, Yoshida S, Mahmoud MMA, Kitamura Y, Abou El Hassan WH (2017) water and salt movement in soils of the Nile Delta. In: Satoh M, Aboulroos S (eds) Irrigated agriculture in Egypt. Springer International Publishing, Switzerland, pp 153–186. https://doi.org/10.1007/978-3-319-30216-4_7

Liniger H, Studer RM, Hauert C, Gurtner M (2011) Sustainable land management in practice. Guidelines and best practices for Sub-Saharan Africa. TerrAfrica, World Overview of Conservation Approaches and Technologies (WOCAT) and Food and Agricultural Organization of the United Nations (FAO), Rome

Mabrouk B, Arafa S, Farahat H, Badr M, Gampe D, Ludwig R (2013b) Impact of climate change on water and agriculture: challenges and possible solutions for the Nile Delta. In: EGU general assembly conference abstracts, vol 15, p 6293

Mabrouk B, Farhat Abd-Elhamid H, Badr M, Ludwig R (2013c) Adaptation to the impact of sea level rise in the Northeastern Nile Delta, Egypt. In: EGU general assembly conference abstracts, vol 15, p 4042. Retrieved from http://adsabs.harvard.edu/abs/2013EGUGA.15.4042M

Mabrouk B, Jonoski A, Solomatine D, Uhlenbrook S (2013a) A review of seawater intrusion in the Nile Delta groundwater system—the basis for assessing impacts due to climate changes and water resources development. Hydrol Earth Syst Sci Discuss 10 (8):10873–10911. https://doi.org/10.5194/hessd-10-10873-2013

Masria A, Negm AM, Iskander M (2017) Assessment of Nile Delta Coastal Zone using remote sensing. In: Negm AM (ed) The Nile Delta, The handbook of environmental chemistry. Springer International Publishing, Switzerland. https://doi.org/10.1007/698_2016_55

Matin NH, Jalali M (2017) The effect of waterlogging on electrochemical properties and soluble nutrients in paddy soils. Paddy Water Environ 15:443–455. https://doi.org/10.1007/s10333-016-0562-y

McBratney A, Field D, Koch A (2014) The dimensions of soil security. Geoderma 213:203–213

McBratney AB, Morgan CLS, Jarrett LE (2017b) The value of soil's contributions to ecosystem services. In: Field DJ et al (eds) Global soil security, Progress in soil science. Springer International Publishing, Switzerland, pp 227–235. https://doi.org/10.1007/978-3-319-43394-3_20

McBratney AB, Field DJ, Jarrett LE (2017a) General concepts of valuing and caring for soil. In: Field DJ et al (eds) Global soil security, Progress in soil science. Springer International Publishing, Switzerland, pp 101–108. https://doi.org/10.1007/978-3-319-43394-3_9

Mohamed ES, Saleh AM, Belal AA (2014) Sustainability indicators for agricultural land use based on GIS spatial modeling in North of Sinai-Egypt. Egypt J Remote Sens Space Sci 17(1):1–15

Mohamed N (2017a) Management of salt-affected soils in the Nile Delta. In: Negm AM (ed) The Nile Delta, The handbook of environmental chemistry. Springer International Publishing AG, Switzerland. https://doi.org/10.1007/698_2016_102

Mohamed N (2017b) Land degradation in the Nile Delta Nader. In: Negm AM (ed) The Nile Delta, The handbook of environmental chemistry. Springer International Publishing AG, Switzerland. https://doi.org/10.1007/698_2016_101

Morgan CLS, McBratney AB, Field DJ, Koch A, Bouma J, Carré F (2017) Synthesis: goals to achieve soil security. In: Field DJ et al (eds) Global soil security, Progress in soil science. Springer International Publishing, Switzerland, pp 457–462. https://doi.org/10.1007/978-3-319-43394-3_43

Murphy B (2017) Testing the links between soil security, sustainable land management practices and land evaluation. In: Field DJ et al (eds) Global soil security, Progress in soil science. Springer International Publishing, Switzerland, pp 87–97. https://doi.org/10.1007/978-3-319-43394-3_8

Negm AM (2017) The Nile Delta. The handbook of environmental chemistry series, vol 55. Springer International Publishing AG, Switzerland. https://doi.org/10.1007/978-3-319-56124-0

Negm AM, Armanuos AM (2017) GIS-based spatial distribution of groundwater quality in the Western Nile Delta, Egypt. In: Negm AM (ed) The Nile Delta, The handbook of environmental chemistry, vol 55. Springer International Publishing, Switzerland, pp 89–120. https://doi.org/10.1007/698_2016_66

Negm AM, Eltarabily MGA (2017) Modeling of fertilizer transport through soil, case study: Nile Delta. In: Negm AM (ed) The Nile Delta, The handbook of environmental chemistry, vol 55. Springer International Publishing, Switzerland, pp 121–158. https://doi.org/10.1007/698_2016_88

Negm A, Elsahabi M, El-Sayed R, Abd El-Basset N, Ghaly S, Ali K (2017b) Morphological variation of the Nile River first and second reaches using RS/GIS techniques. In: Negm AM (ed) The Nile River, The handbook of environmental chemistry. Springer International Publishing AG, Switzerland. https://doi.org/10.1007/698_2016_126

Negm A, Saavedra O, El-Adawy A (2017a) Nile Delta biography: challenges and opportunities. In: Negm AM (ed) The Nile Delta, The handbook of environmental chemistry, vol 55. Springer International Publishing, Switzerland, pp 3–18. https://doi.org/10.1007/698_2016_62

Noaman MN (2017) Country profile. In: Satoh M, Aboulroos S (eds) Irrigated agriculture in Egypt. Springer International Publishing, Switzerland. https://doi.org/10.1007/978-3-319-30216-4_1

Noaman MN, El Quosy D (2017) Hydrology of the Nile and ancient agriculture. In: Satoh M, Aboulroos S (eds) Irrigated agriculture in Egypt. Springer International Publishing, Switzerland. https://doi.org/10.1007/978-3-319-30216-4_2

Omran EE (2017) Will the traditional agriculture pass into oblivion? Adaptive remote sensing approach in support of precision agriculture. In: Rakshit A et al (eds) Adaptive soil management: from theory to practices. Springer Nature Singapore Pte Ltd, Singapore, pp 39–67. https://doi.org/10.1007/978-981-10-3638-5_2

Ouda SAH, Zohry A, Khalifa H (2016) Combating deterioration in salt-affected soil in Egypt by crop rotations. In: Ouda SAH, Zohry A (eds) Management of climate induced drought and water scarcity in Egypt. Springer Briefs in Environmental Science, pp 77–96. https://doi.org/10.1007/978-3-319-33660-2_6

Powell O, Fensham R (2016) The history and fate of the Nubian Sandstone Aquifer springs in the oasis depressions of the Western Desert, Egypt. Hydrogeol J 24:395–406. https://doi.org/10.1007/s10040-015-1335-1

Purakayastha TJ, Mandal A, Kumari S (2017) Phytoremediation of metal- and salt affected soils. In: Arora S et al (eds) Bioremediation of salt affected soils: an Indian perspective. Springer International Publishing AG, Switzerland, pp 211–231. https://doi.org/10.1007/978-3-319-48257-6_11

Qureshi AS (2016) Perspectives for bio-management of salt-affected and waterlogged soils in Pakistan. In: Dagar JC, Minhas PS (eds) Agroforestry for the management of waterlogged saline soils and poor-quality waters, Advances in agroforestry, vol 13. Springer, India, pp 97–108. https://doi.org/10.1007/978-81-322-2659-8_6

Rao GG, Chinchmalatpure AR, Arora S, Khandelwal MK, Sharma DK (2013) Coastal saline soils of Gujarat: problems and their management. CSSRI Technical Bulletin 01/2013, Central soil Salinity Research Institute, Karnal, pp 1–60

Rezapour S, Kalashypour E, Asadzadeh F (2017) Assessment of the quality of salt-affected soils after irrigation and cultivation in semi-arid condition. Int J Environ Res. https://doi.org/10.1007/s41742-017-0028-0

Salem SM (2017) ASD field hyperspectral measurements for discrimination of the ferruginous rocks and the iron ore types at El Gedida-Ghorabi area, Bahariya Oasis, Western Desert, Egypt. Arab J Geosci 10:166. https://doi.org/10.1007/s12517-017-2944-x

Satoh M, Aboulroos S (2017) Irrigated agriculture in Egypt: past, present and future. Springer International Publishing, Switzerland. https://doi.org/10.1007/978-3-319-30216-4

Shaheen SM, Abo-Waly ME, Ali RA (2013) Classification, characterization, and management of some agricultural soils in the North of Egypt. In: Shahid SA et al (eds) Developments in soil classification, land use planning and policy implications: innovative thinking of soil inventory for land use planning and management of land resources. Springer Science + Business Media, Dordrecht, pp 417–448. https://doi.org/10.1007/978-94-007-5332-7_23

Shalaby A, Ali RR, Gad A (2012) Land degradation monitoring in the Nile Delta of Egypt, using remote sensing and GIS. Int J Basic Appl Sci 1(4):283–294

Shankar S, Shikha S (2017) Management and remediation of problem soils, solid waste and soil pollution. In: Singh RL (ed) Principles and applications of environmental biotechnology for a sustainable future. Applied environmental science and engineering for a sustainable future series. Springer Science + Business Media, Singapore, pp 143–171. https://doi.org/10.1007/978-981-10-1866-4

Sharaky AM, El Hasanein AS, Atta SA, Khallaf KM (2017) Nile and groundwater interaction in the western Nile Delta, Egypt. In: Negm AM (ed) The Nile Delta, The handbook of environmental chemistry. Springer International Publishing AG, Switzerland. https://doi.org/10.1007/698_2016_127

Sharma DK, Singh A (2017) Current trends and emerging challenges in sustainable management of salt-affected soils: a critical appraisal. In: Arora S et al (eds) Bioremediation of salt affected soils: an Indian perspective. Springer International Publishing AG, Switzerland, pp 1–40. https://doi.org/10.1007/978-3-319-48257-6_1

Sherif M, Sefelnasr A, Javadi A (2012) Incorporating the concept of equivalent freshwater head in successive horizontal simulations of seawater intrusion in the Nile Delta aquifer, Egypt. J Hydrol 465:186–198. https://doi.org/10.1016/j.jhydrol.2012.07.007

Singh YP (2017a) Multifunctional agroforestry systems for bio-amelioration of salt-affected soils. In: Arora S et al (eds) Bioremediation of salt affected soils: an Indian perspective. Springer International Publishing AG, Switzerland, pp 173–193. https://doi.org/10.1007/978-3-319-48257-6_9

Singh YP (2017b) Crops and cropping sequences for harnessing productivity potential of sodic soils. In: Arora S et al (eds) bioremediation of salt affected soils: an Indian perspective. Springer International Publishing AG, Switzerland, pp 53–70. https://doi.org/10.1007/978-3-319-48257-6_3

Somenahally A (2017) Root-microbe interactions in response to soil conditions. In: Field DJ et al (eds) Global soil security, Progress in soil science. Springer International Publishing, Switzerland, pp 137–144. https://doi.org/10.1007/978-3-319-43394-3_12

Warkentin BP (2006) Footprints in the soil—people and ideas in soil history. Elsevier, Amsterdam

Winiwarter V, Blum WEH (2006) Souls and soils: a survey of worldviews. In: Warketin BP (ed) Footprints in the soil—people and ideas in soil history. Elsevier, Amsterdam/New York, pp 107–122

Yadav RS, Mahatma MK, Thirumalaisamy PP, Meena HN, Bhaduri D, Arora S, Panwar J (2017) Arbuscular Mycorrhizal Fungi (AMF) for sustainable soil and plant health in salt-affected soils. In: Arora S et al (eds) Bioremediation of salt affected soils: an Indian perspective. Springer International Publishing AG, Switzerland, pp 133–156. https://doi.org/10.1007/978-3-319-48257-6_7

Yousif M, van Geldern R, Bubenzer O (2016) Hydrogeological investigation of shallow aquifers in an arid data-scarce coastal region (El Daba'a, northwestern Egypt). Hydrogeol J 24:159–179. https://doi.org/10.1007/s10040-015-1308-4

Youssef AM, El-Shater A, El-Khashab MH, El-Haddad BA (2017) Coupling of field investigations and remote sensing data for karst hazards in Egypt: case study around the Sohag City. Arab J Geosci 10:235. https://doi.org/10.1007/s12517-017-3029-6

Zaghloul SS (2013) Consideration of the agricultural problems as a base of water resources management in Egypt. In: Seventeenth international water technology conference, IWTC 17 Istanbul, 5–7 Nov 2013, Turkey

Zahran MA (2010a) Afro-Asian red sea coastal lands. In: Zahran M (the author) Climate–vegetation, plant and vegetation, vol 4. Springer Science + Business Media B.V., Berlin, pp 105–217. https://doi.org/10.1007/978-90-481-8595-5_2

Zahran M.A. (2010b). Afro-Asian Mediterranean Coastal Lands. In: M. Zahran (the author), Climate–Vegetation, Plant and Vegetation 4, https://doi.org/10.1007/978-90-481-8595-5_1, pp: 1 – 103. Springer Science + Business Media B.V

Zahran M.A. (2010c). Climate – Vegetation: Afro-Asian Mediterranean and Red Sea Coastal Lands. Plant and Vegetation Series Vol. 4, Springer Science + Business Media B.V., https://doi.org/10.1007/978-90-481-8595-5

Zahran MA, Willis AJ (2009) The Western desert. In: Zahran M, Willis AJ (authors) The vegetation of Egypt. Springer Science + Business Media B.V., Berlin, pp 13–100

Zeidan BA (2017) Groundwater degradation and remediation in the Nile Delta Aquifer. In: Negm AM (ed) The Nile Delta, The handbook of environmental chemistry, vol 55. Springer International Publishing AG, Switzerland, pp 159–232. https://doi.org/10.1007/698_2016_128

Land Degradation

9

Elsayed Mohamed, Abdel-Aziz Belal, R. R. Ali, Ahmed Saleh, and Ehab A. Hendawy

Abstract

Soil degradation issue is one of the most important factors that threaten the crop production, thereby reflecting on food security. When discussing this issue, it must take into account the factors of surrounding, such as geographical location and climate in addition to social and economic conditions. This chapter illustrates the factors causing land degradation in Egypt as an example of semi-arid condition. It was noted that, under Egyptian conditions, there are many factors that differ from one place to another. Salinization, alkalinization and waterlogging are the most common factors in North Nile Delta. In addition, the rest of the Egyptian territories suffer from urban encroachment due to the high population density, where it was noted that the area located between the main branches of Nile Delta Rosetta and Damietta has lost 30510 acres during 1984–2014 where it is converted to urban areas. Finally, this chapter reviews the different methods used to assess land degradation using advanced techniques.

Keyword

Salinization • Alkalinization • Waterlogging
Urban encroachment • Soil erosion

9.1 Introduction

Soil is the main source of human food, so scientists and governments around the world are interested in finding solutions to ensure food security. Hence, recently, numerous studies focus on the quality of soil and crops, where they emphazise on factors reducing the productivity of soil such as degradation, pollution and urban encroachment (Field et al. 2017). Approximately 40% of agricultural lands worldwide are moderately degraded and around 9% are highly degraded, reducing global crop yield by as much as 13%. Nearly 1.6 million ha per year (i.e. 16,000 km^2/year) are lost globally due to salinization. Poor management policies for irrigation water management cause water logging and salinization in arid and semi-arid condition, where that represents 10% of the world's irrigated lands. The total area of Egypt is around one million square kilometres which is equivalent to 238 million feddans (1 feddan = 4200 m^2). Arable lands area is estimated to be around 14 million feddans, i.e. 6% of the Egyptian territory. The total agricultural area of Egypt is about 8.5 million feddans, distributed as 5.7 million feddans, occupies the old irrigated land in Nile delta and valley, 2.5 million feddans reclaimed irrigated lands. About 0.3 million feddans rely on rainwater inhabits the northern coast of Egypt in northern Egypt, (CAPMAS 2006; Darwish et al. 2013). This shows the agricultural land in the Nile Delta is gradually decreasing. On the other hand, the population rate is increasing and this is the great challenge facing the decision makers in Egypt; hence, we are interested, in this chapter, to clarify the factors that lead to decline of agricultural lands and their capability for crop production.

E. Mohamed (✉) · A.-A. Belal · A. Saleh · E. A. Hendawy
Division of Agriculture Applications and Soil Science, National Authority for Remote Sensing and Space Sciences (NARSS), Cairo, Egypt
e-mail: salama55@mail.ru; salama55_55@yahoo.com

R. R. Ali
Soils and Water Use Dept, National Research Centre, Cairo, Egypt

9.2 Definition of Land Degradation

Land degradation is set off processes that lead to changes in the values of the biophysical environment and land characteristics to be deleterious. These include changes in soil, water, topography, vegetation cover and climate. Furthermore, those changes causes several impacts such as loss of soil productivity, biodiversity, change in vegetative cover and increased vulnerability of the environment (Stockings and Murnaghan 2000). Conversely, many researchers have

H. El-Ramady et al. (eds.), *The Soils of Egypt*, World Soils Book Series,
https://doi.org/10.1007/978-3-319-95516-2_9

agreed that this phenomenon represents the greatest threat to agricultural development in the world. It is considered the first challenge to humans in terms of its adverse impact on biomass productivity and environmental quality (FAO 2007). Soil degradation is defined as a change in soil health status resulting in a diminished capacity of the ecosystem to provide goods and services for its beneficiaries. Degraded soils reflect health status as they do not provide the normal goods and services of the particular soil in its ecosystem. Universally, soil degradation has been defined as a human-induced phenomenon, and point out that in a general sense soil degradation could be described as the deterioration of soil quality: the partial or entire loss of one or more soil functions (Oldeman et al. 1991). There is another concept of soil degradation that includes many processes that affect the soil in different ways and different extents. The quantity and quality of information on soil degradation is very variable in different regions, in addition, there are great differences between countries in data and data availability on soil resources and soil change information (FAO 2015).

9.3 Land Degradation Assessment

9.3.1 Framework

Mitigation of land degradation impacts depends on understanding the natural cause of degradation. Therefore, scientists around the world, particularly those interested with dry climate, are focusing their works on land degradation appraisal. During the last three decades, many methods and frameworks have been developed to assess the land degradation. Those methods have discussed extent and intensity of soil degradation such as climate, vegetation, soil characteristics, soil management, topography and type of land utilization to identify areas of potential degradation hazard (FAO 1979). Precise and relevant appraisal methods of land degradation in drylands with a flexible scale that, integrating socio-economic parameters, and biophysical aspects are needed to potential national plans and investments to -land degradation, promote socio-economic livelihoods, and conserving dryland ecosystems (Bridges and Oldeman 1999; Snel and Bot 2003). Global Assessment of Human-induced Soil Degradation (GLASOD) provided real information on the world distribution and intensity of erosional, chemical and physical types of degradation also support policymakers and governments for the continuing need for soil conservation (Bridges and Oldeman 1999). Van Lynden (1995) has develop a guidelines for Assessment of the Status of Human-Induced Soil Degradation (ASSOD) based on GLASOD method but more emphasis was placed on the rate of degradation and on the impact of degradation on productivity for three different management levels. Reed and Dougill (2002) provide a report on the experience of implementing the first stages of the proposed Land Degradation Assessment in Drylands (LADA) methodological framework to assessing and mapping land degradation at different spatial scales—small to large and at various levels local to global. The main factors of LADA are soil health, biomass, biodiversity, water quantity and social economic benefits. The modelling of land degradation requires greater use of advanced technologies such as remote sensing and Geographic Information Systems (GIS) that can be used to evaluate spatial and temporal patterns over very large areas. Ponce and Koohafkan (2010) proposed a methodology for land degradation assessment at multiple scales based on indicators of drivers-pressures-state-impacts-responses (DPSIR). The authors augmented and refined their method for field application by integrating several frameworks such as LADA as presented in Fig. 9.1). The framework uses indicators of both, the state of land degradation and its causes (i.e. driving forces and pressures) as the vehicle for the assessment, thus integrating biophysical to socio-economic data on such indicators. DPSIR including twelve factors, i.e. (1) Area and scale definition; (2) Selection of Indicators of DPSIR; (3) Selection of procedures and tools to measure or estimate the values of indicators; (4) Collection of existing data and identification of data gaps; (5) Partition variability through stratification; (6) Design of a statistically reliable sampling scheme based on the stratification and data collection through field forms and surveys of relevant indicators at the scale of the assessment; (7) Data analysis; (8) Integration of results; (9) Identification of 'hot spots' and 'bright spots'; (10) Validation of results and accuracy assessment; (11) Design of mapping legend integrating all DPSIR indicators and mapping of results and (12) Monitoring changes over a time. However, the elements of the legend used for the integration of indicators of states and drivers and pressures into an assessment can be expressed according to DPSIR indicators as follows;

Degradation type:

P—Physical–chemical, B—biological
Processes of physical land degradation (p) :
Processes (on site).
se = soil erosion by water (sheet erosion), sw = soil erosion by wind (sheet erosion)
co = compaction, cr = crusting and sealing
Processes (off-site)
sed = sediment deposition, sec = water contaminated by erosion, seft = flooding, swd = deposition of dust

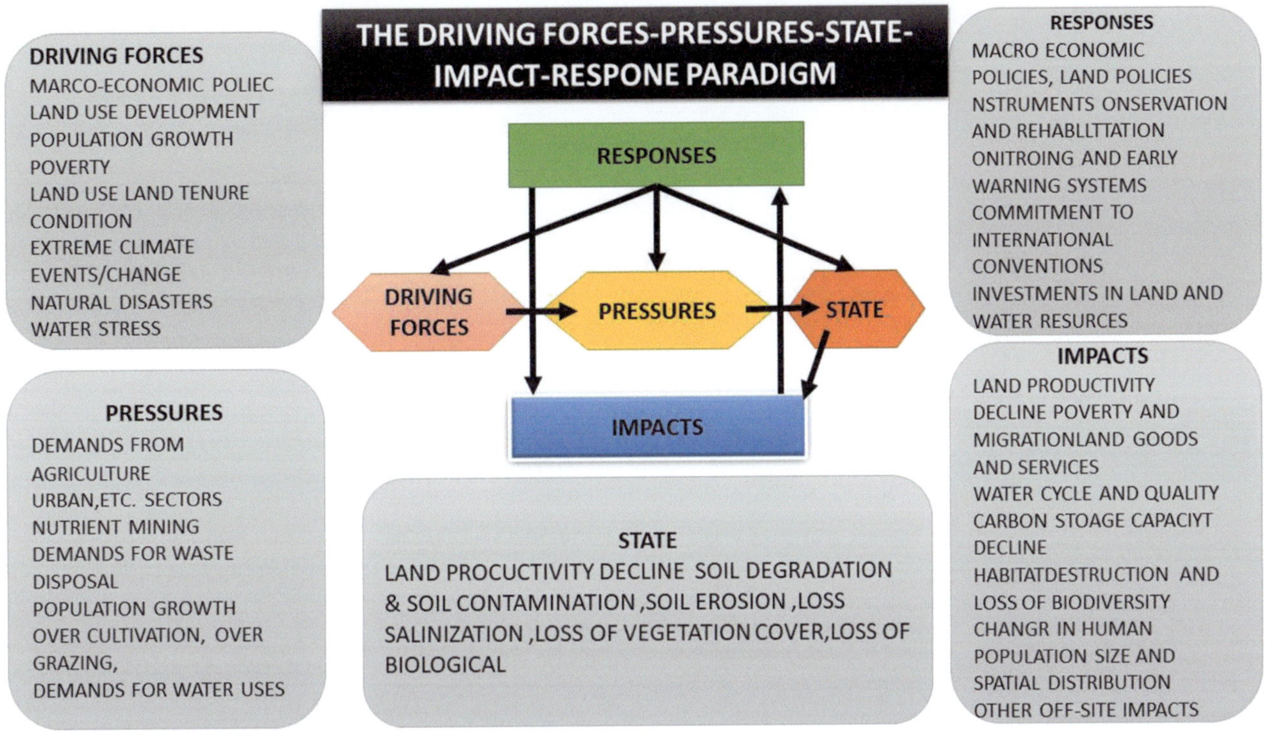

Fig. 9.1 LADA—DPSIR framework. Modified after (FAO 2006)

Processes of land deformation :

seg = soil erosion by water (gully) ser = soil erosion by water (rills)
processes of chemical land degradation (c)
f = soil nutrient and fertility depletion
sa = salinity, na = alkalinity, h = acidity, lx = toxic compounds (pollutants in soil matrix), wt = solid wastes (soil surface)
processes of biological land degradation (b)
ic = loss of lang cover & biomass, om = organic matter depletion,
bio = loss of biological diversity

Degree of intensity:-

1 = very slight, 2 = slight,3 = moderate 4 = intense, 5 = very intense

a. **Calculation of the composite land degradation index**
i. **Unit of measurement and rating**

A common measurement for different processes could be the loss of soil production, in tonnes/ha/year or loss of benefit. Another possibility is to use the cost of soil conservation treatment. Productivity is a function of soil degradation but it is also a function of many other factors. Crop type, climate, cultural practices and so on. Benefits depend not only on the productivity but also on the price, while the cost of soil conservation measures varies widely from place to place. On the other side, the soil degradation classes and rates of each group are shown in Table 9.1. Another difficulty is that the quantitative relationships between soil degradation and soil productivity are never well established. In the light of these difficulties, measurements units specific to each group of processes have been chosen as follows:

Salinization	Increase in (EC) in des/m/year
Alkalinization	increase in ESP%/year
Compaction	increase in apparent density in g/cm^3/year
Waterlogging	decrease in water table level per cm/year

ii. **Natural vulnerability assessment**

Based on soil, topography and climatic factors which are defined and described by using FAO/UNEP (1978) methodology for assessing soil degradation, the natural vulnerability for soils could be evaluated and confirmed with the landforms to produce neutral vulnerability maps. The ratings used are present in Tables 9.2 and 9.3.

Table 9.1 Soil degradation classes and rates

Chemical degradation	Salinization (Cs) increase in (Ec) per dS/m/year	Alkalinization (Ca) increase in ESP %/year
None to slight Moderate High Very high	<0.5 0.5–3 3–5 >5	<0.5 0.5–3 3–7 >7
Physical degradation	Compaction/increase in bulk density per g/cm^3/year	Waterlogging/decrease in water table in cm/year
None to slight Moderate High Very high	<0.1 0.1–0.2 0.2–0.3 >0.3	<1 1–3 3–5 >5

Adapted from FAO (1978)

Table 9.2 Rating for physical vulnerability

Factor	Index	Class			
		Low	Moderate	High	Very high
Climate Soil Topography	$\sum P^2/p$- Silt%/Clay% Slope%	0–50 >0.2 0–2	50–500 0.2–0.3 2–8	500–1000 0.3–0.7 >8	>1000 >0.7 –

Adapted after FAO (1978)
Note P = monthly precipitation, p- = annually precipitation

Table 9.3 Rating for chemical vulnerability

Factor	Index	Class			
		Low	Moderate	High	Very high
Climate Soil Topography	PET/(P + Q) 10 Texture class Slope %	<0.1 Clay 0–2	0.1–0.3 Silt 2–8	0.3–0.5 Sand >8	>0.5 – –

Adapted after FAO (1978)
Note PET = potential evapotranspiration p = precipitation/year Q = irrigation water

Table 9.4 Criteria used to determine the degree of different degradation types (UNEP 1991)

Hazard type	Indicator	Unit	Class			
			Low	Moderate	High	Very high
Salinization	EC	dS/m	4	4–8	8–16	>16
Alkalinization	ESP	%	10	10–15	15–30	>30
Compaction	Bulk density	g/Cm^3	1.2	1.2–1.4	1.4–1.6	>1.6
Waterlogging	Water table	Cm	150	150–100	100–50	<50

iii. **Land degradation status**

The status of soil degradation is an expression of the severity of the process. The severity of the processes is characterized by the degree in which the soil degraded and by the relative extent of the degraded area within a delineated physiographic unit. Degree, relative extent, severity level and causative factors were defined and described by using the GLASOD approach (UNEP 1991) as the following:

(a) **Degree of soil degradation**

The criteria used to assess the degree of land degradation are presented in Table 9.4.

(b) **Relative extent of the degradation type**

It is not possible to separate areas of soil degradation individually on the map. It is, however, possible to estimate

Table 9.5 The severity level of soil degradation (UNEP 1991)

Degree of soil degradation	Relative extent (%)				
	0–5	6–10	11–25	25–50	50–100
Slight	1.1	1.2	1.3	1.4	1.5
Moderate	2.1	2.2	2.3	2.4	2.5
Strong	3.1	3.2	3.3	3.4	3.5
Extreme	4.1	4.2	4.3	4.4	4.5
Severity level	Low	Moderate	High	Very high	

the relative extent of each type of soil degradation within the mapped unit. Five categories are recognized as follows:

- Infrequent: up to 5% of the unit is affected.
- Common: 6–10% of the unit is affected.
- Frequent: 11–25% of the unit is affected.
- Very frequent: 26–50% of the unit is affected.
- Dominant: over 50% of the unit is affected.

(c) **The severity level of soil degradation**

The severity level is indicated by the combination of the degree and the relative extent as shown in Table 9.5.

9.4 Land Degradation in Egypt

Land degradation phenomenon has become one of the most dangerous environmental impacts affecting all countries of the world (poor and industrial countries) as reported by Le et al. (2016). Degradation issue is considered as an important factor for sustainable agricultural. In this chapter, we focus on the factors causing degradation under Egyptian conditions. Topographical Characteristics such (slope and aspect), and climate elements such (temperature, rainfall) and groundwater level are affecting the concentration of salts. The fertile lands of Egypt are located in the Nile Valley and Delta (Darwish et al. 2013). These lands were caused the rising ancient Egyptian civilization, which considered the main source of food production throughout Egyptian history (Kees 1961). The fertile land is not renewable and has been formed in a long period may took up hundreds of thousands time spans as their formation (Lal 1994). Since the Aswan High Dam constructed, the complex ecosystem of the Nile, has been deeply affected. Unfortunately, this Dam has caused big changes to the lives of farmers and soil characteristics downstream Before the constructing of High Dam the flooding have been formed fertile soils from upstream on its banks downstream. The flooding had formed the fertile soil from upstream on its banks downstream. This washed up soil was extremely fertile, and renewed itself every year in time for the crop season. Since the high dam built, the annual flood has been stopped. Thereby, farmers forced to use fertilizer to grow their crops, which is risky and more expensive. Therefore, the management of land resources has a great importance to food security of future generations. Despite the fact that, the Aswan Dam has a significant impact through organizing the crop irrigation and insured continuation of agriculture throughout the year, but also had a negative impact as it was deprived valley and Delta of crucial nutrients and minerals (FAO 2011a, b, c). Decision makers in Medal East region focus their efforts for understanding the causes of land degradation. Hence, several workshops and conferences have been convened to discuss the causes of land degradation and their impacts on agricultural development. Those conferences discussed the challenges of sustainable resource utilization as well as the sustainable societal development goals intend to improve the living conditions of local communities. Moreover, the discussions dealt with advanced technologies that can help researchers to develop new means to mitigate the environmental hazards posed by climate change and degradation of natural resources. Also, it can help to set ambitious targets aimed at addressing environmental degradation (e.g. The Sixth National GIS Symposium 2011 ICRISSM 2014; UNCCD COP12 2015 and AUC 22, 2016). Several authors studied the processes of land degradation in Egypt. Mohamed et al. (2013a, b) studied the land degradation incident during the period from 1997 to 2010, east of Nile delta. The author concluded that the area located east of Nile Delta suffered from several factors of soil degradation (salinity, alkalinity and water table). These factors threaten the ongoing agricultural activities and prohibit further reclamation expansions (Fig. 9.2). These results are consistent with some other researches in different areas of the central and northern Delta, where they reported that salinity, sodicity, compaction, waterlogging, wind and water erosion are the most effective factor of land degradation (El Baroudy 2005; Abdel Rahman 2009). Also, El Fayoum area is classified as a high actual hazard, where 39.14% of these soils are considered as vulnerable to chemical and physical degradation risks salinization process is activated as saline groundwater are rising by capillary action to the surface soil (Fig. 9.3). Here, we must not neglect the important role of

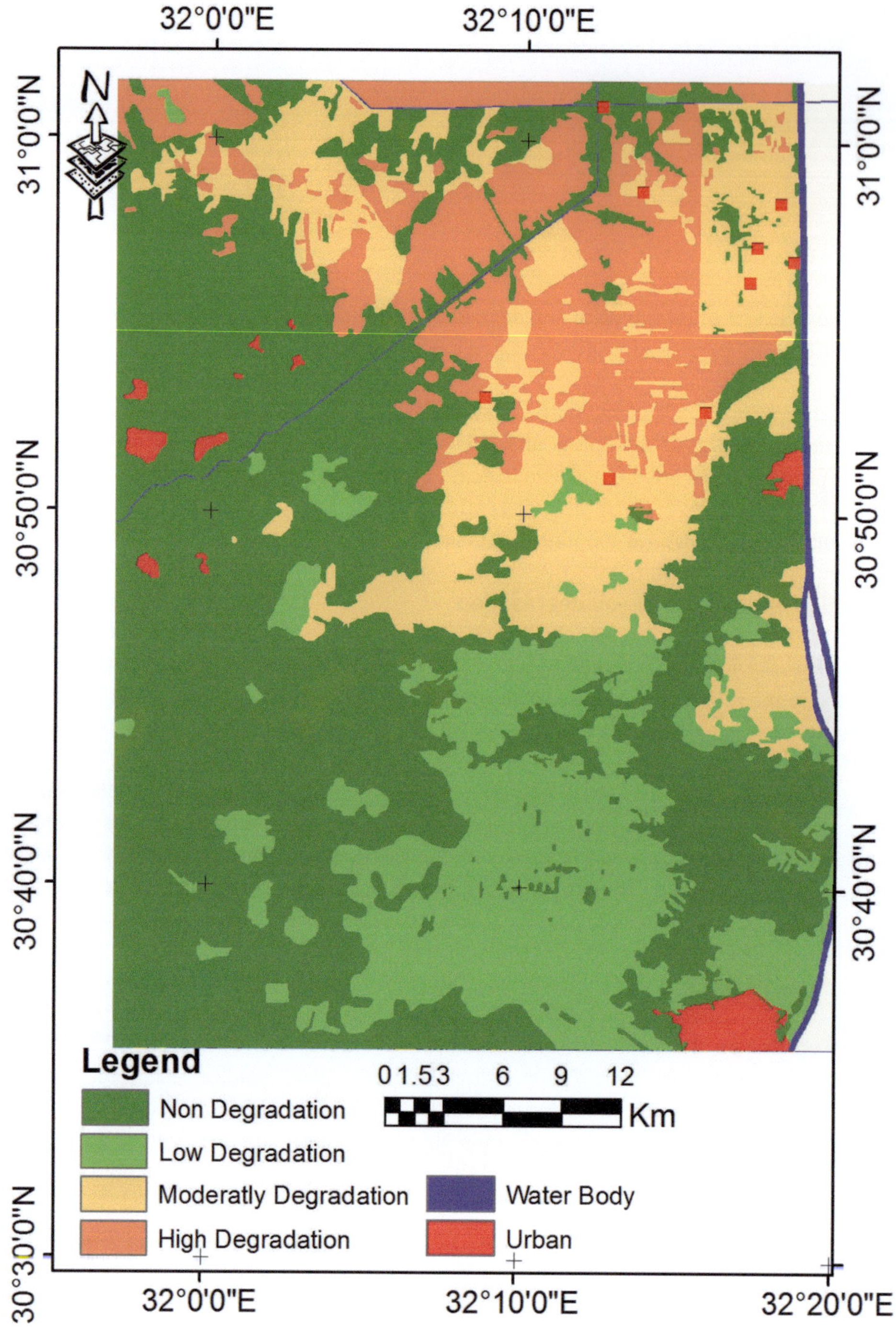

Fig. 9.2 Degradation degree in East Delta. Modified after Mohamed et al. (2013a, b)

arid climate in the emergence of salinization processes and characterizes El Fayoum area, where this area is characterized by high temperature besides evapotranspiration from surface soils (Ali and Abdel Kawy 2013).The authors also illustrated that the human activities are not sufficient to overcome the land degradation processes in El Fayoum depression.

9.5 Causes of Land Degradation in Egypt

The causes of degradation vary from region to another, where the rate of land degradation is associated with several parameters such as topography, climate and human activities. Previous research shows that topographic factors such

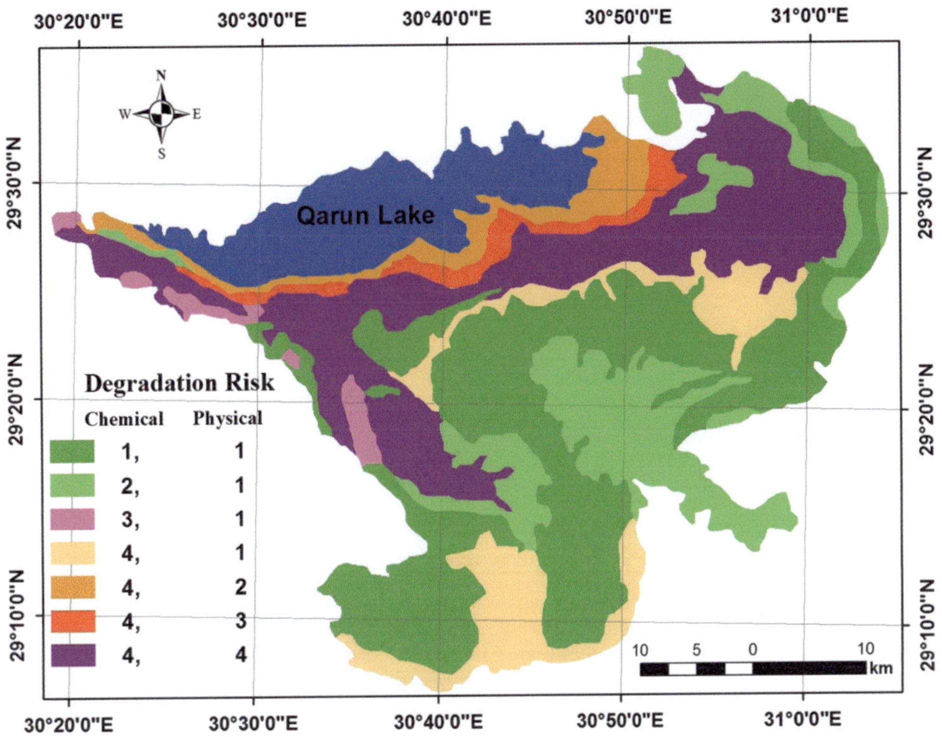

Fig. 9.3 Physical and chemical degradation risks in El Fayoum depression. Modified after Ali et al. (2013)

slope gradient affects directly on increasing degradation. Also, climate factors affect degradation rate as rainfall and temperature that controls the spatial distribution of vegetation density, biological activity, soil erosion, salinity, alkalinity and other factors. On the other hand, the human factor seriously affects the degradation rate, where mismanagement policy and overgrazing. Moreover, the human factor is influential in soil sealing, which is considered the main problem in Egypt (Mohamed et al. 2013a, b; Nkonya et al. 2016). Generally, this chapter throws light on the main causes of land degradation in Egypt as follow:

9.5.1 Soil Sealing and Human Impact

Soil sealing can be defined as changing the state of the land, from agricultural to any other usage, such as urban constructions and layers of completely or partly impermeable artificial material, asphalt, concrete, etc. (European Commission 2012). About 95% of the country's people living are along the banks of the Nile and Delta (CAPMAS 2006). Soil sealing is the most dangerous cause of soil degradation in Egypt. Saleh (2016) monitored the changes in urban sprawl in the area located between the main branches of Nile Delta Rosetta and Damietta during 1984, 1998 to 2014. The author reported that urban area has been increased by 30,510 feddan during that period from, in addition, it increased by 89,751 feddan in the period from 1998 to 2014. Moreover, this result illustrates the magnitude loss of fertile soils has increased dramatically during the period from 1998 to 2014 (Fig. 9.4). Likewise, Mohamed et al. (2015) focused on the magnitude of soil sealing in medial Nile Delta before and after 2011 and he reported that the rate of soil sealing has been highly increased after 2011. The high rate of sealing, during 2011 to 2013, resulted from the weakness of government to control the growth rate of urban sprawl on cultivated lands (Fig. 9.5). However, urban sprawl effects on the variation of vegetation cover where it causes reduction of biomass. The National Authority for Remote Sensing has carried out numerous projects about the monitoring of urban sprawl on some Egyptian governorates (Figs. 9.6a, b; 9.7a, b; 9.8a, b). Those projects highlighted the negative impacts of urban encroachment on cultivated lands where it represents a serious phenomenon that threatens the Egyptian national security from south to north. The total loss of land was recorded 187, 169 and 131 Km^2 in Qena, Minya and Beni Suef Governorate as illustrated in Table 9.6. However, the soil sealing could indicate the high population density where Egypt is the most populous country in the Arab World and the third-most populous on the African continent (after Nigeria and Ethiopia). The area located between the main branches of Nile Delta Rosetta and Damietta has lose 30510 acr during 1984 to 2014 where it converted to urban areas Fig 9.5. While in other places in the governorate of Qena in Upper Egypt, it was noted that the urban size was 54 km^2 and was extended to 240 km^2 in 2017 Fig 9.8. The problem can conclude that soil sealed and lost from the delta by urban sprawl cannot be replaced by the same soil quality

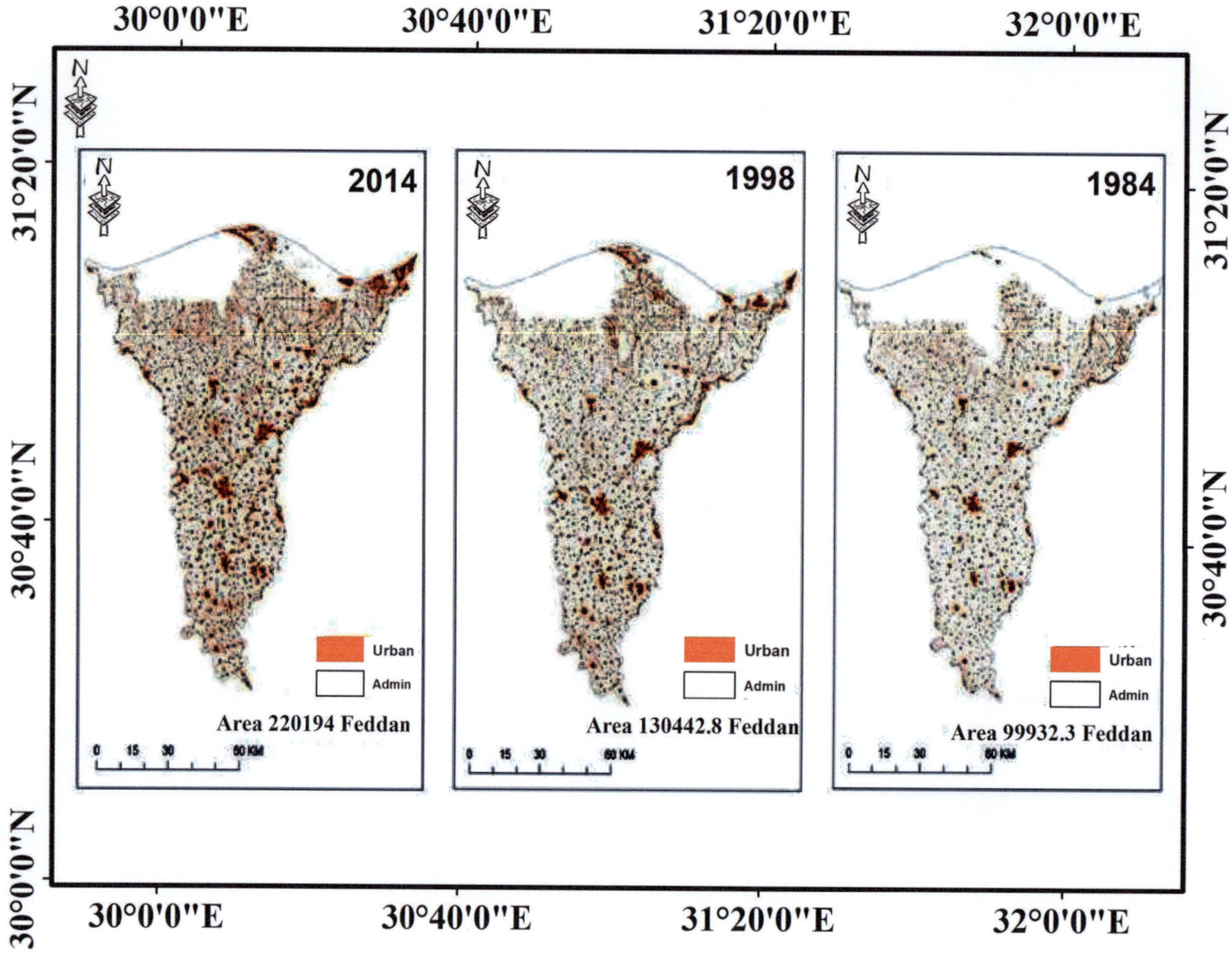

Fig. 9.4 Urban sprawl over the Nile Delta during 1984–2014. Modified after Saleh (2016)

elsewhere, where soils of Egyptian valley and delta are characterized with productivity and fertility, furthermore any other cause such as salinity and soil erosion can be mitigated their impact deterioration and can be maintained, thereby return again to agricultural production, except for the urbanization factor cannot be maintained and return to agricultural productivity. So, soil sealing is the most dangerous factor that threatens the Egyptian food security.

9.5.2 Salinization and Waterlogging

Soil salinity spreads worldwide, and there is no continent which is completely free from salinity phenomena. Recently, global studies revealed that an area about 932.2 Mha worldwide, suffered from various levels of soil salinization (Rengasamy 2006). Soil salinity varies trans-country and even within the country at different locations, it extends in irrigated agriculture regions to farmers' fields in arid and semi-arid. There are many sources of soil salinization in Egypt that was caused by natural processes such as mineral weathering and release salts; these salts are transported off the soil by drainage water according to soil type, amount of water and percent of slope (Fig. 9.9). The occurrence of water-soluble salts in most of the soils is considered to be the result of natural pedogenesis, whereas the high intensity of salinization of agricultural soils is supposed to be anthropogenic. The origins of salts are dependent on the geographic distribution of salinization sources while in agricultural areas are determined by land reclamation technologies (Zinck 2009). Soil salinization is considered the most ecologically risky that is an obstacle in agricultural productivity especially in arid and semi-arid (FAO 2011c). Water evaporation of surface layers leads to increase mineral weathering causing high salt levels in soils of arid and semi-arid regions. Groundwater table is considered as the main source of land degradation in Egypt. In general, it is noted that north Delta suffers from rising of saline groundwater, at surface by capillary action where salts accumulate on the surface soil causing many environmental problems.

Fig. 9.5 Field photos show the urban sprawl on cultivated lands after 2011

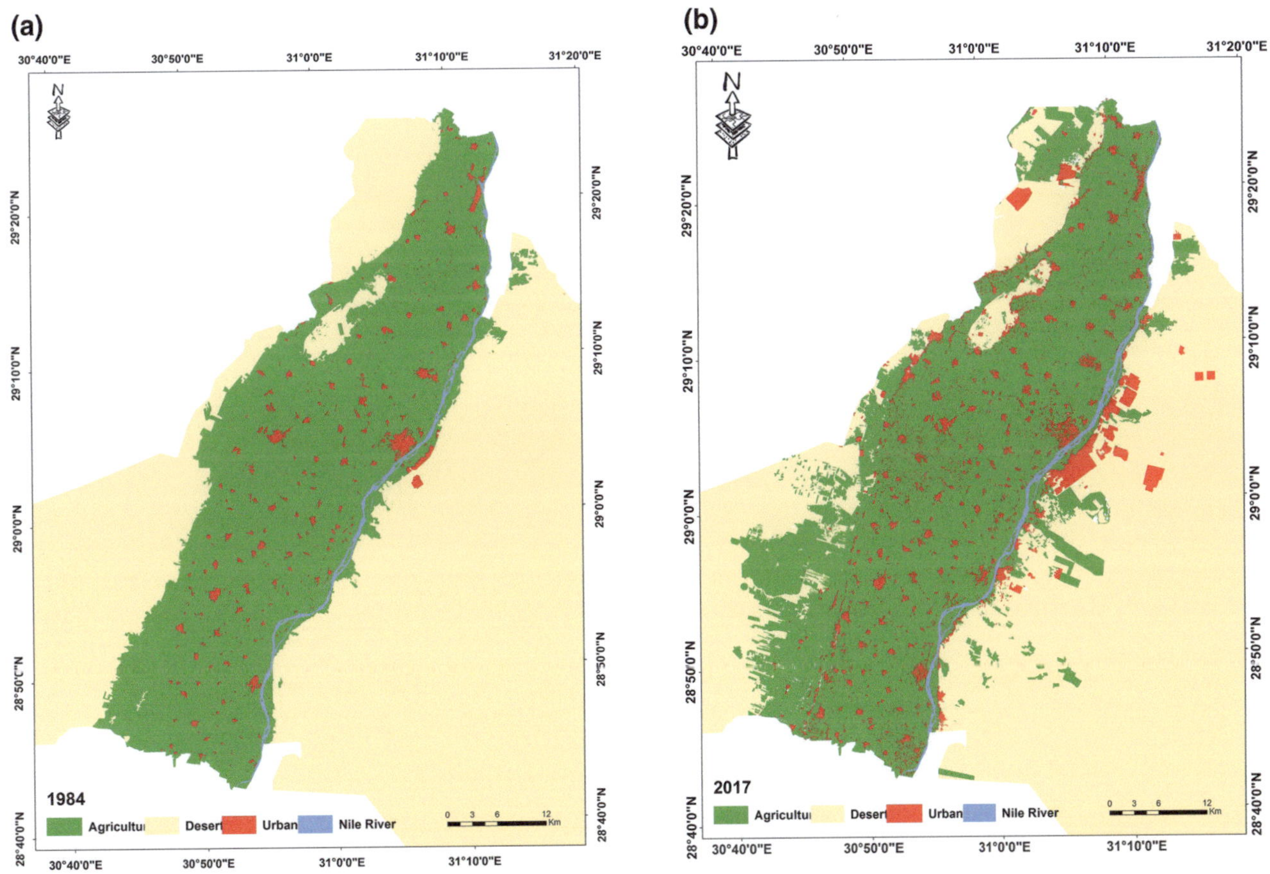

Fig. 9.6 **a** Urban sprawl of Beni Suef Governorate (1984), **b** Urban sprawl of Beni Suef Governorate (2017)

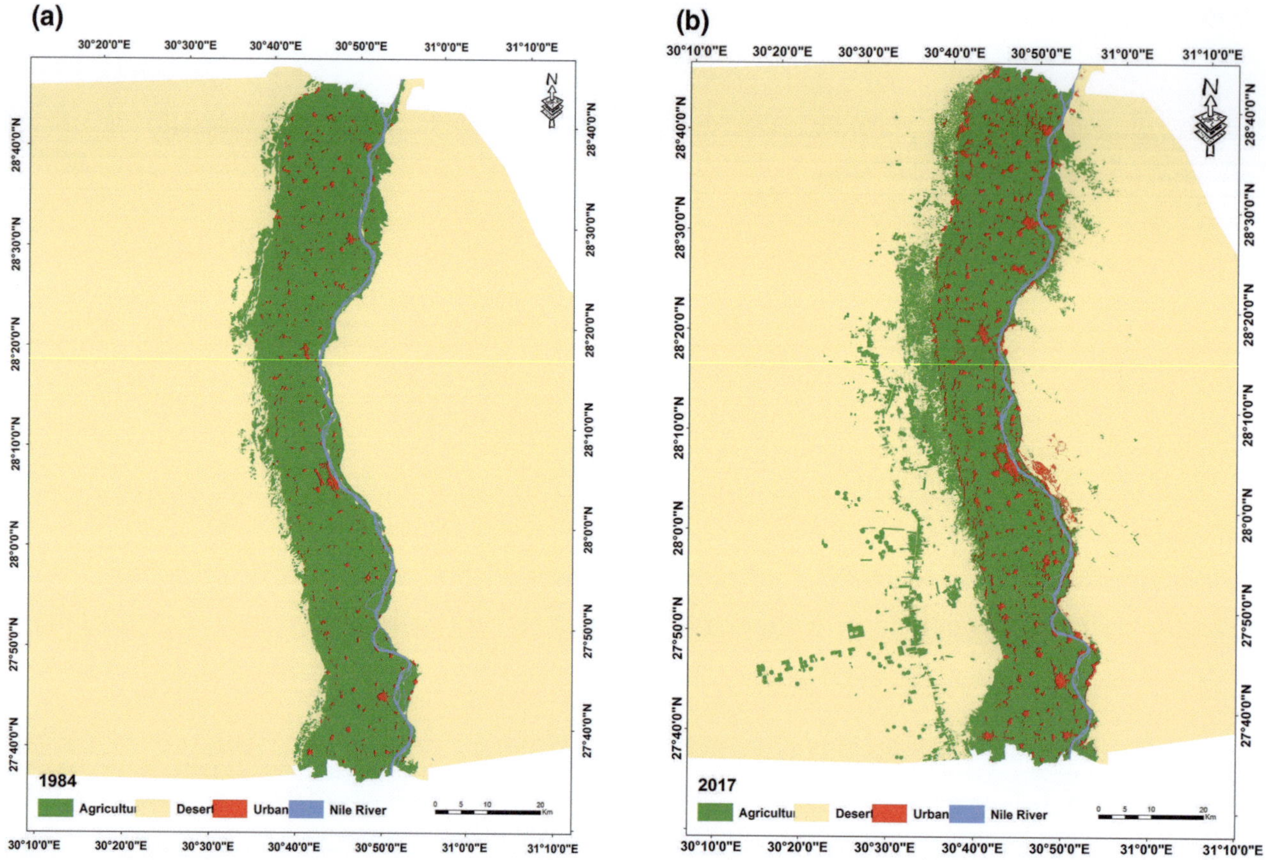

Fig. 9.7 **a** Urban sprawl of El Minya Governorate (1984), **b** Urban sprawl of El Minya Governorate (2017)

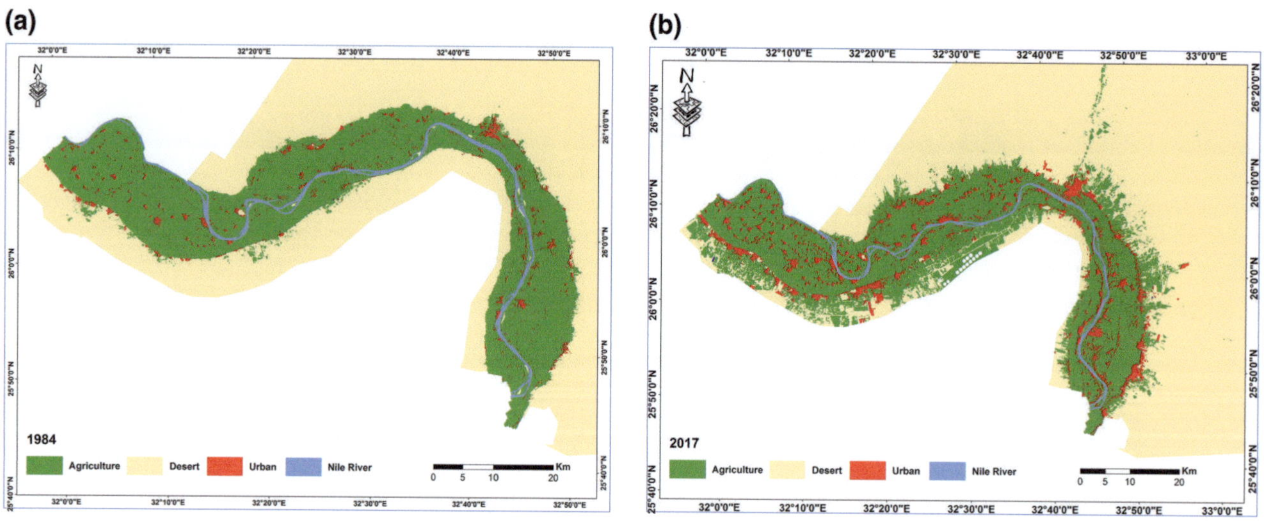

Fig. 9.8 **a** Urban sprawl of Qena Governorate (1984), **b** Urban sprawl of Qena Governorate (2017)

Moreover, the quantities of rainfall in Egypt are scarce, except for some coastal areas. This prevents the washing of salts from upper layers and encourages salt accumulation on the surface. Human factor is one of the most important factors causing the accumulation of salts through mismanagement of soils. Soil salinity is influenced by several factors that can be increased or decreased according to the conditions of each region.

Soil salinity is the foremost factor causing land degradation. The high concentrations of salinity affect crop

Table 9.6 Changes in LULC during 1984 to 2017

Year	Beni Suef		El Minia		Qena	
	1984	2017	1984	2017	1984	2017
	Km^2	Km^2	Km^2	Km^2	Km^2	Km^2
Urban	61.24613	192.5466	90.01	258.90	53.48	240.5351
Agricultural	1367.802	1578.506	2086.23	2419.47	1147.91	1445.406
Desert	18466.47	8903.003	29100.03	28596.11	8102.99	7591.502

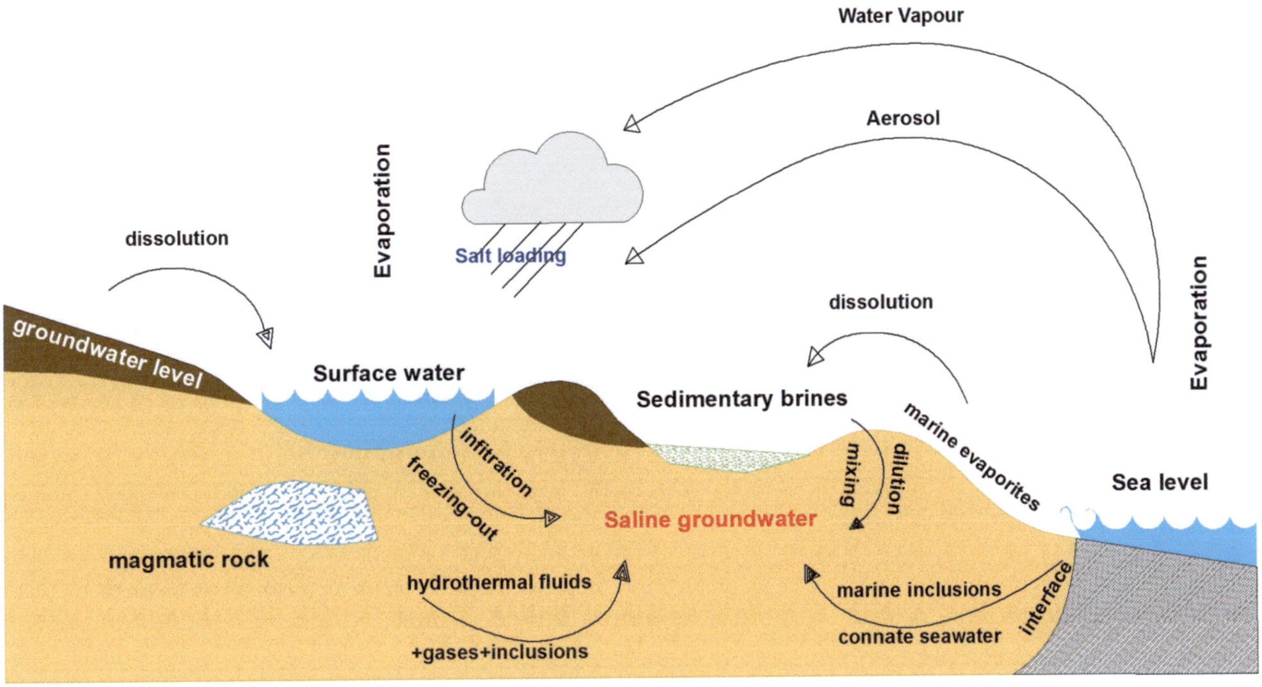

Fig. 9.9 Soil salinization processes. Modified after Daliakopoulos et al. (2016)

growth due to osmotic stress influence which limits the spread of roots to soil solution uptake, thereby causing negatively influences on crop quality, (Shahid et al. (2013). According to Hefny and Shata (2004), the groundwater depth in Egypt vary from location to another where northern parts of Egypt are attributed by shallow depth <2 m. On the other side, the aquifer depth increases towards south of Egypt as it ranges between 60 and 500 m (Fig. 9.10). The northern parts of the Nile Delta suffering from high salinity levels as shown in (Figs. 9.11 and 9.12)., where soil salinity changes according to land use and management methods. Soils located south of Manzala Lake and Idku Lake characterized by high salt content reached to more than 100 dS/m. The high values of salinity is due to water seeping from Manzala Lake. Generally soils of east Nile Delta are characterized by slight to moderate saline. In addition, soil salinity levels decrease towards middle Delta in spite of the high groundwater levels. Groundwater in North Delta is characterized by high concentration of salinity. Thus, contributing to the composition of soil salinity in the northern Nile Delta. It was found that there is a high relationship between the depth of groundwater and soil salinity degrees under arid conditions (Mohamed et al. 2011; Ali and Moghanm 2013). In addition, soils of El-Tina plain east of Suez canal characterized by very high to high saline where it ranges from 8 to 16 dS/m, except for some patches where salinity values reached to 400 dS/m (Sallam et al. 2013).

9.5.3 Soil Erosion

In arid zones, soil erosion is one of the most important factors causing deterioration. Soil erosion has become a seriously threatening problem to the agriculture environment. Soil erosion is traditionally associated with agriculture in tropical and semi-arid areas and some of the associated problems which include loss of fertile topsoil, productivity and threat the sustainable agriculture (Morgan 2005;

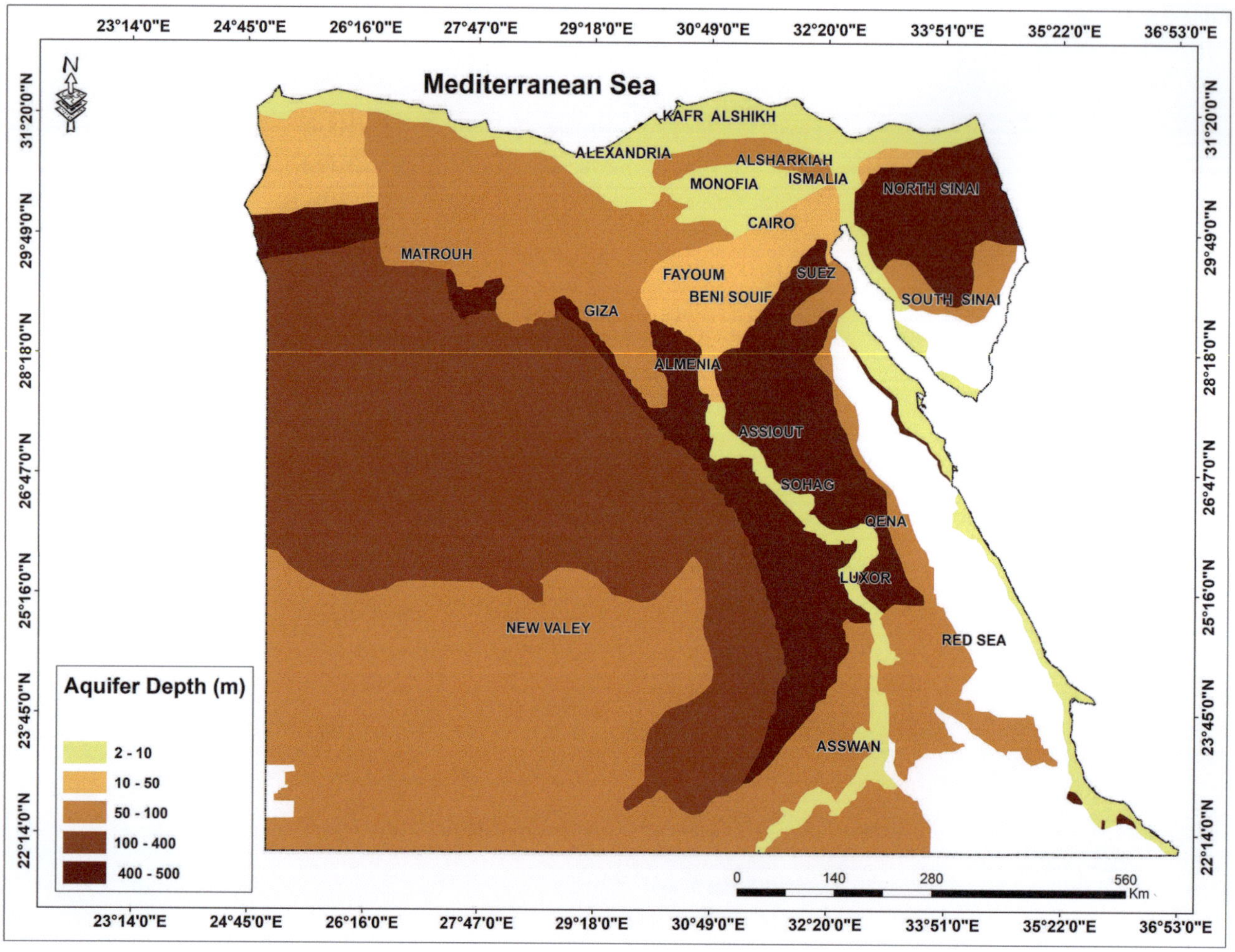

Fig. 9.10 Distribution of aquifer depth in Egypt. Modified after Hefny and Shata (2004)

Pimental 2009). Soil erosion affected by main factors such as amount of rain, besides the soil type and topography. There are two types of soil erosion, i.e. water erosion and wind are the main causes of land degradation (Terrence et al. 2002). Both of them are influenced by soil type. In arid and semi-arid regions, like Egypt, it is noticed that water erosion is seasonal and depends on the amount of rainfall. The rate of water erosion in El-Arish, Rafah located in north coast of Sinai is very high, and vulnerable to erosion risk due to high average annual rainfall where it reaches to 250 mm/year at Rafah. In addition, average annual rainfall at northwest coast reaches 150 mm/years (Mohamed et al. 2013b and Mohamed et al. 2014). In addition, the rate of erosion of the northern west parts of Egypt depends on the soil texture, degree of slope and vegetation cover. Figure 9.13 represents the water erosion risk in the north-western coast of Egypt (Mohamed et al. 2013a, b) the author concluded that water erosion moves salts, nutrients and organic matter from southern to the northern parts of the studied area. On the other hand, although rainfall is low in south-east of Egypt, it has serious impacts. Wahid (2016) evaluated risk of Flash Flood at El-Qaa Plain; he reported that the plain is seasonally susceptible to high risk by water erosion as presented in Fig. 9.14.

9.6 Conclusion

This chapter provides useful information on soil degradation situation under Egyptian conditions which discusses spatial distribution of this phenomena and their affective factors such as natural resources, soil sealing, population, salinity, erosion and desertification. Various definitions of soil degradation, causes and types were briefly discussed. This work provides information with valuable knowledge at the regional scale, pertaining to the assessment of land degradation hazard, status, characteristics, types, distribution of degradation processes in Egypt. Land degradation is complex and varied from place to place. The degradation causes in Nile Delta differs from desert lands. Salinization,

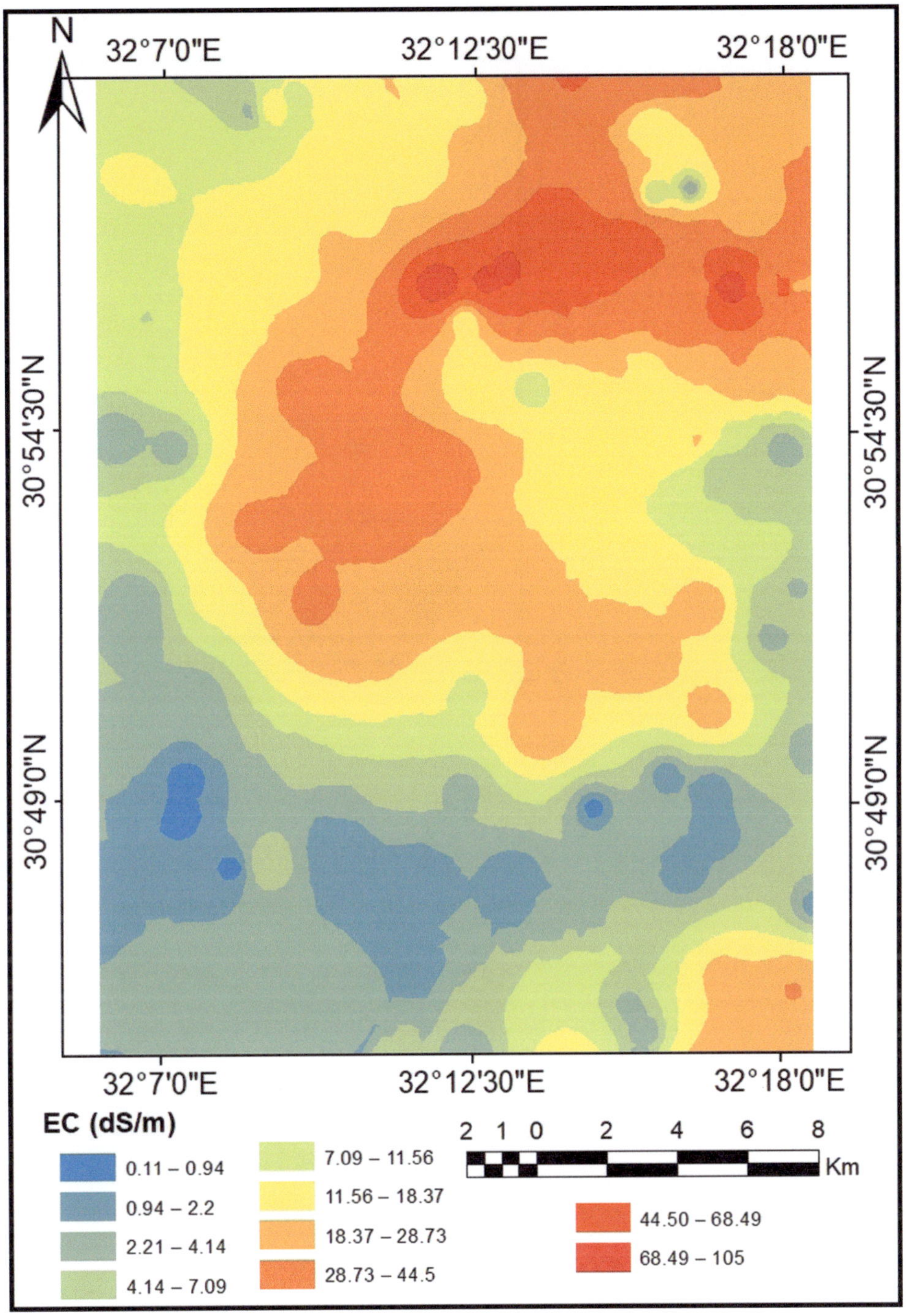

Fig. 9.11 Soil salinity distribution in east of Nile Delta. Modified after Mohamed et al. (2011)

alkalinization and waterlogging are the dominant causes of land degradation in Nile Valley and Delta. Human act has two-sided effect: positive action by right management that leads to mitigate soil degradation, this needs to understand the cause of degradation. Human can also accelerate degradation by mismanagement of land resources. Moreover, social and economic factors have significant effects on reducing or accelerating degradation processes. Meanwhile, desert land is caused by erosion hazards. This chapter highlights the most important methods used to assess degradation. The advanced technologies such as remote sensing with its multi-temporal, multispectral images could offer useful information, real magnitude of land degradation and its changes over time. Therefore, monitoring indicators of degradation processes by means of remote sensing and GIS tools considered as an effective technique to understand the temporal and spatial characteristics of degradation process.

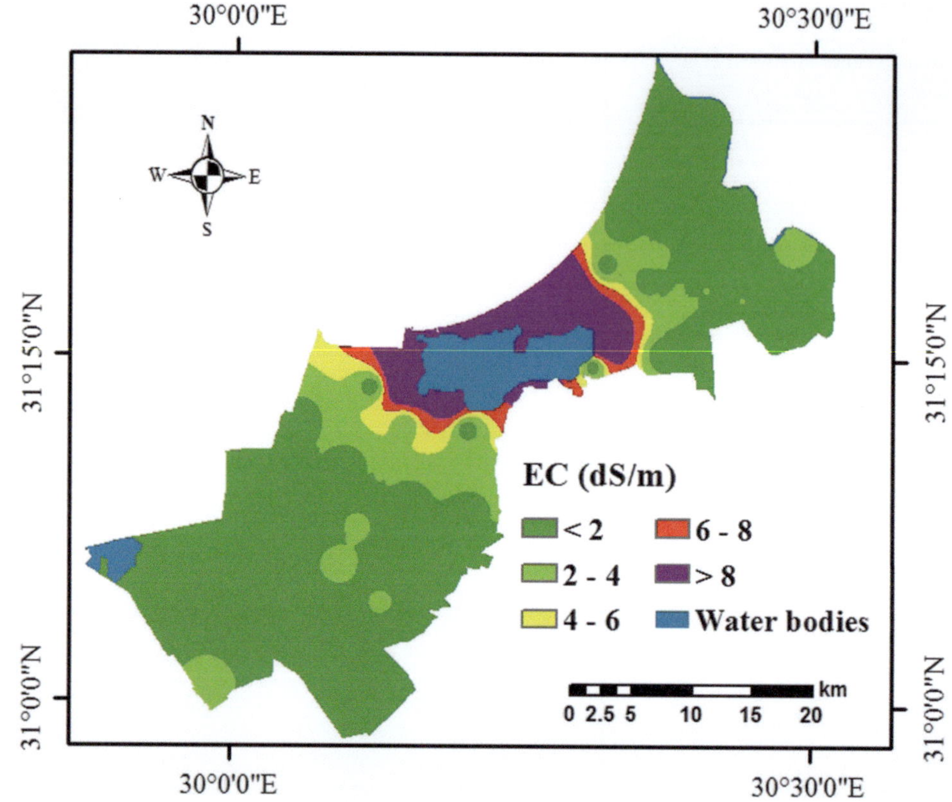

Fig. 9.12 Soil salinity around Idku Lake North West of the Nile Delta. Modified after Ali and Moghnam (2013)

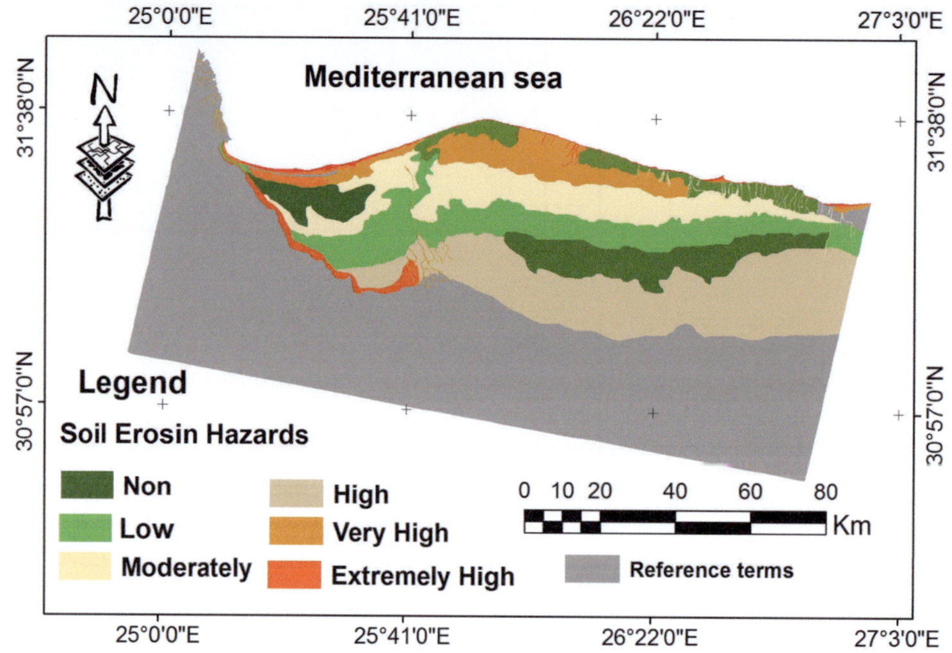

Fig. 9.13 Soil Erosion hazards along the western coast. Modified after Mohamed et al. (2013a, b)

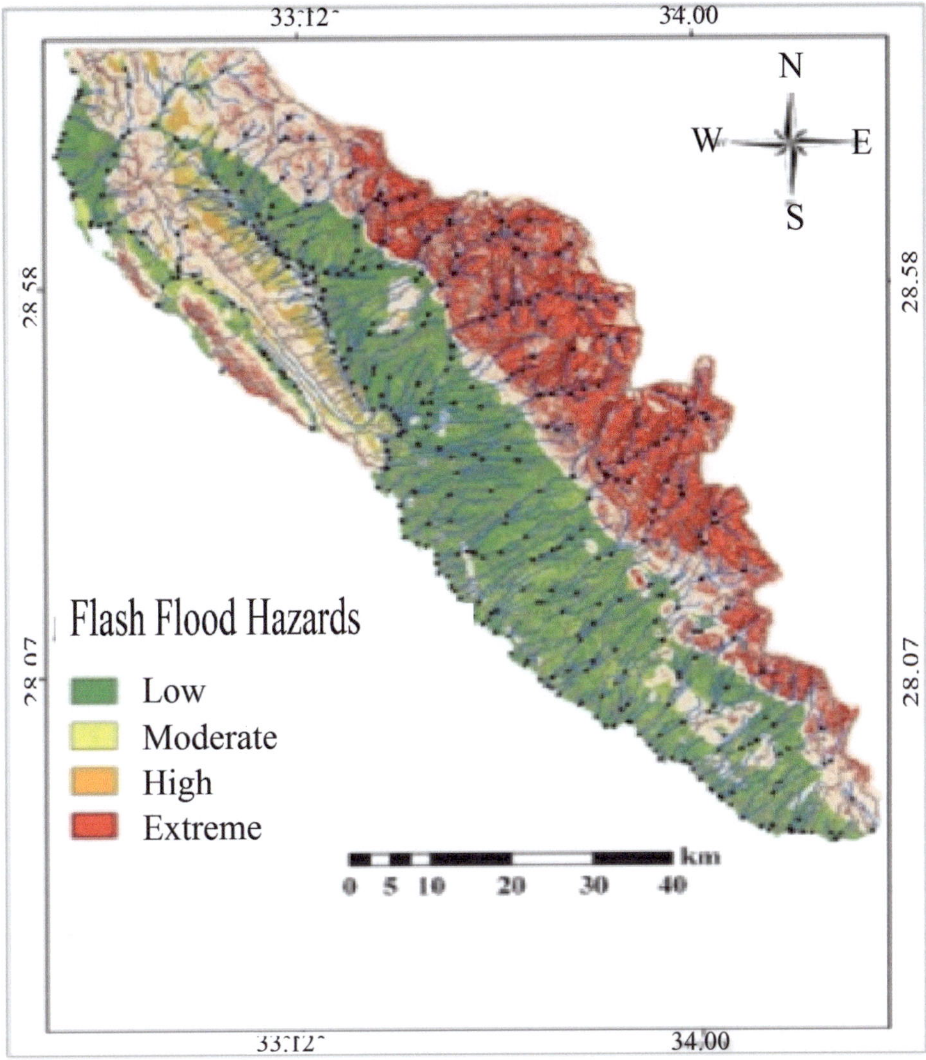

Fig. 9.14 Flash flood hazard in El-Qaa plain. Modified after Wahid et al. (2016)

Acknowledgements The authors acknowledge all staff members in soil department at NARSS for their efforts of maps production and technical support, also acknowledges Dr. Nasser Salim for his permission to use a map from his Ph.D.

References

Abdel Rahman MAE (2009) Quantification of land degradation indicators using remote sensing and GIS techniques. M. Sc. Thesis submitted to Zagazig University, Egypt

Ali RR, Abdel Kawy WAM (2013) Land degradation risk assessment of El Fayoum depression Egypt. Arab J Geosci 6:2767–2776

Ali RR, Moghanm FS (2013) Variation of soil properties over the landforms around Idku lake Egypt. Egypt J Remote Sens Space Sci 2013(16):91–101

AUC's(2016) 22nd annual research conference: cultural survival in Egypt in the 21st Century, May 4–5

Bridges EM, Oldeman LR (1999) Global assessment human-induced soil degradation (GLASOD). Arid Soil Res Rehabil 13:319–325

CAPMAS (2006) Statistical year book central agency for public mobilization and statistics (CAPMAS), annual report July 2006. Egypt, Cairo

Darwish KH, Safaa M, Momou A, Saleh SA (2013) Egypt: land degradation issues with special reference to the impact of climate change. In: Combating desertification in Asia, Africa and the Middle East, Springer Netherlands, pp 113–136

Daliakopoulos IN, Tsanis IK, Koutroulis A, Kourgialas NN, Varouchakis AE, Karatzas GP, Ritsema CJ (2016) The threat of soil salinity: a European scale review. Sci Total Environ 573 (2016):727–739

Egypt Facts. National Geographic. https://en.wikipedia.org/wiki/Demographics_of_Egypt

El Baroudy AA (2005) Using remote sensing and GIS techniques for monitoring land degradation in some areas of Nile delta, Ph.D. Thesis, Fac. of Agric., Tanta Univ., Egypt

European Commission (2012) Guidelines on best practice to limit, mitigate or compensate soil sealing, SWD(2012) 101 final/2

FAO (1979) A provisional methodology for soil degradation assessment M-57 ISBN 92-5-100869-8. FAO, Rome, Italy

FAO (2015) The Food and agriculture organization of the united nations. climate change and food systems: global assessments and implications for food security and trade; FAO: Rome, Italy,.

Available online:http://www.fao.org/3/a-i4332e.pdf.Accessed on 10 Oct 2015

FAO (2007) Global assessment of land degradation and improvement GLADA report 5. FAO, Rome, Italy

FAO (2011a) Country pasture/forage resource profiles EGYPT. Available online http://www.fao.org/ag/agp/agpc/doc/counprof/PDF%20files/Egypt.pdf

FAO (2011b) Land degradation assessment in dry lands (LADA), Manual for local level assessment of land degradation and sustainable land management, Roma

FAO (2011c) The land resource base. Room: FAO (ARC/86/3), Roma, 1986

FAO/UNEP (1978) Methodology for assessing soil degradation Rome, 2527 January 1978 Italy

Field D, Morgan CL, McBratney AB (2017) Global soil security. progress in soil science, Springer International Publishing Switzerland. ISSN 2352–4774, https://doi.org/10.1007/978-3-319-43394-3

Hefny K, Shata A (2004) Underground water in Egypt. Ministry of Water Supplies and Irrigation, Cairo, Egypt

ICRISSM (2014) International conference on research and innovation for sustainable soil management—27th to 29th November 2014 in Hurghada, Egypt

Kees H (1961) Ancient Egypt: a cultural topography. University of Chicago Press, Print, Chicago

Lal R (1994) Sustainable land use systems and soil resilience. In: Greenland DJ, Szabolcs I (eds) Soil resilience and sustainable land use. pp 41–67 Wallingford, UK:CAB International

Le QB, Nkonya E, Mirzabaev A (2016) Biomass productivity-based mapping of global land degradation hotspots. In: Nkonya E, Mirzabaev A, von Braun J (eds) Economics of land degradation and improvement—A global assessment for sustainable development. Springer International Publishing, Cham, pp 55–84

Mohamed ES, Belal A, Saleh A (2013a) Assessment of land degradation east of the Nile Delta, Egypt using remote sensing and GIS techniques. Arab J Geosci 6(8):2843–2853

Mohamed ES, Saleh AM, Belal AA (2014) Sustainability indicators for agricultural land use based on GIS spatial modeling in North Sinai—Egypt. Egypt. J Remote Sens Space Sci 17:1–15

Mohamed ES, Belal A, Shalaby A (2015) Impacts of soil sealing on potential agriculture in Egypt using remote sensing and GIS techniques. Eurasian Soil Science 48(10):1159–1169

Mohamed ES, Morgun EG, Bothina SG (2011) Assessment of soil salinity in the Eastern Nile Delta (Egypt) using geoinformation techniques. Mosc Univ Soil Sci Bull 66(1):11–14

Mohamed ES, Schütt B, Belal A (2013b) Assessment of environmental hazards in the north western coast-Egypt using RS and GIS. Egypt J Remote Sens Space Sci 16(2):219–229

Morgan RPC (2005) Soil Erosion and Conservation, 3rd edn. Blackwell Publishing, p. 316

Nkonya E, Mirzabaev A, von Braun J (2016) Concepts and methods of global assessment of the economics of land degradation and improvement, in the economics of land degradation and improvement—a global assessment for sustainable development, Springer International Publishing, pp 15-32, https://doi.org/10.1007/978-3-319-19168-3_1

Oldeman LR, Hakkeling RTA, Sombroek WG (1991) World map of the status of human-induced soil degradation: an explanatory note. Second Revised Edition, ISRIC/UNEP, p 1991

Pimental D (2009) World soil erosion and conservation. Cambridge University Press, p. 364

Ponce-Hernandez R, Koohafkan P (2010) A methodology for land degradation assessment at multiple scales based on the DPSIR approach: experiences from applications to drylands. In: Zdruli P, Pagliai M, Kapur S, Faz Cano A (eds) Land degradation and desertification: assessment. Springer, Mitigation and Remediation

Reed MS, Dougill AJ (2002) Participatory selection process for indicators of rangeland condition in the Kalahari. Geog J 168:224–234

Rengasamy P (2006) World salinization with emphasis on Australia. J Exp Bot 57:1017–1023. https://doi.org/10.1093/jxb/erj108

Saleh NHS (2016) Monitoring the agriculture land change and its impact on water consumption between branches damietta and rosetta, Egypt using Remote Sensing and GIS Techniques. thesis Ph.D. Faculty of Women for Arts, Ain Shams university

Sallam GAH, Nasralla MR, Magdy AR (2013)Water use Ef fi ciency for leaching saline-sodic clayey soils: case study of tina plain area of Egypt; in Developments in soil salinity assessment and reclamatio. https://doi.org/10.1007/978-94-007-5684-7_6, © Springer Science + Business Media Dordrecht 2013

Shahid SA, Abdelfattah MA, Taha FK (2013) Soil salinity modeling, approaches, and key issues developments in soil salinity assessment and reclamation: innovative thinking and use of marginal soil and water resources in irrigated agriculture, Springer Science + Business Media Dordrecht, pp 45–60. https://doi.org/10.1007/978-94-007-5684-7_2

Snel M, Bot A (2003) Some suggested indicators for land degradation assessment of drylands (draft paper). Food and Agriculture Organization, Rome

Stockings M, Murnaghan N (2000) Land degradation—guidelines for field assessment (PDF). Norwich, UK, pp 7–15

Terrence TJ, Foster GR, Renard KG (2002) Soil erosion: processes, prediction, measurement, and control. John Wiley & Sons. p 1. ISBN 978-0-471-38369-7

The Sixth National GIS Symposium (2011) Al-Khobar, Eastern Province, Saudi Arabia April 24–26

UNEP (1991) Global assessment of soil degradation, United Nations Environment Programme (UNEP), United Nations, GLASOG Project

Van Lynden GWJ (1995) Guidelines for the assessment of the status of human-induced soil degradation in south and southeast Asia. ISRIC WP 95/2, ISRIC, Wageningen

Wahid A, Madden M, Khalaf F, Fathy I (2016) Geospatial analysis for the determination of hydro-morphological characteristics and assessment of flash flood potentiality in arid coastal plains: a case in southwestern Sinai, Egypt Earth Sci Res J 20(1) ISSN 1794–6190

Zinck JA (2009) Remote sensing of soil salinization impact on land management, New York

Soil Health and Its Biology

10

Tarek Alshaal, Hassan El-Ramady, Nevien Elhawat, Sahar El-Nahrawy, Alaa El-Dein Omara, Tamer Elsakhawy, Azza Ghazi, Mohamed H. Abbas, Ihab M. Farid, Neama Abdalla, Miklós Fári, and Éva Domokos-Szabolcsy

Abstract

It is well known that, the health of humans is mainly depends on the physical, social and nutrient factors. The health of humans requires safe, proper and sufficient supply of nutrients through the foods. This quality of foods mainly depends on the biology, fertility and quality of soils, in which crops were cultivated. Thus, the biology of soils has direct and/or indirect effects on the health of humans through the exposure to soils and their microorganisms, extract different medicines from soils and deficiency of nutrients through food chain. The biology of soils also has a close relationship with the nutrition of both plants and humans. Therefore, soil quality and its biology is a crucial issue towards sustainable agriculture. Thus, the main purpose of this book chapter is to emphasize the link between soil quality through its biology and human health. The great roles of soils in our life also will be highlighted.

Keywords

Soil health • Human health • Plant nutrition
Soil biology • Sustainable agriculture

10.1 Introduction

It is well known that, soils are a dynamic, complex and open system. This system supports all different soil microbial communities including bacteria, actinomycetes, algae, fungi, viruses, protozoa, nematodes, etc. These microbial or biological activities have a great role in cycling of nutrient elements in the biosphere, maintaining the fertility of soils, formation of humus, biological conversions, geochemical cycling of elements, ecosystem sustenance, etc., as well as supporting plant life and its productivity (Bharti et al. 2017; Finkel et al. 2017; Vimal et al. 2017). Some soil microbes have the ability to increase the tolerance of plants to root/soilborne plant pathogens like arbuscular mycorrhizal fungi, thereby acting as a biocontrol agent. Several reports have been stated that, mycorrhizal fungi have the ability to reduce damage caused by soilborne plant pathogens through enhancing plant vigor due to increased nutrients uptake and then increasing the plant resistance to pathogen infection (Finkel et al. 2017).

Soil and its relationships with human health were and still one of the most important issues in soil sciences. These relationships have been built on basis that soil is a crucial source for our life, supporting us with safe and enough foods, different medicines, etc. The biological properties of soils were and still the limit factor in nutrition of both plants and humans (Singh et al. 2017). These biological characteristics represent the most important indicators for soil quality. This soil quality and its biology become an essential approach for the sustainable agriculture. It is well documented that, soil biology and its ecology was and still one of the most important sustainable management of soils (Eisenhauer et al. 2017). Therefore, soil ecology and its

T. Alshaal (✉) · H. El-Ramady
Faculty of Agriculture, Soil and Water Department, Kafrelsheikh University, Kafr el-Sheikh, Egypt
e-mail: alshaaltarek@gmail.com

N. Elhawat
Department of Biological and Environmental Sciences, Al-Azhar University, Cairo, Egypt

S. El-Nahrawy · A. E.-D. Omara · T. Elsakhawy · A. Ghazi
Soils, Water and Environment Research Institute (SWERI), ARC, Sakha, Kafr el-Sheikh, Egypt

M. H. Abbas · I. M. Farid
Soils and Water Department, Faculty of Agriculture, Moshtohor, Benha University, Benha, Egypt

N. Abdalla
Genetic Engineering Division, Plant Biotechnology Department, National Research Center, Giza, Egypt

M. Fári · É. Domokos-Szabolcsy
Agricultural Botany, Plant Physiology and Biotechnology Department, University of Debrecen, Debrecen, Hungary

H. El-Ramady et al. (eds.), *The Soils of Egypt*, World Soils Book Series,
https://doi.org/10.1007/978-3-319-95516-2_10

biology is the main key for the protection and conservation of soils, for sustainable environmental management and for the nutrition and health of humans. As a consequence, soil ecology and its biology will remain an extremely important field of research into the future and requires a coordinated global effort to address the most important issues facing the sustainability of soils and gaps in soil ecological knowledge (Eisenhauer et al. 2017).

The biology, ecology and quality of Egyptian soils are in a crucial need to a great concern at all levels. That means there is no any sustainable agriculture in Egypt without a real guarantee for protection the Egyptian soils from all forms of degradation and careless. Therefore, this review will focus on the soil biology and its impact on health of humans and plants as well as the soils as a main source of our life needs.

10.2 Soil Biology: As a Treasure

Soil biology is a branch of soil sciences dealing with all forms and types of life in soil including soil micro- and macro-organisms. So, there is crucial need for more understanding of the effect of soil biological components on chemical and physical fertility. The maintaining of the fertility and productivity of soils is mainly depends on the desirable levels of numbers, activity and diversity of soil organisms. The biology of soil has serious roles in different issues such as soil fertility and cycling of nutrients, decomposition of organic matter by soil organisms, soil structure, plant growth and soil carbon storage. Regarding soil biota, soil life and soil fauna could be considered a term for all organisms including nematodes, earthworms, protozoa, bacteria, fungi and different arthropods. Therefore, the biology of soils is a real and valuable treasure for plant nutrition and its growth. There is no productive soil without the biology or biological activities in soils and the soil health starts from the biology of this soil (Bharti et al. 2017; Hüberli 2017; Xiao et al. 2017).

The treasure of soil biology means that, all processes and factors controlling the fate and bioavailability of nutrients in soil rhizosphere as well as the biological activities (i.e., soil enzyme activities and soil microbial counts) give the soil its vitality and reflect the role of this treasure in plant nutrition, soil biochemistry, soil quality, soil microbiology, soil biogeochemistry, soil geomicrobiology, soil ecology, etc. Therefore, more investigations should be carried out regarding the biology of soils including studies on different levels in particular molecular and genetic level. The most important applications of soil biology related to human health including both the direct contact of human with soils (several human health problems) and discovery of many antibiotics produced by soil organisms. So, the relationship between soils (their biology) and human health was and still a great natural resource for the sustainable development. Furthermore, several investigations regarding the scientific approach of soils and human health are needed.

In Egypt, the biology of soil as a great treasure was well known for ancient Pharaohs several thousand years ago. They expressed about this treasure through different paintings in their Temples (e.g., Karnak) including many agricultural practices like ploughing with a yoke of horned cattle, and other agricultural practices on a large scale. This civilization of Ancient Egypt was indebted to the Nile River and its dependable seasonal flooding. Due to the fertile soil and the Nile River, the Egyptians have been built an empire on the basis of great agricultural wealth. The farming practices of Egyptian s allowed them to grow staple food crops (i.e., grains such as wheat and barley) and industrial crops such as papyrus and flax. From their paintings in different temples and tombs, the Egyptians have been proved about the significance of soils in their entire life including production of food, storing these foods, manufacturing these foods, production some medicines from the soils, manufacturing some cosmetic from soils, etc.

Therefore, soils as complex, dynamic and open systems are very important for the entire of human life. The humans are created from this soil, their natural necessaries from this soil (food, feed, fiber and fuel) are supported, a lot of his medicines from soil are manufactured, a lot of vaccinations against human diseases are produced from soil, the essential nutrients in human diets are depleted from the soil, nearly all nutrient deficiencies on human are definitely recovered from soil, etc.

10.3 Soil Biology and Human Health

The biology of soils refers to soils encompasses a very wide range of different organisms including different species of both plants and animals. These organisms even in macro- or micro-forms give the real biological face for soils. These organisms generally have several activities in soils including the soil enzyme activities and the soil microbial activities, which called as soil biological indicators. Therefore, these soil biological indicators for evaluating the quality of soils are more recently interest and very important indicators, which could serve as early and more sensitive indicators of agro-ecosystems as a response to soil management (Bhaduri et al. 2017; Plekhanova et al. 2017). The soil organisms have a great role in both soil fertility and its quality from one side and human health from the other side (Shao et al. 2017).

In 1952, Selman Waksman was Nobel Prize Winner, and his students cultured *Streptomycin* from a soil borne microbe. He produced the first chemical that was a cure for tuberculosis, as bactericidal antibiotic and on the World

Health Organization's List of Essential Medicines as well as also as a pesticide for crops and a bactericide for animals (SSSA 2017). The impact of soils on human health generally includes (1) its role in soilborne human pathogens or diseases from soils, (2) its role in vaccinations and medicines produced from soils and (3) its role in recovering of deficiency of nutrients through the food chain. These three issues could be presented in the following sections in more details.

10.3.1 Soils and Medicine in Ancient Egypt

Nowadays, the using of soils in producing and manufacturing of different medicines, cosmetics and some personal care products is a profitable industry. This industry has several advantages and disadvantages and related to medical geology. The medical geology (or geomedicine) is a science of studying the health problems related to the place (or environment) and these studies should go well beyond simplistic comparisons of geochemical and epidemiological data (Selinus 2013). Therefore, it could use soil organisms or soil materials in isolating some antibiotics and then manufacturing some medicines. These soil medicines include antibiotics, anti-diarrheal medications, cancer drugs, emollient and drying agents, which could be used to treat poison ivy, poison oak and poison sumac cases and others (Brevik et al. 2017).

It is worth to mention that, ancient Egypt has been treated with seaweed (as a good source for iodine) in dealing with goiter disease (severe iodine deficiency) several thousand years ago. Several descriptions also of lead poisoning have been found in texts from Egyptian civilization further corroborate the heavy uses of lead. Clay tablets have been provided for lead-poisoning symptoms from the middle and late Assyrian periods (1550–600 BC) as do ancient Egyptian medical papyri and Sanskrit texts dating from over 3000 years ago. More details about the role of ancient Egyptian civilization in the history of medicine, science and technology, it could be referred to *Encyclopedia of the History of Science, Technology, and Medicine in Non-Western Cultures*, which published by Springer (Selin 2016). This encyclopedia includes several topics regarding ancient Egypt such as the medicine in ancient Egypt, meat preservation, fishing, surveying, gardens, basketry, maps and mapmaking, tombs, temples and archaeoastronomy, ship- and boatbuilding, obelisks, and origins of agriculture in Egypt. A comprehensive history of the origins of agriculture in Egypt could be found in this encyclopedia edited by Janick (2016).

Concerning the medicine in ancient Egypt, they concluded their life in this sentence "*Magic is effective together with medicine. Medicine is effective with magic*". They have been recognized these beliefs and used the prosthetic medicine for living persons as well as some mummies are outfitted with prosthetic devices. The ancient Egyptian have been achieved a great progress in medicine more than the modern progressive world nowadays especially in case of mummification. Ancient Egyptians also had advanced their physiological and anatomical knowledge. It is reported that, the known medical papyri originate beginning with the Middle Kingdom and the last we hear about are mentioned by Clement of Alexandria (second half of the second century AD). He knew about six Egyptian handbooks devoted to specific aspects of medicine such as illnesses, anatomy, surgical instruments, drugs, eye diseases and gynaecology (Gyóry 2016).

Therefore, it is well stated that, ancient Egypt would be the preferred place if anyone had to be ill in antiquity due to their medications were mainly empirical and aimed at the relief of symptoms. The medical knowledge of the physicians in ancient Egypt was without equal at that time; their formal, structured and logical approach to the patient is unchanged in modern medicine; and their techniques and treatments are often comparable to today's medical profession. About 300 diseases are mentioned in ancient Egyptian texts (Gyóry 2016). The soils have a lot of roles in medicine in ancient Egypt as painted in many walls of temples. These roles are still in concern in modern Egypt but needs more management and protection.

10.3.2 Soils and Human Diseases

The complete physical, mental and social state of human is defined as the health. This health of humans has been linked directly and/or indirectly with soils (Brevik and Sauer 2015). It is reported that, soils have some important roles on human health including disposal of wastes, food availability with high quality, human diseases from contacting with some chemicals and/or harmful microorganisms in soils, and production of some medicines manufacturing from soils (Brevik et al. 2017). Therefore, several studies have been published about the effects of soils on the health of humans through the exposure of humans to radioactive elements (Wang et al. 2017) or some chemicals like heavy metals in soils (Zhang et al. 2017) or exposure to some microbial pathogens in soils (Brevik et al. 2017), and others. It is worth to mention that, soil are teeming with life as well as being a home for a huge array of living organisms mainly microorganisms. The most common soil-dwelling microbes pathogenic for humans include bacteria, fungi, protozoa and viruses, both of which require a plant or animal host for their survival. These microorganisms could be found in different environments including salt and fresh water, all soils, different climatic zones, in deep-sea hydrothermal vents, throughout the atmosphere and deep below the surface of the Earth in oil wells (Vimal et al. 2017).

It is well known that, some elements have the ability to emit radiations (like alpha, beta and gamma rays) due to the radioactive behavior of a nuclide (e.g., $^{89-90}Sr$, ^{131}I, ^{140}Ba, ^{137}Cs). A lot of modifications will cause in cells and/or death of cells, when human or plant cells interact with these radiations. Under high radiation level, several problems could be created such as nausea, diarrhoea, vomiting, leukaemia, etc. The relative effect of different radionuclides depends on management practices such as fertilization and its types, organic matter application and cultivation method (Bharti et al. 2017). These radionuclides could be resulted from the soil, fertilizer application, nanoparticles or nanomaterial application, pollution with radiation, nuclear medicine, etc. Therefore, the universe faces nowadays the over-use of nanomaterials in almost all fields in our life, which may allow to the radioactive of some elements like gold and its nanotechnology to destroy the human health (Jain 2017).

Concerning the soil biology and human diseases in Egypt, it is reported that, several helminthes have the ability to transmit into the soils of Egypt (Farghly et al. 2016). Soils could be considered, as well known, an important source for several infections such as protozoal, bacterial and helminthic ones. Many parasites also could be transmitted through soils as well as several parasitic infective stages and their development. Indeed, soil could transmit different protozoal cysts such as *Entamoeba histolytica*, *Balantidium coli* and *Giardia lamblia* (Farghly et al. 2016). In Egypt, the spread of parasites through soils is a little bit common and different parasites could transmit through soils (e.g., Etewa et al. 2016; Farghly et al. 2016) and waters (e.g., Elbana et al. 2017a, b). The most vulnerable Egyptian groups could host or infect with the large numbers of intestinal worms include school-aged children and preschool children. Further studies with holistic survey are needed to know exactly the current situation in Egypt about the most common soil-borne human pathogens as well as the role of soils in qualitative and quantitative spread of these pathogens.

10.3.3 Soils, Deficiency of Nutrients and Human Health

Human must consume a variety of foods in modest quantities to protect his body from mineral nutrient deficiencies as well as these foods should include low fat dairy, different whole grains and proper quantities of vegetables, fruits and meats. Concerning the macro-mineral nutrients like potassium and phosphorus, human is required them in amounts of up to $10\ g\ d^{-1}$. Regarding the daily requirement of secondary and micro-mineral nutrients, they will be requested by humans in ranges from 400 to 1 500 $mg\ d^{-1}$ and 45 $\mu g\ d^{-1}$ to 11 $mg\ d^{-1}$, respectively (Gupta and Gupta 2014). Therefore, an increase in published investigations has been recorded day by day regarding the interrelations between essential metal ions and human diseases (Cakmak et al. 2017; Davison 2017; Liu et al. 2017; Marles 2017; Palzer 2017).

It is well documented that, mineral nutrients are the main key and the essential engine for many functions in the human body. That means, mineral nutrients are required for many biological processes in human in very particular amounts and a specific minerals. For example, there are no vitamins could be absorbed by human or can carry out its intended function without their specific mineral nutrients (Gupta and Gupta 2014). Therefore, many ways should be followed to increase the level of these mineral nutrients in human diets like the biofortification approach. This biofortification is considered one of the most important strategies in supporting humans with enough and essential mineral nutrients such as copper, iodine, iron, selenium, zinc, etc. This approach has been already used in enriching some strategic crops with these previous mineral nutrients such as wheat, rice, maize, sweet potato, etc. Several books (e.g., de Pee et al. 2017; Field et al. 2017; Rakshit et al. 2017; Pearce 2017; Pilon-Smits et al. 2017; Zhao 2017) and other publications have been issued this year focusing on the human health and soil mineral nutrients (e.g., dos Reis et al. 2017; Kim et al. 2017; Bharti et al. 2017).

Therefore, it could be listed some common human diseases related to deficiency of mineral nutrients or some trace elements as follows (Davison 2017):

(1) **Calcium**: Ca-functions in human include bone and tooth structure, nerve transmission, muscle contraction, blood clotting, blood pressure regulation, hormone secretion. Ca-deficiency cause bone loss disease or osteoporosis.
(2) **Copper**: Cu-functions include its role in lipid metabolism, iron absorption, collagen synthesis, antioxidant protection, nerve and immune function. Cu-deficiency cause anemia, poor growth and bone abnormalities.
(3) **Iron**: Fe-functions include part of hemoglobin, electron carriers in electron transport chain and immune function. Fe-deficiency causes anemia, pale red blood cells, low hemoglobin, weakness and reduced immunity.
(4) **Magnesium**: Mg-functions include ATP stabilization, bone structure, enzyme activity, nerve and muscle function. Mg-deficiency cause vomiting, nausea, weakness, confusion, muscle pain and depressed pancreatic hormone secretion.

(5) **Phosphorus**: P-functions include acid–base balance, structure of bones, teeth, membranes, ATP, DNA. P-deficiency causes bone loss, weakness.
(6) **Potassium**: K-functions include fluid balance, protein synthesis, nerve transmission and muscle contraction. K-deficiency cause muscle weakness, paralysis, confusion, can cause death and accompanies dehydration.
(7) **Sodium**: Na-functions include fluid and acid–base balance and nerve impulse transmission. Na-deficiency cause mental apathy, muscle cramps and loss of appetite.
(8) **Zinc**: Zn-functions include regulation protein synthesis, growth development, antioxidant protection, wound healing, immunity, vitamin A transport and making sperm. Zn-deficiency includes poor growth and development, skin rashes, decreased immune function, loss of taste and poor wound healing.
(9) **Fluoride**: F-functions include formation of bones and teeth; resistance to tooth decay. F-deficiency causes susceptibility to tooth decay.
(10) **Chromium**: Cr-functions include its association with glucose metabolism and enhancing insulin action. Cr-deficiency causes abnormal glucose metabolism and high blood glucose.
(11) **Iodine**: I-functions include its need for synthesis of thyroid hormones and cellular energy. I-deficiency causes goiter disease, intellectual disability, growth and developmental abnormalities.
(12) **Selenium**: Se-functions include its role in fat metabolism, spares vitamin E, synthesis of thyroid hormones and it is an antioxidant as part of glutathione peroxidase. Se-deficiency causes muscle pain, weakness or Keshan disease.

Regarding the health care system in Egypt, it is quite complex. It is involved with a large number of public entities in financing, the management and provision of care. The public health expenditure is low and has pluralistic and complex financing mechanisms including health insurance, tax-based financing and fee for service through out-of-pocket expenditures. It is estimated that Egypt produces more than 90% of the pharmaceuticals consumed and pharmaceuticals account for more one-third of all health spending, which around 85% is private expenditure (Elshamy 2016). In the Nile Delta, the diseases of both liver and kidney have been recorded the highest number of patients and ranking worldwide. Several studies have been proved these figures and confirmed that about 38% of all newly diagnosed cancer cases recorded in Egypt (e.g., Elshamy 2016; Bogan et al. 2017). Studies also suggested that, cancer is an increasing problem in Egypt recording for liver (23.8%), breast (15.4%) and bladder (6.9%) for both sexes (Elshamy 2016). There is also some evidence suggesting a positive association between particular pesticides and liver cancer or hepatocellular carcinoma (VoPham et al. 2017). Concerning the prevalence of the hepatitis C virus (HCV) worldwide, it is recorded by 2.8%, whereas Egypt has the highest HCV prevalence in the world with 14.7% (Youssef et al. 2017). Serious fungal infections also have been reported in Egypt (Zaki and Denning 2017). Egypt also is considered an endemic country according to several European reports as well as other outbreaks. It is estimated that hepatitis A virus (HAV) has a high prevalence and circulating in Egypt (Hamza et al. 2017a). This virus is transmitted primarily by the ingestion of contaminated food or water, the fecal-oral route, person-to person contact, or on rare occasions transmission via blood has also been also reported (Hamza et al. 2017b).

Therefore, great challenges face Egypt in frame of the safety and protection of health for the Egyptian people. This urgently needs an effective system in health insurance to include all Egyptian citizens with supporting the nursery, doctors and patients. Health care systems should be strengthened focusing on patient centered care to optimize outcomes for patients. Making real improvements in health management will require the proactive efforts of many organizations with stress on education as a cornerstone to improve health care in general and pain relief and palliative care specifically.

10.4 Soil Health and Its Biology Towards Sustainable Agriculture

In the near future, the world will face great and serious challenges regarding issues of climate changes, food security, energy security, freshwater scarcity, and biodiversity losses. Therefore, the UN has been published a fascinating perspective for soil science called the sustainable development goals. These goals included the major soil-related environmental issues, which are presented in a societal context and soil expertise are needed to reach these goals. These goals also covered many soil functions such as food production, water availability, climate change, biodiversity losses and energy, which are covered in goal no. 2, 1, 13, 15 and 7, respectively.

The soil health or quality is a measure of the ability of soil to fulfill its requirements for plant growth and/or human health. In other words, soil health is the ability of soils in providing social and ecosystem with proper and enough services through its capacities to perform its functions under different changing conditions. Soil quality also is the capacity of soil to fulfill its functions including management of ecosystem boundaries, sustaining the plant and animal productivity, maintaining or enhancing both water and air quality and supporting both human health and its habitation. Therefore, the quality of soils reflects how soils perform their functions including the maintenance of soil biodiversity and its productivity, filtering and buffering, partitioning water and solute flow, cycling of nutrients and providing support for plants and other structures (Shivlata and Satyanarayana 2017).

The health or quality of soils may be declined under many soil practices such as excessive use of chemicals like fertilizers or pesticides, intensive cropping system, pollution with different sources and its forms, reducing the recycling of organics and other waste generated from the farm, etc (Bharti et al. 2017). It is worth to mention that, soil quality has several indicators including physical, chemical and biological indicators. Regarding soil physical indicators, they are very important in determining the characteristics and micro-habitat utilization. These soil physical indicators for example include permeability, porosity, texture, water-holding capacity and temperature. Therefore, all of these aspects of the soil environment are important for crop production and thereby for human health through the supply of nutritious and adequate crop products (Shivlata and Satyanarayana 2017).

Concerning soil health and its biology towards sustainable agriculture, many studies have been published focusing on this relationship including the following issues:

(1) Role of soils in biodegradation of plastics in frame the sustainable modern agriculture.
(2) Role of soil health, soil biology, soilborne diseases in sustainable agriculture (e.g., Hüberli 2017).
(3) Different adaptive practices in soil management such as digital soil mapping for best management of soil resources, adaptive remote sensing approach in supporting of precision agriculture, site-specific nutrient management approach in maintaining soil health, technological conservation of soil resources using sustainable soil health management, changes in soil–plant–microbes interactions under climatic change conditions, etc. (Rakshit et al. 2017).
(4) The role of biofertilizers microbial inoculants in soil quality and plant health (e.g., Alori et al. 2017).
(5) Sustainable management of salt-affected soils using nanotechnology and more (e.g., Sharma and Singh 2017).
(6) Different approaches for sustainable agriculture through plant-microbe interaction (e.g., Choudhary et al. 2016).
(7) Study of soil pollution as an emerging threat to agriculture and its effects on soil and crop quality as well as soil fertility (Saha et al. 2017).

In Egypt, almost all soil studies have been involved the soil health and different management practices. These studies generally include physical, chemical and biological soil properties and different effects of soil management practices on these soil properties as soil quality indicators. These investigations included sustainable agriculture and climate changes in Egypt (e.g., El-Ramady et al. 2013; Froehlich and Al-Saidi 2017), sustainable irrigation management and conservation of soil and water (e.g., Elbana et al. 2017a, b), sustainable land use management using GIS and remote sensing (e.g., Kamel and Abu El Ella 2016; Elbasiouny et al. 2017a), effects of irrigated agriculture in Egypt on soils in past, present and future (e.g., Satoh and Aboulroos 2017), the sustainable management for the soils of Nile Delta as well as the remediation of agricultural drainage water for sustainable reuse (e.g., Fleifle and Allam 2017), management of salt-affected soils in the Nile Delta (Mohamed 2017b), the expected degradation of some Egyptian soils after the construction of Aswan High Dam (Khalifa and Moussa 2017), etc. It could mention some examples for some investigations carried out under Egyptian conditions:

(1) The effects of recommended soil fertilization as an important factor controlling the biology of soil and its health such as the study of recommended soil NPK fertilization and selenite foliar application (up to 20 mg kg^{-1}) on spinach plants significantly increased the yield, Se accumulation in leaves as well as the activity of nitrate reductase enzyme under soil salinity and acidity or pH to 3.89 dS m^{-1} and 7.67, respectively (Fouda 2016).
(2) Distribution of nutrients in soils is totally controlled by soil properties and its climatic and pedogenic properties like a clear relation in field study in different Egyptian areas (i.e., Matrouh, El-Arish, El-Hesynia Plain, El-Tina Plain, El-Mansoura, El-Gabal EL-Asfar, El-Fayoum and Toshki). The obtained correlation between indigenous soil parameters and these nutrients (e.g., selenium) showed that soil parameters including clay content, cation exchange capacity, soil pH and sulfur content are the domain factors (Abd El-Razik et al. 2013; Table 10.1).
(3) A comparison between organic and conventional farming systems has been evaluated using soil quality indicators (e.g., chemical and fertility). It was found that, available N, P, K, Fe, and Mn, soil organic carbon, total N, and CEC were significantly higher in the organic farming system comparing with the conventional system. The results also showed that, the soil nutritional status in organic farming was significantly better than the conventional system regardless the periods of farming practice (Ahmed et al. 2015).

Table 10.1 Soil characterization of surface layer (0–30 cm) of coastal, middle Delta and new reclaimed areas of Egypt as well as their parent materials for different lacations

Parent material	Soil texture	pH (1:2.5)	EC (dS m^{-1})	$CaCO_3$ (%)	OM (g kg^{-1})	CEC (cmol kg^{-1})	Available Se (mg kg^{-1})
Matrouh (Mediterranean coastal region)							
Alluvial deposits	Silty loam	7.69	10.33	78.94	3.5	15.67	0.13
El-Arish (Mediterranean coastal region)							
Alluvial deposits	Silty loam	7.87	2.71	4.36	3.9	8.80	0.13
El-Mansoura (a middle Delta region)							
Alluvial deposits	Clay loam	7.78	2.17	4.33	3.7	8.08	0.27
Tina Plain (a new reclaimed area)							
Fluvio marine deposits	Silty loam	7.89	10.25	76.95	3.7	17.77	0.36
El-Hesynia Plain (a new reclaimed area)							
Alluvial deposits	Silty loam	7.88	2.37	4.53	3.7	8.08	0.36
Qaroun Lake (El-Fayoum region)							
Fluvio lacustrine deposits	Clay loam	7.99	27.27	20.08	10.9	23.82	0.49
Kom Oshim (El-Fayoum region)							
Fluvio lacustrine deposits	Sandy clay loam	7.86	20.80	5.70	15.4	23.7	0.32
Toshki region (uncultivated soils)							
Alluvial deposits	Sandy loam	7.85	15.32	1.94	2.4	11.76	0.19
Toshki region (cultivated soils)							
Alluvial deposits	Sandy loam	7.78	1.03	3.55	3.2	4.33	0.41
EL-Gabal EL-Asfar area (normal soil without irrigation with sewage effluent)							
Aeolian deposits	Sandy	8.77	3.31	3.89	0.9	2.41	Nil
EL-Gabal EL-Asfar area (Soil irrigated for 15 years with sewage effluent)							
Aeolian deposits	Sandy clay loam	6.58	9.17	3.00	37.9	4.85	Nil
EL-Gabal EL-Asfar area (Soil irrigated for 100 years with sewage effluent)							
Aeolian deposits	Sandy clay loam	6.08	14.16	1.57	61.5	8.76	0.03

Adapted from Abd El-Razik et al. (2013)
Abbreviations Soil EC = soil salinity, CEC = cation exchange capacity, OM = organic matter

(4) Study the effects of silicon foliar application and biofertilization (*Azotobacter chroococcum* and *Bacillus megatherium*) on productivity of some crops like sunflower (*Helianthus annuus* L.) under New Valley conditions. It was found that, the microbial activity in rhizosphere and some enzymatic activities (dehydrogenase, nitrogenase and phosphatase) have a positive response in all treatments comparing with control (Abd El-Gwad and Salem 2013).

(5) Survey of several locations, monitoring of soil fertility, land use change and sprawl urban using Remote Sensing and GIS using and measuring the change in soil parameters (e.g., Elbasiouny et al. 2017b; Ibrahim et al. 2017).

(6) Effects of different organic acids (i.e., tartaric, salicylic, oxalic, humic and fulvic acid) on nutrients availability from natural alternative fertilizers (e.g., rock phosphate and feldspar). The results confirmed that rock phosphate and feldspar as alternative fertilizers under acidulation with different previous acids could enhance release of available nutrients (e.g., N, P and K) improving some soil chemical properties reflecting on the soil productivity (Seddik et al. 2016).

(7) The soil biology and its health in Egypt suffer from different sources of pollution like pollutants resulting from automobile exhausts. These exhausts are source for airborne pollutants, which precipitate on soils

surrounding highways (e.g., Cairo-Alexandria agricultural highway) causing serious ecological hazards). The results revealed that the investigated soils were contaminated with some potentially toxic elements such as Pb, Ni and Cd (Hashim et al. 2017).

(8) Abiotic stresses including salinity of soil or water and/or drought have a serious impact on soil biology in Egypt. Several studies confirmed that these stresses represent a real threat for the health of Egyptian soils (e.g., Abdel-Aziz and Sadik 2017; Nossier et al. 2017; Mousa 2017).

(9) Soil erosion could consider a serious problem in Egypt including different kinds of erosion (e.g., chemical, water and wind erosion) as presented in many studies (e.g., El-Nady and Shoman 2017).

(10) The competition between energy and food crops under conventional and stress conditions will influence and effect on the health of Egyptian soils (e.g., Hokam and Abo El-Soud 2017).

(11) The using of municipal wastewater in agriculture may change some soil properties or effects on soil health playing an important role in the availability of nutrients present in the applied wastewater (e.g., Elgharably and Mohamed 2016; Farrag et al. 2017; Elbana et al. 2017a, b).

(12) The intensive use of agro-chemicals including pesticides (e.g., El-Kady et al. 2017), mineral fertilizers (e.g., Negm and Eltarabily 2017) and nanomaterials (e.g., Belal and El-Ramady 2016; Shalaby et al. 2016) may influence on soil biology and aquatic ecosystem in Egypt.

(13) Impact of soil or land degradation on soil biology has been involved in many investigations in Egypt. Land degradation could be resulted from many processes such as seawater rise, soil sodicity, salinization, deterioration of soil structure, urban sprawl, land compaction and nutrient depletion as well as water logging (e.g., Rashed 2016; Kotb et al. 2017; Mohamed 2017a).

(14) Modern agricultural management should be handled to overcome the conservation the biology of soil such as precision farming (e.g., Saleh et al. 2017b), nanotechnology (e.g., El-Ramady et al. 2017a, b) and magnetization of irrigated water (e.g., Amer 2016; Alsaeedi et al. 2017; Fanous et al. 2017; Abd-Elrahman and Shalaby 2017).

(15) The main cause for depleting natural resources in arid lands is desertification. This desertification process also is considered the major environmental threat affecting the soil biology of about 40% of the world dry lands. Several applications of satellite or GIS and remote sensing have been documented in monitoring desertification in arid lands (e.g., Mohamed 2013; Saleh et al. 2017a).

It could be concluded that, there is no sustainable agriculture without conservation of soil health and its biology. The health of soil and its biology mainly depends on soil characterization as well as climatic and environmental conditions. The biology and health of Egyptian soils need more concern and further studies and these studies should be linked to the human health.

10.5 Conclusion

No doubt that, formidable challenges face the universe regarding the security and safety of foods including the scarcity of freshwater, the degradation of lands and climate changes. These challenges or problems will be aggravated in the future according to the available evidences. The current food production and its distribution systems could not be able to ensure the global food and its nutritional requirements. Therefore, the agricultural production must be increased and should be produced from existing agricultural lands as well as the marginal or abandoned lands using non-conventional and stable crops. Indeed, the properties of soil (i.e., physical, chemical and biological ones) are main factor controlling the productivity of these soils. Thus, soil biology and its fertility have a close relationship with nutrition of plants and humans. Therefore, new challenges should be considered regarding the feeding of soils in order to feed the humans and soil quality and its biology will be included within these challenges towards sustainable agriculture.

Acknowledgements The research was financed by the Higher Education Institutional Excellence Program of the Ministry of Human Capacities in Hungary, within the framework of the biotechnology thematic program of the University of Debrecen".

References

Abd El-Gwad AM, EMM Salem (2013). Effect of biofertilization and silicon foliar application on productivity of sunflower (*Helianthus annuus* L.) under new valley conditions. Egypt J Soil Sci 53(4): 509–536

Abd El-Razik EM, Soaud AA, ElKilani R, Habashy NR (2013) Distribution of selenium in some Egyptian soils. Egypt J Soil Sci 53 (4):475–488

Abd-Elrahman SH, Shalaby OA (2017) The response of wheat plants to magnetized water irrigation under Egyptian soil conditions. Egypt. J. Soil Sci. (in press). https://doi.org/10.21608/EJSS.2017.891.1122

Abdel-Aziz A, Sadik A (2017) Effect of water and salt stresses on productivity of cantaloupe in Ismailia soil. Egypt J Soil Sci 57 (2):201–214. https://doi.org/10.21608/ejss.2017.1529

Ahmed MM, Daif MA, Khater HA, Abdul Aleem M (2015) Chemical and fertility indicators of soil quality in organic and conventional farming systems in Egypt. Egypt J Soil Sci 55(2):215–230

Alori ET, Dare MO, Babalola OO (2017). Microbial Inoculants for soil quality and plant health. In: Lichtfouse E (ed) Sustainable agriculture reviews, sustainable agriculture reviews, vol 22. Springer International Publishing, Switzerland, https://doi.org/10.1007/978-3-319-48006-0_9, pp 281–307

Alsaeedi A, El-Ramady H, Alshaal TA, Almohsen M (2017) Enhancing seed germination and seedlings development of common bean (*Phaseolus vulgaris*) by SiO_2 nanoparticles. Egypt J Soil Sci (in press). https://doi.org/10.21608/EJSS.2017.891.1098

Amer MM (2016) Effect of biochar, compost tea and magnetic iron ore application on some soil properties and productivity of some field crops under saline soils conditions at north nile delta. Egypt J Soil Sci 56(1):169–186

Belal E, El-Ramady H (2016) Nanoparticles in water, soils and agriculture. In: Ranjan et al S (eds) Nanoscience in food and agriculture 2, sustainable agriculture reviews, vol 21, Springer International Publishing Switzerland, https://doi.org/10.1007/978-3-319-39306-3_10

Bhaduri D, Purakayastha TJ, Patra AK, Singh M, Wilson BR (2017) Biological indicators of soil quality in a long-term rice–wheat system on the Indo-Gangetic plain: combined effect of tillage–water–nutrient management. Environ Earth Sci 76:202. https://doi.org/10.1007/s12665-017-6513-0

Bharti VS, ML Dotaniya, SP Shukla, VK Yadav (2017) Managing soil fertility through microbes: prospects, challenges and future strategies. In Singh JS, Seneviratne G (eds) Agro-Environmental sustainability, Springer International Publishing AG, https://doi.org/10.1007/978-3-319-49724-2_5, pp 81–111

Bogan D, Meile L, El Bastawisy A, Yousef HF, Zekri AN, Bahnassy AA, ElShamy WM (2017) The role of BRCA1-IRIS in the development and progression of triple negative breast cancers in Egypt: possible link to disease early lesion. BMC Cancer 17:329. https://doi.org/10.1186/s12885-017-3283-8

Brevik EC, Sauer TJ (2015) The past, present, and future of soils and human health studies. Soil 1:35–46. https://doi.org/10.5194/soil-1-35-2015

Brevik EC, Steffan JJ, Burgess LC, Cerdà A (2017) Links between soil security and the influence of soil on human health. In: Field et al (eds) Global soil security, progress in soil science, Springer International Publishing, Switzerland, https://doi.org/10.1007/978-3-319-43394-3_24, pp 261–274

Cakmak I, McLaughlin MJ, White P (2017) Zinc for better crop production and human health. Plant Soil 411:1–4. https://doi.org/10.1007/s11104-016-3166-9

Choudhary DK, Varma A, Tuteja N (2016) Plant-microbe interaction: an approach to sustainable agriculture, Springer Nature Singapore Pte Ltd, https://doi.org/10.1007/978-981-10-2854-0

Davison KM (2017). Mineral nutrients: from macro-level to ultra trace. In: Temple et al NJ (eds) Nutrition guide for physicians and related healthcare professionals, nutrition and health, Springer International Publishing AG, https://doi.org/10.1007/978-3-319-49929-1_26, pp 261–272

de Pee S, Taren D, Bloem MW (2017) Nutrition and health in a developing world. Nutrition and Health Series, Springer Science + Business Media New York, https://doi.org/10.1007/978-3-319-43739-2

dos Reis AR, El-Ramady H, Santos EF, Gratão PL, Schomburg L (2017) Overview of selenium deficiency and toxicity worldwide: affected areas, selenium-related health issues, and case studies. In: Pilon-Smits et al (eds) Selenium in plants, Plant ecophysiology, vol 11. Springer International Publishing AG, https://doi.org/10.1007/978-3-319-56249-0_13, pp 209–230

Eisenhauer N, Antunes PM, Bennett AE, Birkhofer K, Bissett A, Bowker MA, Caruso T, Chen B, Coleman DC, de Boer W, de Ruiter P, DeLuca TH, Frati F, Griffiths BS, Hart MM, Hättenschwiler S, Haimi J, Heethoff M, Kaneko N, Kelly LC, Leinaas HP, Lindo Z, Macdonald C, Rillig MC, Ruess L, Scheu S, Schmidt O, Seastedt TR, van Straalen NM, Tiunov AV, Zimmer M, Powell JR (2017) Priorities for research in soil ecology. Pedobiologia. J Soil Ecol 63:1–7. https://doi.org/10.1016/j.pedobi.2017.05.003

Elbana TA, N Bakr, M Elbana (2017b). Reuse of treated wastewater in Egypt: challenges and opportunities. In Negm AM (ed) Unconventional water resources and agriculture in Egypt, Hdb Env Chem, Springer International Publishing AG. https://doi.org/10.1007/698_2017_46

Elbana TA, Bakr N, George B, Elbana M (2017a) Assessment of marginal quality water for sustainable irrigation management: case study of Bahr El-Baqar area Egypt. Water Air Soil Pollut 228:214. https://doi.org/10.1007/s11270-017-3397-2

Elbasiouny H, Abowaly M, Gad A, Abu_Alkheir A, Elbehiry F (2017a) Restoration and sequestration of carbon and nitrogen in the degraded northern coastal area in Nile Delta, Egypt for climate change mitigation. J Coast Conserv 21:105–114. https://doi.org/10.1007/s11852-016-0475-3

Elbasiouny H, Elbehiry F, Abowaly M (2017b) Soil quality indices-special focus on salt-affected soil: review and case study in northern Egypt. Env. Biodiv. Soil Security 1:85–100

Elgharably A, Mohamed HM (2016) Heavy metals uptake by wheat, bean and onion and characterization of microorganisms in a long-term sewage wastewater treated soil. Egypt J Soil Sci 56 (4):605–620

El-Kady AA, Wade TL, Sweet ST, Sericano JL (2017) Distribution and residue profile of organochlorine pesticides and polychlorinated biphenyls in sediment and fish of Lake Manzala Egypt. Environ Sci Pollut Res 24:10301–10312. https://doi.org/10.1007/s11356-017-8714-1

El-Nady MA, Shoman MM (2017) Assessment of soil erosion risk in the basin of wadi maged in northern west coast of Egypt using corine model and gis techniques. Egypt J Soil Sci 57(1):31–45. https://doi.org/10.21608/EJSS.2017.3587

El-Ramady H, Alshaal T, El-Henawy A, Abdalla N, Taha HS, Elmahrouk M, Shalaby T, Elsakhawy T, Omara AE-D, El-Marsafawy S, Elhawat N, Shehata S, Domokos-Szabolcsy É (2017a) Environmental nanoremediation under changing climate. Env Biodiv Soil Secur 1:109–128. https://doi.org/10.21608/jenvbs.2017.1550.1009

El-Ramady H, Alshaal T, Abowaly M, Abdalla N, Taha HS, Al-Saeedi AH, Shalaby T, Amer M, Fári M, Domokos-Szabolcsy É, Sztrik A, Prokisch J, Selmar D, Pilon-Smits EAH, Pilon M (2017a) Nanoremediation for sustainable crop production. In Ranjan et al S (eds) Nanoscience in food and agriculture, sustainable agriculture reviews, vol 26. pp 335–363, Springer International Publishing AG. https://doi.org/10.1007/978-3-319-58496-6_12

El-Ramady HR, El-Marsafawy SM, Lewis LN (2013) Sustainable agriculture and climate changes in Egypt. In Lichtfouse E (ed) Sustainable agriculture reviews, sustainable agriculture reviews, vol 12. Springer Science + Business Media Dordrecht, pp 41–95, https://doi.org/10.1007/978-94-007-5961-9_2

Elshamy K (2016). Challenges and future trends for cancer care in Egypt. In: Silbermann M (ed) Cancer care in countries and societies in transition, Springer International Publishing Switzerland, pp 117–146, https://doi.org/10.1007/978-3-319-22912-6_9

Etewa SE, Abdel-Rahman SA, Abd El-Aal NF, Fathy GM, El-Shafey MA, Ewis AMG (2016) Geohelminths distribution as affected by soil properties, physicochemical factors and climate in Sharkyia governorate Egypt. J Parasit Dis (Apr-June 2016) 40(2):496–504. https://doi.org/10.1007/s12639-014-0532-5

Fanous NE, Mohamed AA, Shaban KA (2017) Effect of magnetic treatment for irrigation ground water on soil salinity, nutrients, water productivity and yield fruit trees at sandy soil. Egypt J Soil Sci 57(1):113–123. https://doi.org/10.21608/EJSS.2017.1528

Farghly AM, Mohamed SMA, Abdel-Rahman SA, Mohammed FE, El-Bahaie ES, El-Shafey MA (2016) The relation between the prevalence of soil transmitted parasites in the soil and among school children in Zagazig district, Sharkyia Governorate, Egypt. J Parasit Dis (July-Sept 2016) 40(3):1021–1029. https://doi.org/10.1007/s12639-014-0627-z

Farrag HM, El-Desoky MA, Basha AAA, Roshdi NMK (2017) Long-term impact of treated sewage water on some soil properties and nutrients status in luxor governorate Egypt. Egypt J Soil Sci 57 (1):1–14. https://doi.org/10.21608/ejss.2017.3358

Field DJ, Morgan CLS, McBratney AB (2017) Global soil security: progress in soil science series, Springer International Publishing Switzerland, https://doi.org/10.1007/978-3-319-43394-3

Finkel OM, Castrillo G, Paredes SH, González IS, Dangl JL (2017) Understanding and exploiting plant beneficial microbes. Curr Opin Plant Biol 38:155–163

Fleifle A, Allam A (2017) Remediation of agricultural drainage water for sustainable reuse. In: Negm AM (ed) The nile delta, Hdb Env Chem, Springer International Publishing AG, pp 297–324, https://doi.org/10.1007/698_2016_119

Fouda KF (2016) Quality parameter and chemical composition of spinash plant as affected by mineral fertilization and selenite foliar application. Egypt J Soil Sci 56(1):149–167

Froehlich P, Al-Saidi M (2017) Community-based adaptation to climate change in Egypt—status quo and future policies. In: Leal Filho W (ed) Climate change research at universities, Springer International Publishing AG, pp 235–250, https://doi.org/10.1007/978-3-319-58214-6_14

Gupta UC, Gupta SC (2014) Sources and deficiency diseases of mineral nutrients in human health and nutrition: a review. Pedosphere 24:13–38

Gyóry H (2016) Medicine in ancient Egypt. In: Selin H (ed) Encyclopaedia of the history of science, technology, and medicine in non-western cultures. Springer Science + Business Media Dordrecht, pp 2957–2970, https://doi.org/10.1007/978-94-007-7747-7

Hamza H, Leifels M, Wilhelm M, Hamza IA (2017a) Relative abundance of human bocaviruses in urban sewage in Greater Cairo, Egypt. Food Environ Virol, 1–10. https://doi.org/10.1007/s12560-017-9287-3

Hamza H, Abd-Elshafy DN, Fayed SA, Bahgat MM, El-Esnawy NA, Abdel-Mobdy E (2017b) Detection and characterization of hepatitis a virus circulating in Egypt. Arch Virol 162:1921–1931. https://doi.org/10.1007/s00705-017-3294-4

Hashim T, Abbas HH, Farid IM, El-Husseiny O, Abbas MH (2017) Accumulation of some heavy metals in plants and soils adjacent to cairo—alexandria agricultural highway. Egypt J Soil Sci 57(2):215–232. https://doi.org/10.21608/EJSS.2016.281.1047

Hokam EM, Abo El-Soud IH (2017). Evaluation of growth parameters for *Jatropha curcas* L. (Biofuel Plant) under salinity and water stress using tensiometer for irrigation guiding. Egypt J Soil Sci 57 (1):47–60. https://doi.org/10.21608/ejss.2017.3589

Hüberli D (2017) Soil health, soil biology, soilborne diseases and sustainable agriculture: a guide Graham. Australasian Plant Pathol. https://doi.org/10.1007/s13313-017-0493-0

Ibrahim MM, Wahab MA, Ali RR, EL-Baroudy AA, Hussein AA (2017) Physiographic and soil mapping of Qena and El-Qarn wadis by using remote sensing and GIS. Egypt J Soil Sci 57(2):137–141

Jain KK (2017) the handbook of nanomedicine, Springer Science + Business Media LLC, https://doi.org/10.1007/978-1-4939-6966-1

Janick J (2016) Origins of agriculture in Egypt. In: Selin H (ed) Encyclopaedia of the history of science, technology, and medicine in non-western cultures, Springer Science + Business Media Dordrecht, https://doi.org/10.1007/978-94-007-7747-7, pp 3460–3478

Kamel M, Abu El Ella EM (2016) Integration of Remote Sensing & GIS to Manage the Sustainable Development in the Nile Valley desert fringes of assiut-sohag governorates, upper Egypt. J Indian Soc Remote Sens 44(5):759–774. https://doi.org/10.1007/s12524-015-0529-2

Khalifa HE, Moussa HA (2017) Soil and agriculture after the Aswan High Dam. In: Satoh M, Aboulroos S (eds) Irrigated agriculture in Egypt, Springer International Publishing Switzerland, https://doi.org/10.1007/978-3-319-30216-4_5, pp 81–124

Kim SC, Lim KJ, Yang JE (2017) The measurement of soil security in terms of human health: examples and ideas. In: Field et al DJ (eds) Global soil security, progress in soil science, Springer International Publishing Switzerland, https://doi.org/10.1007/978-3-319-43394-3_26, pp 297–303

Kotb MM, Ali RR, El Semary MA (2017) Use of remote sensing and GIS for land degradation assessment of qarun lake coastal area, el-fayoum, Egypt. In: Ouessar et al M (eds) Water and land security in drylands, Springer International Publishing AG, https://doi.org/10.1007/978-3-319-54021-4_22, pp 251–267

Liu Y, Zhang C, Zhao Y, Sun S, Liu Z (2017) Effects of growing seasons and genotypes on the accumulation of cadmium and mineral nutrients in rice grown in cadmium contaminated soil. Sci Total Environ 579:1282–1288

Marles RJ (2017) Mineral nutrient composition of vegetables, fruits and grains: the context of reports of apparent historical declines. J Food Compos Anal 56:93–103

Mohamed ES (2013) Spatial assessment of desertification in north Sinai using modified MEDLAUS model. Arab J Geosci 6:4647–4659. https://doi.org/10.1007/s12517-012-0723-2

Mohamed NN (2017a). Land degradation in the Nile Delta. In: Negm AM (ed) The Nile Delta, Hdb Env Chem, Springer International Publishing AG, https://doi.org/10.1007/698_2016_101, pp 235–264

Mohamed NN (2017b) Management of salt-affected soils in the Nile Delta. In Negm AM (ed) The Nile Delta, Hdb Env Chem, Springer International Publishing AG, https://doi.org/10.1007/698_2016_102, pp 265–295

Mousa AA (2017) Effect of using some soil conditioners on salt affected soil properties and its productivity at El-tina plain area, north Sinai. Egypt Egypt J Soil Sci 57(1):101–111. https://doi.org/10.21608/EJSS.2017.1526

Negm AM, Eltarabily MGA (2017) Modeling of fertilizer transport through soil, case study: nile delta. In Negm AM (ed) The Nile Delta, vol 55, Springer International Publishing Switzerland, https://doi.org/10.1007/698_2016_88, pp 121–157

Nossier MI, Gawish SM, Taha TA, Mubarak M (2017) Response of wheat plants to application of selenium and humic acid under salt stress conditions. Egypt J Soil Sci 57(2):175–187. https://doi.org/10.21608/EJSS.2017.3715

Palzer S (2017) Technological solutions for reducing impact and content of health sensitive nutrients in food. Trends Food Sci Technol 62:170–176

Pearce EN (2017). Iodine deficiency disorders and their elimination. Springer International Publishing AG, https://doi.org/10.1007/978-3-319-49505-7

Pilon-Smits EAH, Winkel LHE, Lin ZQ (2017) Selenium in plants: molecular, physiological, ecological and evolutionary aspects. In: Plant ecophysiology series, vol 11, Springer International Publishing AG, https://doi.org/10.1007/978-3-319-56249-0

Plekhanova IO, Zarubina AP, Plekhanov SE (2017) Ecotoxicological assessment of nickel pollution of soil and water environments

adjacent to soddy-podzolic soil. Moscow Univ Soil Sci Bulletin 72 (2):71–77. https://doi.org/10.3103/S0147687417020065
Rakshit A, Abhilash PC, Singh HB, Ghosh S (2017) Adaptive soil management: from theory to practices, Springer Nature Singapore Pte Ltd, https://doi.org/10.1007/978-981-10-3638-5
Rashed HSA (2016) Change detection in land degradation and environmental hazards sensitivity in some soils of siwa oasis. Egypt J Soil Sci 56(3):433–451
Saha JK, Selladurai R, Coumar MV, Dotaniya ML, Kundu S, Patra AK (2017) Soil pollution—an emerging threat to agriculture. Springer Nature Singapore Pte Ltd, https://doi.org/10.1007/978-981-10-4274-4
Saleh A, Belal A, Jalhoum M (2017a) Quantitative assessment of environmental sensitivity to desertification in Sidi Abdel-Rahman area, Northern West Coast of Egypt. Egypt J Soil Sci (in press)
Saleh AM, Belal AB, Mohamed ES (2017b) Mapping of soil salinity using electromagnetic induction: a case study of east Nile Delta. Egypt Egypt J Soil Sci 57(2):167–174. https://doi.org/10.21608/EJSS.2017.3705
Satoh M, Aboulroos S (2017) Irrigated agriculture in Egypt: past, present and future. Springer International Publishing Switzerland, https://doi.org/10.1007/978-3-319-30216-4
Seddik WMA, Hassan WZ, Zein El-Abdeen HA (2016) Influence of organic acids on nutrients availability from natural alternative fertilizers. Egypt J Soil Sci 56(3):503–517
Selin H (2016). Encyclopaedia of the history of science, technology, and medicine in non-western cultures. Springer Science + Business Media Dordrecht, https://doi.org/10.1007/978-94-007-7747-7
Selinus O (2013) Essentials of medical geology. Revised Edition. Springer Science + Business Media Dordrecht, https://doi.org/10.1007/978-94-007-4375-5
Shalaby T, Bayoumi Y, Abdalla N, Taha H, Alshaal T, Shehata S, Amer M, Domokos-Szabolcsy É, El-Ramady H (2016) Nanoparticles, soils, plants and sustainable agriculture. In: Ranjan S et al (eds) Nanoscience in food and agriculture 1, Sustainable Agriculture Reviews, vol 20, Springer International Publishing Switzerland, https://doi.org/10.1007/978-3-319-39303-2_10
Shao H, Lu H, Xu G, Marian B (2017) Integration into plant biology and soil science has provided insights into the total environment. Sci Total Environ 579:928–929
Sharma DK, Singh A (2017) Current trends and emerging challenges in sustainable management of salt-affected soils: a critical appraisal. In: Arora S et al (eds) Bioremediation of salt affected soils: an indian perspective, Springer International Publishing AG, https://doi.org/10.1007/978-3-319-48257-6_1, pp 1–40
Shivlata L, Satyanarayana T (2017) Actinobacteria in agricultural and environmental sustainability. In Singh JS, Seneviratne G (eds) Agro-environmental sustainability, Springer International Publishing AG, https://doi.org/10.1007/978-3-319-49724-2_9, pp 173–218
Singh R, Kumar M, Mittal A, Mehta PK (2017) Microbial metabolites in nutrition, healthcare and agriculture. 3. Biotech 7:15. https://doi.org/10.1007/s13205-016-0586-4
SSSA, Soil Science Society of America (2017). When soil won the nobel prize in medicine. Accessed on https://www.soils.org/files/sssa/.../when-soil-won-the-nobel-prize-in-medicine-kb.pptx
Vimal SR, Singh JS, Arora NK, Singh S (2017) Soil-plant-microbe interactions in stressed agriculture management: a review. Pedosphere 27(2):177–192
VoPham T, Bertrand KA, Hart JE, Laden F, Brooks MM, Yuan J-M, Talbott EO, Ruddell D, Chang CCH, Weissfeld JL (2017) Pesticide exposure and liver cancer: a review. Cancer Causes Control 28:177–190. https://doi.org/10.1007/s10552-017-0854-6
Wang S, Shang Z, Wang P, Wei G, Dany Y (2017) Study on radioactive contaminated soil remediation technologies and selection principles. In: Jiang H (ed) Proceedings of the 20th pacific basin nuclear conference, Springer Science + Business Media Singapore, https://doi.org/10.1007/978-981-10-2314-9_49, pp 547–554
Xiao L, Sun Q, Yuan H, Lian B (2017) A practical soil management to improve soil quality by applying mineral organic fertilizer. Acta Geochim. https://doi.org/10.1007/s11631-017-0139-5
Youssef NFA, El Kassas M, Farag A, Shepherd A (2017) Health-related quality of Life in patients with chronic hepatitis C receiving Sofosbuvir-based treatment, with and without Interferon: a prospective observational study in Egypt. BMC Gastroenterol 17:18. https://doi.org/10.1186/s12876-017-0581-1
Zaki SM, Denning DW (2017) Serious fungal infections in Egypt. Eur J Clin Microbiol Infect Dis 36:971–974. https://doi.org/10.1007/s10096-017-2929-4
Zhang H, Huang B, Dong L, Hu W, Akhtar MS, Qu M (2017) Accumulation, sources and health risks of trace metals in elevated geochemical background soils used for greenhouse vegetable production in southwestern China. Ecotoxicol Environ Saf 137:233–239
Zhao G (2017). Mineral containing proteins: roles in nutrition, Springer Nature Singapore Pte Ltd, https://doi.org/10.1007/978-981-10-3596-8

11 Soil Chemical Pollution and Sustainable Agriculture

Tamer Elbana, Hesham M. Gaber, and Fawzy M. Kishk

Abstract

Sources of the agricultural soil pollution in Egypt include the utilization of agrochemicals, the reuse of agricultural drainage water, the recycling of the partially treated wastewater, the contaminated air, and the improper disposing of solid wastes. These various sources pose a real threat to the sustainable agricultural system and human health. In this chapter, we reviewed the sources and fate of the organic and inorganic pollutants in Egyptian soils. In brief, we explained sorption, mobility, and degradation properties of contaminants and their rules for realizing proper management of contaminated soils. Also, we reviewed the status of soils contamination with heavy metals and pesticides. Notably, the fate of the contaminants in soils depends on its characteristics and soil properties as well as the surrounding environmental condition. Therefore, the management of contaminated soils is a site-specific process. Such management can be ensured through prevention, remediation, and adaptation approaches. The literature review revealed the need to develop and implement a national plan for preventing soil pollution in the newly reclaimed areas. Besides, the necessity for providing a well-defined Egyptian guideline for managing contaminated sites in old agricultural lands.

Keyword

Pollution • Heavy metals • Pesticides • Sorption Degradation • Mobility

T. Elbana (✉)
Soils and Water Use Department, National Research Centre (NRC), Cairo, Egypt
e-mail: tamerelbana@yahoo.com

H. M. Gaber · F. M. Kishk
Soil and Water Sciences Department, Faculty of Agriculture, Alexandria University, Alexandria, Egypt

H. El-Ramady et al. (eds.), *The Soils of Egypt*, World Soils Book Series,
https://doi.org/10.1007/978-3-319-95516-2_11

11.1 Introduction

Pollution problem is a global issue that is directly related to human health and environmental deterioration. The recently published environmental reports confirmed the relation between human death and pollution. The pollution-related diseases caused 16% of the worldwide deaths during 2015 (Landrigan et al. 2017). The third sustainable development goal (SDG) of the United Nations is related to reducing the pollution in order to achieve good health and enhance community well-being. Generally, the term "environmental pollution" indicates the occurrence of deterioration in a natural resource(s) that is negatively affecting the living organisms.

Soil pollution is defined, according to encyclopedia of public health, as "*physical, chemical, biological, or radiological modification of the surface layer of the earth's crust by accumulation of a large quantity of natural materials or occurrence of new synthetic materials that disturb the composition of the soil, influence the natural balance of the ecological system, and disable the purification process (self-cleaning) of the soil*" (Backović 2008). However, this definition emphasizes the forms and the consequences of soil pollution; it provides a qualitative description for the cause of pollution as a substantial buildup of a pollutant material in soil. For instance, disposing of large amount of plastic and glass wastes can be considered as a physical modification that could prevent plant growth on the polluted soil. The physical modification is visible, whereas chemical or biological alteration such as the impact of disposing wastewater into soil is invisible pollution. Soil contamination is another term that is applied to express the occurrence of soil deterioration due to the accumulation of a specific substance in soil at a higher level than its initial background value. However, soil contamination is commonly applied as an interchangeable term for soil pollution.

Obviously, pollution did not respect geographical boundaries. Pollutants can transfer from one soil to another through wind blows or water runoff among other carriers. Soil pollution is associated mainly with human activities such as urbanization, agricultural, and industrial developments. For example, in the adjacent surrounding urban areas, continuous disposing of untreated wastewater into the soil is an obvious example of the accumulation of toxic chemical and biological contaminants in soil (Elbana et al. 2017a). In the agricultural sector, the use of fertilizers causes the accumulation of some potentially toxic elements that adversely affect plants and soil biota in the long term. Polluted air, wastewater, and the applied agrochemicals in agriculture can be considered as sources of soil pollution. These different sources can carry numerous organic, inorganic, and biological pollutants. The fate of different pollutants in a soil depends on the physical, chemical, and biological characteristics of the soil as well as pollutant characteristics. Reactivity and mobility of chemical pollutants are controlled by many factors including soil reaction (pH), soil organic matter (OM), ion exchange capacity, and calcium carbonate contents, oxides and hydroxides among other (Han 2007; Shaheen et al. 2009; Kabata-Pendias 2011; Kodešová et al. 2011). Consequentially, the different soils exhibit varied potentiality to retain or release certain contaminant. Moreover, the retention and release of a specific contaminant are regulated by adding another reactive chemical into the soil. Elbana and Selim (2013) showed that the release of Cd was influenced by introducing Cu into a soil column, whereas the release of Pb was not affected by introducing Cd or Cu into a soil that has 2.8% $CaCO_3$.

Vulnerability of surface and ground waters for contamination is directly related to polluted soil. This vulnerability depends on soil properties that control contaminant sorption and mobility in soil. The contaminants are generally more mobile in sandy soils that have low $CaCO_3$, OM, and amorphous oxides contents than its mobility in the clay soils that usually exhibit higher CEC. Under intensive agricultural system, Muhammetoğlu et al. (2002) explained that in addition to soil properties, the depth of the groundwater and the quality of irrigation water are key factors for groundwater vulnerability. Ceplecha et al. (2004) applied vulnerability matrix as a field assessment tool to evaluate the groundwater vulnerability to nitrate contamination. They considered soil texture, irrigation efficiency, nitrogen application rate/timing, and best management practices for quantifying vulnerability rating. Aquifer contamination with N and P due to chemical fertilization in the Nile valley, Egypt was explored by Shamrukh et al. (2001). They concluded the implementation of the best management practices to control nitrate leaching especially for the newly reclaimed areas where there is an absence of clay–silt cap. In addition, they related the contamination of shallow depth with nitrate because of the nearness to the agricultural regions. However, their study emphasized that P fertilization was not the source of groundwater contamination. Moreover, the pesticides vulnerability of the Quaternary aquifer in Sohag area, Upper Egypt, was high to very high because of the high hydraulic conductivity of this aquifer (Ahmed 2009).

In this chapter, sources of agricultural soil pollution in Egypt are discussed with giving a focus on the chemical pollutants types. Moreover, the processes of sorption, mobility, and degradation of soil contaminants are briefly discussed to understand the fate of contaminates in soils.

11.2 Sources of Agricultural Soil Pollution in Egypt

The pollution sources of agricultural soils include all inputs that contain elevated levels of chemicals or biological matter into soils. For example, these inputs include phosphatase fertilizers that are borne trace elements (Nziguheba and Smolders 2008). Also, irrigation with low water quality is another example of soil pollution source (Elbana et al. 2017a). Specifically, the main sources of soil pollution in Egypt are categorized by Hussein (2011) into: (i) discharging of industrial effluents into irrigation waterways, (ii) discharging of domestic wastewater in rural areas to drains that are reused in agricultural irrigation, and (iii) the misuse of the applied fertilizers and pesticides during the intensive Egyptian cropping system. Additionally, we can consider the contaminated wet and dry atmospheric deposits as a source of soil pollution in Egypt.

11.2.1 Applied Agrochemicals in the Agricultural Sector

Crop intensification in Egypt is essential to cope with the continuous food demand due to the rapid population expansion. Application of different agrochemicals such as pesticides and fertilizers to the soil is commonly practiced to enhance agricultural production. Heavy use of pesticides and fertilizers as a consequence of implementing an intensive cropping system cause the accumulation of agrochemicals in soils. The overuse of the agrochemicals elevates its contents as well as its residues in soils. Moreover, accidental spills of those agrochemicals represent a serious threat of soil pollution with heavy metals and organic pesticides (Osman 2014). Safe storage practices and good handling of agrochemicals are a human safety necessity as well as prerequisites for environmental protection. For this purpose, considering the integrated crop management (ICM), good agricultural practices (GAP), and the proper hygiene procedures reduce environmental and human health risks of the applied agrochemicals (Damalas and Eleftherohorinos 2011).

Egyptian farmers apply excessive amounts of pesticides to control different pests where the average annual consumption reached 12 thousand metric tons during the years from 2000 to 2006 (Mansour 2008). Also, the increase of the applied fertilizers, especially after High dam construction, is taken into consideration because the dam sequestrated the fertile clay to reach Nile delta soils. In fact, the total required fertilizers were increased from 2.3×10^6 ton in 2000 to 6.4×10^6 ton in 2012 (Central Agency for Public Mobilization and Statistics 2014). The main types of the applied fertilizers in Egypt include urea, ammonium nitrate, ammonium sulphate, calcium nitrate, single and concentrated superphosphate, potassium sulphate, and potassium chloride (FAO 2005). Varied concentrations of heavy metals in fertilizers were reported worldwide (Atafar et al. 2010; Carnelo et al. 1997; Mortvedt 1995; Sabiha-Javied et al. 2009). The addition of such fertilizer is building up heavy metal concentrations in soil, and consequently increases their availability and uptake by the growing plants.

11.2.2 Reuse of Agricultural Drainage Water

Reusing of the agricultural drainage water as a nonconventional water resource is a necessity to satisfy water demands that exceed the available conventional Egyptian water resources. According to Central Agency for Public Mobilization and Statistics (2017), the recycling of agricultural drainage water is estimated at 11.9×10^9 m^3. This water source is mostly mixed or consequently applied with fresh water. Zidan and Dawoud (2013) classified the Nile Delta region to three zones, where the south of Delta exhibited the lowest drainage rate (1.0 mm day^{-1}) and total dissolved solids (TDS) of less than 1000 mg L^{-1}, whereas north of Delta had the highest drainage rate (>3.0 mm day^{-1}) and TDS of more than 2000 mg L^{-1}. Actually, this classification is an indication of the lower water quality of the drainage agricultural water in the direction of northward. For example, the agricultural reclamation of Sahl El-Hessania area in the northeast of Egypt depends on the reuse of agricultural drainage water for irrigation, where the source of irrigation water is Elsalam canal that carries 1:1 Nile to agricultural drainage waters. Abdelhafez et al. (2015) found the soils characterized by high values of pollution load index of As, Cd, Co, Ni, and Pb. They reported the averages of Ni (46.9 mg kg^{-1}), Co (19.5 mg kg^{-1}), and Pb (21.8 mg kg^{-1}) that were considered higher than the natural levels on the earth crust.

Drainage water transmits elevated levels of contaminates into the soils through irrigation processes. However, the concentrations of trace elements in agricultural drainage water are generally less than the allowable level for crops irrigation (El-Bana et al. 2006; Salman et al. 2017). Monitoring drainage water for pesticides residue and trace elements concentrations should be considered before discharging or reusing this water. El-Kabbany et al. (2000) emphasized the difficulty of quantifying all pesticides in water sources. However, they detected the contamination of drainage agriculture water with organophosphorus and carbamate pesticides because of the industrial and sewage pollution of the studied drain.

Avoiding the reuse of drainage water at the late stages of the crop growing season and alternating between fresh and drainage waters for irrigation is recommended to sustain soil productivity (Wahba 2017). The monitoring of the drainage water quality is a prerequisite for the rational reuse planning. It is important to prevent the reuse of the drainage water if it is polluted with municipal or industrial wastewaters for crops irrigation to protect and sustain agro-ecosystems as well as protect human health (Abdel-Azim and Allam 2005; Zidan and Dawoud 2013).

11.2.3 Reuse of Treated Wastewater

A significant reuse of treated and partially treated wastewater in agricultural sector is commonly practiced in Egypt due to water scarcity. Moreover, disposal of sewage water into the surface streams in rural areas of Egypt is common due to the absence of sewer systems in most villages. Elgabal Elasfar farm, 25 km northeast of Cairo is an obvious case of the effect of irrigation with untreated/partially treated wastewater since 1911. Elbana et al. (2013) quantified the accumulated available and total Cd, Cu, Ni, and Pb concentrations in Elgabal Elasfar surface soils. They reported that the long-term use of partially treated wastewater raised the total concentrations of Pb and Cu above the permissible levels, whereas the total Cd concentrations (0.8–3.0 mg kg^{-1}) was on a critical level for soil contamination. Moreover, the percolation of this wastewater to the shallow groundwater represents an environmental risk in Elgabal Elasfar sandy soils (Gemail 2012). Another example, El-Alfy et al. (2017) quantified the trace elements concentration in soils that are irrigated from El-Gharbia main drain, in the Middle Nile Delta. In fact, 75% of this drain is composed of agricultural drainage, 23 and 2% from partially treated domestic and industrial wastewater, respectively. The results showed that the irrigated soils are enriched with Cr, Pb, and Cd (El-Alfy et al. 2017). Moreover, the impact assessment of wastewater disposal into the desert is necessary to sustain natural resources of the desert ecosystems. The use of well-controlled disposal systems and the monitoring of the contaminant releases in disposal sites are highly recommended to mitigate the contamination risk of the aquifer (Ahmed 2009).

11.2.4 Contaminated Air

The soil's productivity is highly affected with air contamination especially in the industrial areas. Air pollution provides an extensive contamination with heavy metals and metalloids to the industrial regions (Alloway 2013). Atmospheric deposition includes wet deposition that is associated with precipitation, dry deposition settling of atmospheric contaminated particles, and cloud deposition (Osman 2014). A comprehensive air particulates analyses of dust and haze storms as well as straw rice combustions days in Giza, Egypt indicated that SO_4^{2-}, NO_3^-, and NH_4^+ were the dominant ions in the three episodes (Hassan and Khoder 2017). Additionally, this research revealed that the higher concentrations of Fe, Mn, Al, Zn, Pb, Ni, Cd, Cr, Cu, and Co in particulate matters were realized during the dust storm days.

Air pollution can directly reduce quality and quantity of agricultural production. For instance, Ali (1993) showed that contaminated air in the industrial region of Shoubra Elkheima, north of Cairo, Egypt, significantly reduced the yield as well as the quality of clover and Egyptian mallow. That deterioration was ascribed to the accumulation of Pb and Cd by the cultivated plants in such area. Also, Abou-Arab et al. (2015) pointed out the rule of the atmospheric pollution due to the traffic and the industrial activities in urban area as a source for Pb in soils as well as cultivated plants. In agreement with that, Feleafel and Mirdad (2013) reported the elevated levels of Pb in soil and vegetable for the Egyptian's areas that exposed to atmospheric pollution. Moreover, Abdel-latif (2003) reviewed the impact of air pollution on vegetation in Egypt and emphasized that the deposit of the particulate matter in industrial areas on the surface of plants leaves causing yield and quality reduction. Additionally, the areas neighboring cement industry are highly susceptible to be polluted. Abdel-Halim et al. (2003) showed that the dust particulates at Helwan and Tebbin industrial areas, south of Cairo, carried elevated levels of chemical pollutants. In concurring, an environmental risk assessment of Tebbin area revealed high soil contents of Ba, Cd, Cr, Cu, Pb, Ni, Sn, and among other due to anthropogenic contamination in more than 65% of Tebbin area (Melegy et al. 2010).

11.2.5 Solid Wastes

Solid wastes including plastic, glass, metals, textiles, and among other represent an environmental burden and pollution source in residential areas. Dumping these solid wastes into or nearby agricultural soil is a trigger for soil deterioration. Especially in Egypt, 56% of the total solid wastes, around 75 million tons, are organic wastes (Ibrahim and Mohamed 2016). Despite that, municipal solid waste is a renewable energy source; the landfill site is considered a source of various contaminants to the adjacent agricultural soils. Rashad and Shalaby (2007) quantified heavy metals contents in the surface agricultural soils around the main municipal solid waste landfill sites at Alexandria. They reported the closest the landfill site is the highest polluted soils where the ranges for Cd (0.25–5.10 mg kg^{-1}), Cu (38–97 mg kg^{-1}), Ni (6.7–12.4 mg kg^{-1}), Cr (7.1–11.2 mg kg^{-1}), and Zn (70–110 mg kg^{-1}) were detected. Additionally, Abd El-Salam and Abu-Zuid (2015) quantified the heavy metal contents in the leachate samples from Alexandria's landfills that revealed a considerable contamination with Zn (0.749 ± 0.235 mg L^{-1}) and Mn (0.839 ± 0.165 mg L^{-1}).

Proper management and controlled operating of solid waste treatments are highly recommended for such activity to prevent soil pollution. Furthermore, monitoring leachate to the groundwater and continuous improvement of treatment processes are essential elements for protecting environmental resources.

11.3 Chemical Pollutants

Two categories of soil chemical pollutants are considered for the purpose of this chapter. These include inorganic chemicals (e.g., Cd, Pb, and Hg) and organic chemicals (e.g., organophosphorus, and organochlorine pesticides) which inversely affect the living organisms in the soil. Each pollutant has specific reactivity in a particular soil as well as varied toxic effect on the different living organisms. In agricultural soils, these two types of chemicals occur mainly in all at once. Certainly, these contaminants have different sources as discussed through the earlier sections of this chapter.

11.3.1 Inorganic Pollutants

Inorganic soil pollutant can be any chemical element or inorganic compound that occurs in higher concentration than the permissible level and adversely affects living organisms. Examples of inorganic soil pollutants include the elevated concentrations of trace elements such as As, Cd, and Pb. In addition to the inorganic form, these heavy metals also establish complexes with organic compounds. For example, Sn exhibited very limited mobility in soil when it exists as inorganic chemical form (Elbana et al. 2014; Hou et al. 2005). Whereas, organotin fungicide exhibited relatively more toxicity and mobility in soil (Loch et al. 1990; Rüdel 2003). Generally, natural metalliferous soils are those naturally containing high concentration of heavy metals.

On the other hand, there are numerous anthropogenic sources of trace elements into soil environment. Osman (2014) listed different sources that borne certain elements such as pesticides contain As; battery recycling as a source for Pb, Cd, Ni, Cu, and Zn; paints encompasses Pb, Cr, Cd, and Hg.

In Egyptian soils, Rinklebe and Shaheen (2017) calculated the contamination factor for Co, Cu, Ni, and Zn for different soils that revealed moderate contamination by Co, Ni, and Cu in eutric and sodic fluvisols soils. Where, the contamination factor reached 3.3, 1.5, 2.7, and 1.3 respectively for Co, Cu, Ni, and Zn in the eutric and sodic fluvisols soils. In the middle of the Nile delta (El-Gharbia Governorate), Shokr et al. (2016) showed the spatial distribution of Cr, Cu, Ni, V, and Zn. The study revealed that the area has a considerable to a high degree of contamination where the total XRF concentrations of all the studied elements were above the average global contents for the upper earth crust. Badawy et al. (2017) reported the distribution and the enrichment factors of 32 chemical elements in soils that are extended from middle Egypt at Asyut to the northward at Cairo. The results of this research concluded that no serious pollution occurred along the river Nile in the studied area. Also, the heavy metals-enrichment factors of 2.39, 1.27, 1.06, and 1.01 for Cr, Co, V, and Mn, respectively were observed whereas enrichment values were less than one for Ni, Zn, As, Mo, Sb, and Ba. In the south of Egypt, Darwish and Pöllmann (2015) assessed the geochemical characteristics of various trace elements at Aswan agricultural and desert areas where the average enrichment factors for Cd and Zn of 87.3 and 19.7, respectively for the agricultural soils, whereas the corresponding values for desert soils were of 36.1 and 6.2.

The soil pollution by Pb is commonly associated with industrial activities and automobile emissions. In a comprehensive research on Pb status in the Egyptian soils, Shetaya et al. (2018) found that Pb concentration ranged between 1.95 and 29,857 mg kg^{-1} (2.9%) in different Egyptian soils. Elnazer et al. (2015) quantified the contamination levels of Pb, Cd, and Zn in roadside soils. They collected calcareous soil ($CaCO_3$ of 25–91%) samples from 1 to 30 m along the Alexandria-Marsa Matruh Highway, north of Egypt. The results indicated soil contamination with Cd (1.25–3.15 mg kg^{-1}) and Pb (29.2–50.6 mg kg^{-1}), whereas Zn (26.6–67.1 mg kg^{-1}) was found less than the worldwide average level. Authors concluded that the contamination of Pb and Zn was related to motorway influence, whereas Cd was associated to agricultural activities.

Moreover, in the mining activities areas, pollution with heavy metals is a critical environmental dilemma. Rashed (2010) explored the pollution with As, Cr, Ag, Ni, Au, Mo, Zn, Mn, and Cu from gold mining at Wadi El-Allaqi, south of Egypt. The calculations of contamination factor indicated a very strong As and Hg pollution in the adjacent areas of the mining activity, where the most of the area has moderate to strong pollution with As and Hg. Also, the high enrichment factors of 5.9 and 4.9 for Mn and Mo, respectively is an indication of the anthropogenic pollution source in the area. Table 11.1 shows examples of different ranges for various trace element concentrations in Egyptian soils. The data in Table 11.1 revealed the large variation in the contamination with trace elements in the Egyptian soils. Total Cd in soils varied from 0.1 to 33 mg kg^{-1}. This high Cd level was reported for the affected soils by industrial wastes or mining activities. Whereas, the wide variation of Pb from 0.01 mg kg^{-1} in non-polluted soil to the value of 237.0 mg kg^{-1} in soil that is irrigated with partially treated domestic wastewater for long time (>100 years). On the other hand, few studies recently quantified Hg in Egyptian soils. Rashed (2010) reported a range of Hg concentrations in soils of 0.03–2.7 mg kg^{-1} nearby mining area. This range is higher than the Egyptian soils range that is reported by Elsokkary (1982), where the range was 0.035 mg kg^{-1} in non-polluted soil to 0.495 mg kg^{-1} in polluted soil.

On the other hand, nitrogen and phosphorus are essential nutrients for plants. The nutrient elements cannot be considered as soil pollutants until elevated concentrations were released into water bodies. Excessive fertilizer use and mismanagement cause nutrient pollution. For instance, high concentrations of NO_3, and PO_4 in water cause environmental problems such as eutrophication. The enrichment of NO_3 and PO_4 trigger a speedy growth of algae that deplete oxygen and reduce light penetration into water body. Also, leaching of NO_3 and potentially toxic trace elements such as Cd into groundwater is a serious environmental pollution. In a study of evaluating the impact of sewage fertilization on the quality of groundwater at Elgabal Elasfar farm, Abo el Abas (2004) showed that groundwater was polluted with NO_3 and NO_2. Moreover, contamination of the groundwater by Fe, Mn, and Zn was observed but with permissible concentration levels.

11.3.2 Organic Pollutants

The synthetic organic pesticides including herbicides, insecticides, fungicides, and among other represent dangerous soil pollutants whenever it exceeds the safe recommended level of use. Numerous environmental challenges are associated with the application of the pesticides. Human and animal health problems owing to the pesticides exposure and the chemical accumulation in the environment are familiar cases of the pesticides use (Mansour 2008). Moreover, Barakat (2004) stressed the high environmental and health risk of the long lifetime of the persistent toxic pesticides such as organochlorine compounds (e.g., polychlorinated biphenyl (PCBs) and dichlorodiphenyltrichloroethane (DDT). Farag et al. (2011) monitored the pesticide residues in Egyptian local markets of 132 fruits, vegetables, herbs, and spices

Table 11.1 Different ranges of various trace element concentrations in Egyptian soils

Soil or location	Total (mg kg^{-1})									References
	Cd	Co	Cr	Cu	Ni	Pb	Zn	As	Hg	
Eutric fluvisols	–[a]	24.7–29.8	–	47.3–49.0	63.5–70.5	–	77.2–95.6	–	–	Rinklebe and Shaheen (2017)
Sodic fluvisols	–	22.7–24.6	–	32.7–40.2	55.1–59.1	–	70.4–74.1	–	–	
Haplic calcisols	–	7.7–9.1	–	6.0–10.7	15.7–19.8	–	31.3–33.7	–	–	
Aswan agricultural soil	8.3–28.3	16.6–54.9	60.0–218.2	20.2–77.5	23.1–98.4	15.9–42.7	628–2224	–	–	Darwish and Pöllmann (2015)
Aswan desert soil	0.47–2.45	0.0–5.6	1.4–46.5	0.01–9.49	0.0–6.9	0.01–11.33	0.01–444.0	–	–	
Elgabal Elasfar farm	0.5–3.5	–	–	1.6–120.5	0.2–31.3	7.0–237.0		–	–	Elbana et al. (2013)
El-Gharbia governorate	–	–	140–519	0–289	60.6–267.3	–	90.2–377.6	–	–	Shokr et al. (2016)
Wadi El-Allaqi (nearby mining area)	0.1–0.2	12–22	140–530	20–60	68–260	4–29	47–92	4–700	0.03–2.7	Rashed (2010)
Middle Egypt (Asyut to Cairo)	–	5–36	33–308	–	8–84	–	13–165	0.0–4.6	–	Badawy et al. (2017)
Eutric Fluvisols	–	–	43–89	–	–	–	–	25–67.5	–	Shaheen et al. (2017)
El-Mahla El-Kobra area	11–33	–	–	60–386	31–164	48–92	54–449	–	–	Mahmoud and Ghoneim (2016)

[a]Data is not reported for this element within the referenced study

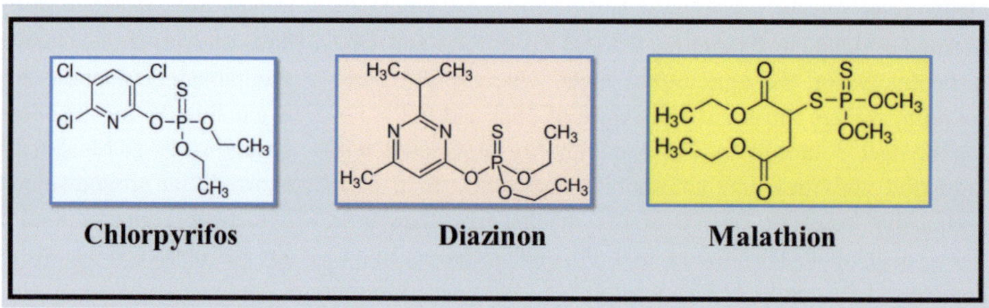

Fig. 11.1 Examples of the pesticides chemical structure. Created by authors

samples. The study reported 17 pesticides residues of 241 analyzed ones were detected in 55% of the total samples. Additionally, the study emphasized that only one sample exceeded the recommended maximum residue limits by carbendazim. Also, Gad Alla et al. (2015) evaluated pesticide residues in 177 fruit samples collected during 2010 from local markets where the residues of 17.5% of the samples were above the safe limit with only 2.3% of the samples exhibited violation of maximum residue limit. The detected residues were mainly for pyrethroids, organophosphates, and benzimidazoles pesticides. Figure (11.1) shows the main detected pesticides as examples of the pesticides chemical structure.

No doubt, rational use of pesticides and implementation of integrated pest management are the key factors to sustain agricultural production and maintain human health (Mansour 2004). Also, awareness of the associated pesticides benefits and risk as well as implementation of effective monitoring appropriated corrections, and proper regulatory restrictions will help to reduce organic pollutants in Egyptian soils.

11.4 Fate and Behavior of Soil Pollutants

Chemical and biological properties of soil pollutant as well as the contaminated soil characteristics influence the contaminant fate in the environment. As shown in Fig. (11.2), variou contaminant fates can take place such as plant uptake, volatilizing to the surrounding air, deep leaching into soil, transport with water runoff and/or by wind erosion. Moreover, numerous chemical and biological transformations

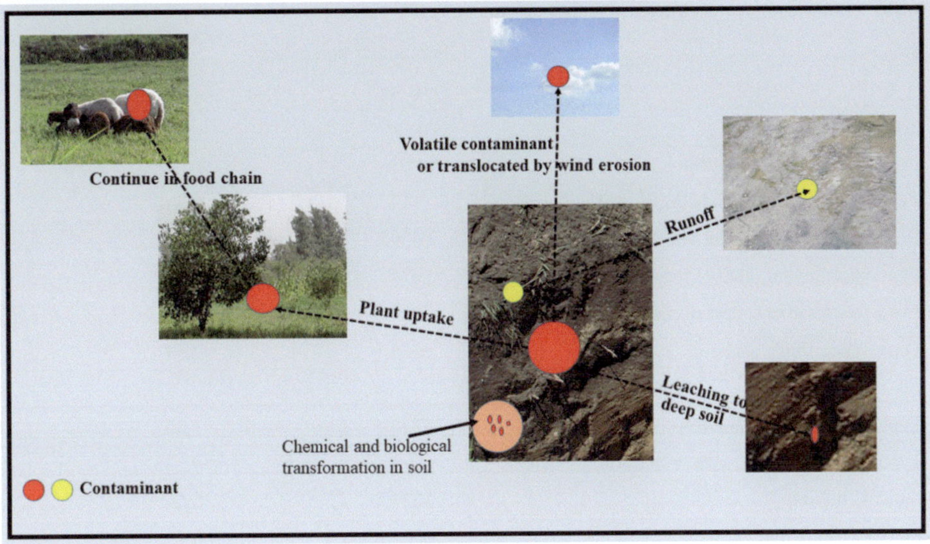

Fig. 11.2 Fate of soil contaminant. Created by authors

happen in soil such as sorption and degradation processes. Soil pollution has direct consequences on water, air, and food quality that consequently adversely affects human health. The polluted soil is a source of surface waterbodies pollution through runoff and seepage to water streams and the deep leaching trigger off groundwater contamination.

Specifically, plants have different potentialities to uptake chemicals from soil. For instance, some plants are hyperaccumulators of certain chemical element, whereas other plants are very sensitive and unable to continue its growth with even low soil contaminant concentration. The metallophytes that exhibit high tolerance and hyperaccumulation of metals can be applied for heavy metals phytoremediation (Babula et al. 2008). Examples of hyperaccumulators include *Pteriscretica* for As; Ecotypes of *T. caerulescens* for Cd; *Betula* and *Salix* trees for Cr; *Salixnigra* for Cd and Cu; *Amanitamuscaria* for Hg; *Alyssum lesbiacum* for Ni (Peer et al. 2005). On the other hand, most of the plants are suffering yield reduction due to the contamination with trace elements. Uncontaminated soil has Cd contents of 0.22–0.51 mg kg^{-1}, whereas the relatively high concentration of 100–500 mg kg^{-1} is an excessive for soil Pb levels (Kabata-Pendias and Mukherjee 2007). This indicates the role of the chemical element properties itself in evaluating the toxicity. There are many factors that control pollutant uptake by plants. For example, Shahid et al. (2016) concluded that Cd concentration and speciation, soil properties, exposure duration, and physiological plant properties are the main factors controlling Cd uptake by plant. As shown in Fig. (11.2), chemical and biological transformations of contaminant control its fate in soil. Recognizing the contaminant sorption, mobility, and degradation in soil is an essential aspect to understand its fate in environment and to quantify its availability for the plants.

11.4.1 Sorption

Quantifying the relation between the sorbed contaminant concentration (*S*) on soil particles and its concentration (*C*) in soil solution provide valuable environmental information. Where,

$$S = f(C) \tag{11.1}$$

Numerous mathematical models can be applied to quantify Eq. 11.1 such as the simple linear model ($S = k_d C$, where k_d is the linear distribution coefficient), Freundlich model ($S = k_f C^n$, where k_f is Freundlich distribution coefficient and n is the reaction order), and Langmuir model ($\frac{S_{\max} k_l C}{1 + k_l C}$, where $S_{\max}$ is the maximum sorption capacity, k_l is the Langmuir coefficient).

Sorption of different contaminants is studied on various Egyptian soils. High sorption of cationic trace element occurred in the Egyptian soils. This elevated affinity can be attributed to the high pH and the occurrence of $CaCO_3$ and oxides in most of Egyptian soils. Table (11.2) shows examples of Freundlich model parameters that are recently published for different Egyptian soils. The data reveals the variation of trace element affinities to be sorbed on different soils. Generally, trace elements exhibited high sorption and a nonlinear behavior in the Egyptian soils, where $n \neq 1$. Based on the literature of this chapter, Pb exhibited the highest sorption followed by relatively lower affinity for Cu then, Zn; whereas Cd and Ni exhibited the lowest sorption on the Egyptian soils.

The chemical behavior of one element is varied from soil to soil based on the soil properties and environmental conditions such as oxidation/reduction circumstances. Moreover, competition between the chemicals on the available

Table 11.2 Examples of different Freundlich parameters of various trace element concentrations in Egyptian soils

Soil	Freundlich parameter		References
	k_f (L kg^{-1})	n	
Cd			
Typic torripsamment, surface soil	203	0.276	Elbana and Selim (2010)
Typic torripsamment, subsurface soil	117	0.346	
Typic ustifluvent, entisol fluvial	639.1	0.53	Shaheen (2009)
Typic xerofluvent, entisol lacustrine	1074	0.64	
Typic xeropsament, entisol marine	245.5	0.36	
Typic calcorthids, aridisol	628.1	0.54	
Cu			
Typic torripsamment, surface soil	5226.4	1.255	Elbana and Selim (2011)
Typic torripsamment, subsurface soil	755.5	0.233	
Entisol (fluvial)	3531.8	0.64	Shaheen et al. (2009)
Entisol (lacustrine)	6760.8	0.93	
Entisol (marine)	912.0	0.19	
Aridisol	2398.8	0.72	
Pb			
Typic ustifluvent, Entisol FLUVIAL	7128.5	0.29	Shaheen (2009)
Typic xerofluvent, entisol lacustrine	7998.3	0.28	
Typic xeropsament, entisol marine	2398.8	0.14	
Typic calcorthids, aridisol	13,031.70	0.67	
Bahr El Baqar region, clay soil	1238.9	0.31	Abd El Razek (2014)
Zn			
Fluvial	1165.7	0.35	Shaheen et al. (2015)
Lacustrine	1970.2	0.44	
Marine	390.1	0.25	
Aridisols	653.9	0.39	
Ni			
Typic torrifluvent	0.312	1.360	Elkhatib et al. (2013)
Typic torripsamment	0.201	1.260	
Typic calciorthids	0.250	1.538	

sites of reaction is another factor that controls chemical behavior in soils. Usman (2008) showed the relative selectivity of various heavy metals by New Valley soils, south of Egypt. Author applied a combative sorption batch experiments for Cd, Cu, Ni, Pb, and Zn. The results showed that Pb and Cu exhibited high sorption on soils, whereas soils have lower selectivity for Cd, Ni, and Zn. The Freundlich parameters (k_f, n) reported by Usman (2008) for the competitive elements sorption were relatively lower than the reported values in Table 11.2 for the single element sorption. This is an indication of the effect of chemical competition on the available reaction sites. Thus, we can conclude that the specific chemical behavior of trace elements in soil depends on multiple interconnected factors including soil properties, element characteristics, and environmental conditions as well as the occurrence of other competitive trace elements in soil.

On the other hand, sorption of organic contaminants in soil is more complicated than the retention of trace elements. This is probably because of the degradation and the residue sorption processes that are involved. In addition to linear, Freundlich, and Langmuir model, scientists commonly calculate the carbon normalized sorption coefficient, which is the product of k_d/OC, where OC is the soil organic carbon. The main sorption mechanisms include Van der Waals attraction, hydrophobic bonding, hydrogen bonding, charge transfer, ion exchange, and ligand exchange (Sadegh-Zadeh et al. 2017). Quantifying sorption of different organic contaminants in soil is a prerequisite to understand contaminant fate in the environment. Sorption of the organic pesticides controls its availability in soil and its leaching to groundwater (Pignatello 2006). There are numerous studies on organic contaminants in the Egyptian soils that revealed each contaminant exhibited certain behavior in different soils and in the presence of other chemicals that can increase the soil sorption or reduce it. For instance, ElShafei et al. (2009) explored the sorption of cadusafos, organophosphate nematicide, on two Egyptian soils. The sorption data was well fitted to the linear model with the k_d of 4.20 and 2.74 L g^{-1} for sandy and clay soils, respectively. The nonlinear sorption behavior is also observed on the Egyptian soils. Kandil et al. (2015) showed the sorption nonlinearity of imidacloprid, insecticide, on a lacustrine sandy clay loam soil. They emphasized the rule of humic acid and clay fractions in increasing the sorption of imidacloprid.

Investigating the sorption of organic contaminants provides valuable information regarding the appropriated materials to remove certain chemicals from soils or water. For example, Osman et al (2017) found that the surfactant-modified bentonite clay removed 66% of atrazine from aqueous solution, whereas natural bentonite recovered 11%. Also, a mix of natural clays, kaolinite (58%), and montmorillonite (42%), removed 28–33% of methomyl insecticide from contaminated solution (El-Geundi et al. 2012). Use of the natural local materials provides an environmental solution to remove organic contaminants and prevent soil pollution.

In fact, few published research on the chemical behavior of antibiotics in the Egyptian soils were found. However, it is important to study their sorption reactivity and behavior in soils. ElSayed and Prasher (2014) quantified the sorption of oxytetracycline which is borne to the soil with the additions of manure as an organic fertilizer on a sandy soil. Sorption data was fitted well to the nonlinear Freundlich model with k_f of 25.46 and 23.55 mL g^{-1} in presence and absence of nonionic surfactant, respectively. The applications of animal manure and the reuse of treated wastewater in agricultural irrigation are sources of various antibiotics that should be quantified and monitored in the Egyptian soils.

11.4.2 Mobility

Quantifying the contaminant transport and the sorption process are equally important to understand the fate of chemicals in soil. Solute transport equations can be applied to model the contaminant mobility in soils. The transport of any chemical in soil depends on advection and dispersion phenomena. In other words, the mobility is controlled by the movement of contaminants that is related to current velocity and the process of mixing this contaminant within the water phase in soil. Mathematically, the simplest advection–dispersion equation (ADE) for nonreactive chemical is written as follows:

$$\frac{\partial C}{\partial t} = -v\frac{\partial C}{\partial z} + \frac{\partial}{\partial z}\left(D\frac{\partial C}{\partial z}\right), \qquad (11.2)$$

where C is the contaminant concentration at time "t" and depth "z". v is the average pore water velocity, D is the dispersion coefficient. Therefore, the rate of any contaminant concentration change can be calculated by solving the ADE by counting for chemical reactivity and the degradation terms. Sorption is included in mobility equation through the calculation of the linear retardation factor (R)

$$R = 1 + \rho K_d/\theta, \qquad (11.3)$$

where ρ is the soil bulk density and θ is the soil moisture content. Also, the degradation rate (μ) can be simply incorporated by calculating the term of μC. Simulation of the contaminants transport through leaching and runoff is mandatory for implementing pollution control (Arias-Estévez et al. 2008).

Numerous studies indicated the limited mobility of trace elements in the Egyptian alkaline calcareous soils. Mobility of Cd, Cu, and Pb in soil columns for a newly reclaimed area in the northwestern desert of Egypt was studied (Elbana and Selim 2013). The authors reported that the average recoveries of <1, 6, and 47% of the total applied Pb, Cu, and Cd, respectively were obtained in the effluent solution. Additionally, the removal of $CaCO_3$ from these soils greatly increased the mobility of Cu. In a large pulse (60 pore volumes) column experiments, Elbana and Selim (2012) showed the increase of Cu recovery in the effluent from 27 to 87% of the applied Cu due to the removal of $CaCO_3$.

Field research in sandy soils that were irrigated with wastewater for a period of 100 years revealed a larger relative mobility of Cd compared with Pb and Cu that exhibited higher retention in the surface and subsurface soil layers (Elbana et al. 2013). The mobility of contaminants in the Nile delta area is a real environmental problem due to the shallow groundwater phenomenon in such area. However, Nile delta soils exhibit higher clay and OM contents that can increase sorption of the inorganic contaminants compared with the new reclaimed Egyptian sandy soils. Abu Khatita (2011) found that Cu was accumulated on the surface with an obvious potential to move downward in the Nile Delta cultivated soils. A comprehensive assessment of groundwater quality of the shallow aquifers in the west of the Nile Delta revealed the contamination with Fe, Mn, Zn, Cu, and Ni were the concentrations that exceeded the safe WHO-limit for drinking water (Masoud 2014). In a study of the aquifer contamination with pesticides and inorganic chemicals in the south of Egypt, Ahmed (2009) found that the vulnerability to pesticides contamination was higher than contamination vulnerability with inorganic chemicals, especially in the new reclaimed sandy soils.

In addition to soil and pollutant chemical characteristics, numerous factors affect the mobility of contaminants such as soil temperature, aeration statues, among other. Therefore, site-specific examinations for contaminant mobility are important to be considered for individual cases of studies. For instance, Fakhry et al. (2016) emphasized the rule of soil organic matter in heavy metals mobility in soil profile. They found that the humic acid formed metal–carboxylate complex on the surface soil layer whereas the reverse reaction occurred at the subsoil where the released metals take place.

11.4.3 Degradation

In addition to the value of understanding sorption and mobility of organic contaminants, the degradation process is in such importance for complete interpretation of pollutants fate in soils. Breakdown of organic contaminants through biological-, chemical- and photo-degradation produce new chemicals that can exhibit different toxicity than the original contaminant. First- (Eq. 11.4) and second- (Eq. 11.5) order models are commonly applied to quantify organic contaminants degradation:

$$\frac{\mathrm{d}c}{\mathrm{d}t} = k_1(C_o - C_t), \tag{11.4}$$

$$\frac{\mathrm{d}C}{\mathrm{d}t} = k_2(C_o - C_t)^2, \tag{11.5}$$

where C_o and C_t are the contaminant concentrations at initial and t times, respectively. $k_1$and k_2 are the rate constants of pseudo-first-order and second-order reactions, respectively.

The biodegradation process requires certain environmental conditions to support the breakdown activity. The required conditions for optimal microbial degradation include soil moisture, pH, O_2 level, temperature, soil nutrients, particle size distribution, and contaminant concentration (Vidali 2001). Contaminant degradation requires the presence of particular microorganism under appropriate environmental conditions. It is worthy to mention that not all biodegradation processes are bioremediation techniques. Thus, degradation of some chemicals can produce more toxic compound to the environment. However, bioremediation process should produce less toxic substances into the environment. The failures of bioaugmentation on large scale can be ascribed to the lack of attention to the principles of degradation of polycyclic aromatic hydrocarbons (Johnsen et al. 2005). Bioremediation will continue as a promising cleanup soil technique in view of the fact that it is an environmental friendly remediation approach.

11.5 Management of Polluted Soils

Agricultural soil pollution is a natural continuous process that will keep on occurring as human exist. Pollutants are substances that are transferred from a matter to another and adversely affect the productivity or functionality of the hosted matter. According to the law of mass conservation, the mass can neither be created nor destroyed and so pollutants. Humans can prevent the soil pollution via removing pollutants from the soil and/or adapting with the pollution. Locating and mapping contaminated sites are key roles for managing pollution problem. It is important to delineate polluted site and quantify the amount of the pollutant. Also, identifying the pollution sources and the vulnerable living target should be considered for proper management of contaminated sites. Furthermore, monitoring and calculating soil pollution indices are basic for soil pollution control. The management of polluted soils can be ensured through three ways (prevention, remediation, and adaptation). Certainly, the management is a specific site that counts for the pollutant properties, soil characteristics, and contiguous environmental conditions.

Prevention of soil pollution can be achieved through minimizing the utilization of contaminated agro-inputs such as low water quality for irrigation, high impurities borne fertilizers, low degradable pesticides, and among other. This is not an easy option; many social and economic factors control the elimination of contaminated agro-inputs. For example, the prevention of soil contamination due to atmospheric deposition of contaminated particles cannot be achieved without bending down the source of air pollution in the adjacent areas. It is a recommended strategy to prevent the source of pollution. Applications of the affordable environmental friendly materials are eternally recommended. For example, using biochar materials from rice straw to remove and reduce pesticides from polluted water would prevent soil contamination (Taha et al. 2014). The prevention of soil pollution is a multidisciplinary management that requires the efforts of a whole society.

Implementation of best management practices (BMP) program is a substantial way to sustain our natural resources. Source control of the agrochemicals applications is the preferred approach to manage pollutant runoff from agricultural soils (Heathwaite et al. 2005). Combination of BMP scenarios is needed to reduce pollution load. Özcan et al. (2017) found that counting for optimizing amount of fertilizers, no tillage, and implementing parallel terraces BMPs in a semiarid watershed significantly decreased pollution load. Moreover, preventing soil pollution requires well-defined standards and regulations to be effective in the management strategies. Chen et al. (2017) emphasized the rule of regulations, pollution tax, and agricultural subsidy to control pollution. They concluded that innovation subsidy can prevent agricultural pollution than quantity subsidy.

In addition to pollution prevention, the remediation of polluted soil is an additional soil management option. Recently, numerous remediation techniques were developed including thermal, physical, chemical, and biological remediation. Soil remediation depends on several factors such as pollutant nature, degree of pollution, number of pollutants, soil properties, and environmental conditions. Morillo and Villaverde (2017) reviewed the innovative technologies for remediating pesticides-polluted soils. They categorized remediation techniques as containment-immobilization, separation, and destruction. The latter category includes chemical and biological remediation. They addressed the advantages and disadvantages of each remediation methods that were related mainly to cost and time of remediation treatment. In agreement with that, Xu et al. (2015) reviewed the Hg remediation techniques such as soil washing, stabilization/solidification, thermal treatment, and phytoremediation. They emphasized the thermal treatment as a timely effective one and the phytoremediation as an environmentally friendly technique for remediating Hg-contaminated soils.

Adaptation with contaminated soil is a proposed term that means the use of the contaminated soil without adversely affecting human health or other environmental resources. Adaptation can be considered as an intermediate management strategy to cope with the current contamination conditions within the available remediation approaches. Immobilization of contaminant can be a kind of adaption. Also, during the long period of phytoremediation and before achieving the complete cleanup can be considered as adaption with the contaminated soils. For example, wood or nonedible plants plantation of the contaminated site can prevent the spreading of contaminated soil particles by wind erosion or runoff. Another example is the planting of biodiesel production crops such as jatropha and jojoba that can provide an opportunity for safe using of contaminated soil. Rational management of the contaminated soil is critical to avoid contamination of surroundings water and air (Wong 2003).

The adaptation with the contaminated soil needs guidelines and well-defined policies. The choice of a tolerant plant to be grown in a contaminated soil is a complex issue where the final product should be safe for human as well as wildlife animals. Here, plant-based phytoremediation should be considered for such a choice. It is important to be sure that the grown plant for the phytoremediation will not increase the mobility of contaminant to the groundwater or adversely affect the ecosystem. Dickinson et al. (2000) provided a review of the available guidelines for planting trees on contaminated soils and how to interpret the related information to achieve appropriate management decisions in the contaminated soils. Also, the authors proposed trace elements-tolerated values for planting trees in contaminated soil. Handling agricultural products and practicing cultivation in the contaminated soil should be accomplished with specific guidelines and regulations. In Egypt, there is a well-established guideline for the safe use of treated wastewater in agriculture (ECP 501 2015). In this code, the agricultural practices and irrigation methods as well as hygiene requirements are specified for specific allowable cultivation plants. Our literature review indicated that a detailed Egyptian guideline for managing contaminated soils is needed.

11.6 Threats and Challenges of Soil Pollution on Sustainable Agriculture

Climate change and water scarcity represent the main threats to agricultural production in Egypt. These two interconnected issues are directly affecting soil pollution in Egypt as drivers for using low water quality and excessive amounts of agrochemicals to cope with the continuous demands of agricultural production. This nexus of the limited resources and soil pollution represent a real challenge for sustainable agriculture in Egypt. Especially, soil deterioration via pollution directly affects the limited agricultural area in Egypt that threatens the food security. Moreover, the safe reuse of treated wastewater in agricultural irrigation is a mandatory option to face the gap between Egyptian water resources and demands. The implementation of the proper and affordable wastewater treatment as well as considering the public awareness of the benefits/risks of reusing the treated wastewater in agricultural production are highly recommended to sustain natural resources (Elbana et al. 2017b). In fact, long-term soil pollution causes loss of a productive soil and/or crop yield reduction. This challenge could adversely affect the social and economic conditions in Egypt. Soil degradation and desertification could lead to malnutrition, hunger, and poverty (Lal 2008).

Soil pollution in the old Nile Delta land negatively affects the quality of the shallow groundwater that is being recycled in agricultural production. On the other hand, soil pollution in the newly reclaimed lands in Egyptian deserts could affect the deep groundwater in the long term, and consequently will adversely affect the agricultural sustainability. The use of the groundwater in the Nile Valley and Delta for irrigation has doubled in recent years and reached around 6.2×10^9 m^3 in 2010 (Omran 2017). Prevention of soil pollution should be the first step in implementing new agricultural land reclamation projects, whereas national soil remediation programs should be considered for the old alluvial agricultural soils in the Nile valley and Delta of Egypt. Literature review revealed the high vulnerability for groundwater pollution in the newly reclaimed sandy soils. This implies the need to develop and implement a national plan for soil pollution prevention in the newly reclaimed areas. Pollution prevention will avoid resorting to remediation processes that are time-consuming and very costly activity. Also, it is recommended to promote policies of soil conservation incentives to reduce the risk of pollution and sustain a healthy agricultural production (Elnemr 2017).

References

Abd El Razek AA (2014) The mobility and speciation of lead and cadmium in Bahr El Baqar region, Egypt. J Environ Chem Eng 2 (1):685–691. https://doi.org/10.1016/j.jece.2013.11.006

Abd El-Salam MM, Abu-Zuid GI (2015) Impact of landfill leachate on the groundwater quality: a case study in Egypt. J Adv Res 6:579–586. https://doi.org/10.1016/j.jare.2014.02.003

Abdel-Azim R, Allam MN (2005) Agricultural drainage water reuse in Egypt: strategic issues and mitigation measures. In: Hamdy A, El Gamal F, Lamaddalena N, Bogliotti C, Guelloubi R (eds) Non-conventional water use: WASAMED project. CIHEAM/EU DG Research, Bari, pp 105–117

Abdelhafez AA, Abbas MHH, Attia TMS (2015) Environmental monitoring of heavy-metals status and human health risk assessment in the soil of Sahl El-Hessania area, Egypt. Pol J Environ Stud 24(2):459–467

Abdel-Halim AS, Metwally E, El-Dessouky MM (2003) Environmental pollution study around a large industrial area near Cairo, Egypt. J Radioanal Nucl Chem 257(1):123–124. https://doi.org/10.1023/A:1024709729606

Abdel-latif NM (2003) Air pollution and vegetation in Egypt: a review. In: Emberson L, Ashmore M, Murray F (eds) Air pollution impacts on crops and forest a global assessment. Air pollution reviews, vol 4. Imperial College Press, UK, pp 215–235

Abo el Abas MY (2004) Distribution pattern of some inorganic pollutants in groundwater and soil of a sewage farm at NE-Cairo/Egypt. In: Zereini F, Jaeschke W (eds) Water in the middle east and in North Africa. Springer, Berlin. https://doi.org/10.1007/978-3-662-10866-6_17

Abou-Arab AAK, AbouDonia MA, Mohamed SR, Enab AK (2015) Risk assessment of lead in Egyptian vegetables and fruits from different environments. Int J Nutr Food Eng 9(3):335–341

Abu Khatita AM (2011) Assessment of soil and sediment contamination in the middle Nile Delta area (Egypt)- geo-environmental study using combined sedimentological, geophysical and geochemical methods. Doctoral Dissertation, Friedrich-Alexander University, Erlangen-Nürnberg, Germany, p 214

Ahmed AA (2009) Using generic and pesticide DRASTIC GIS-based models for vulnerability assessment of the quaternary aquifer at Sohag, Egypt. Hydrogeol J 17:1203–1217. https://doi.org/10.1007/s10040-009-0433-3

Ali EA (1993) Damage to plants due to industrial pollution and their use as bioindicators in Egypt. Environ Pollut 81(3):251–255. https://doi.org/10.1016/0269-7491(93)90207-5

Alloway BJ (2013) Sources of heavy metals and metalloids in soils. In: Alloway B (ed) Heavy metals in soils. Environmental pollution, vol 22. Springer, Dordrecht. https://doi.org/10.1007/978-94-007-4470-7_2

Arias-Estévez M, López-Periago E, Martínez-Carballo E, Simal-Gándara J, Mejuto J, García-Río L (2008) The mobility and degradation of pesticides in soils and the pollution of groundwater resources. Agric Ecosyst Environ 123:247–260. https://doi.org/10.1016/j.agee.2007.07.011

Atafar Z, Mesdaghinia A, Nouri J, Homaee M, Yunesian M, Ahmadimoghaddam M, Mahvi AH (2010) Effect of fertilizer application on soil heavy metal concentration. Environ Monit Assess 160:83–89. https://doi.org/10.1007/s10661-008-0659-x

Babula P, Adam V, Opatrilova R, Zehnalek J, Havel L, Kizek R (2008) Uncommon heavy metals, metalloids and their plant toxicity: a review. Environ Chem Lett 6:189–213. https://doi.org/10.1007/s10311-008-0159-9

Backović D (2008) Soil pollution. In: Kirch W (ed) Encyclopedia of Public Health. Springer, Dordrecht, pp 1322–1324. http://dx.doi.org/10.1007/978-1-4020-5614-7_3280

Badawy WM, Ghanim EH, El Duliu OG, Samman H, Frontasyeva MV (2017) Major and trace element distribution in soil and sediments from the Egyptian central Nile Valley. J Afr Earth Sci 131:53–61. https://doi.org/10.1016/j.jafrearsci.2017.03.029

Barakat AO (2004) Assessment of persistent toxic substances in the environment of Egypt. Environ Int 30:309–322. https://doi.org/10.1016/S0160-4120(03)00181-8

Carnelo LGL, de Miguez SR, Marbán L (1997) Heavy metals input with phosphate fertilizers used in Argentina. Sci Total Environ 204(3):245–250. https://doi.org/10.1016/S0048-9697(97)00187-3

Central Agency for Public Mobilization and Statistics (2014) Economic study of chemical fertilizers industry in Egypt. Cairo. Ref. No. 80-23411-2014. http://www.capmas.gov.eg. Accessed 26 Dec 2017 (in Arabic)

Central Agency for Public Mobilization and Statistics (2017) Statistical yearbook. Cairo. Ref. No. 71-01111-2017. http://www.capmas.gov.eg. Accessed 22 Dec 2017 (in Arabic)

Ceplecha ZL, Waskom RM, Bauder TA, Sharkoff JL, Khosla R (2004) Vulnerability assessments of Colorado groundwater to nitrate contamination. Water Air Soil Pollut 159:373–394. https://doi.org/10.1023/B:WATE.0000049188.73506.c9

Chen Y, Wen X, Wang B, Nie P (2017) Agricultural pollution and regulation: How to subsidize agriculture? J Cleaner Prod 164:258–264. https://doi.org/10.1016/j.jclepro.2017.06.216

Damalas CA, Eleftherohorinos IG (2011) Pesticide exposure, safety issues, and risk assessment indicators. Int J Environ Res Public Health 8:1402–1419. https://doi.org/10.3390/ijerph8051402

Darwish MAG, Pöllmann H (2015) Trace elements assessment in agricultural and desert soils of Aswan area, south Egypt: geochemical characteristics and environmental impacts. J Afr Earth Sci 112:358–373. https://doi.org/10.1016/j.jafrearsci.2015.06.018

Dickinson NM, Mackay JM, Goodman A, Putwain P (2000) Planting trees on contaminated soils: issues and guidelines. Land Contam Reclam 8(2):87–101. https://doi.org/10.2462/09670513.561

ECP 501 (2015) Egyptian code of practice for the use of treated municipal wastewater for agricultural purposes. The Ministry of Housing Utilities and Urban Communities (in Arabic)

El-Alfy MA, El-Amier YA, Abd El-Hamid HT (2017) Soil quality and health risk assessment of heavy metals in agricultural areas irrigated with wastewater from Kitchener Drain, Nile Delta, Egypt. J Sci Agric 1:158–170. http://dx.doi.org/10.25081/jsa.2017.v1.50

Elbana TA, Selim HM (2010) Cadmium transport in alkaline and acidic soils: miscible displacement experiments. Soil Sci Soc Am J 74:1956–1966. https://doi.org/10.2136/sssaj2010.0146

Elbana TA, Selim HM (2011) Copper mobility in acidic and alkaline soils: miscible displacement experiments. Soil Sci Soc Am J 75:2101–2110. https://doi.org/10.2136/sssaj2011.0185

Elbana TA, Selim HM (2012) Copper transport in calcareous soils: miscible displacement experiments and second-order modeling. Vadose Zone J. https://doi.org/10.2136/vzj2011.0110

Elbana TA, Selim HM (2013) Lead mobility in calcareous soils: influence of cadmium and copper. Soil Sci 178:417–424. https://doi.org/10.1097/SS.0000000000000000

Elbana TA, Ramadan MA, Gaber HM, Bahnassy MH, Kishk FM, Selim HM (2013) Heavy metals accumulation and spatial distribution in long term wastewater irrigated soils. J Environ Chem Eng 1:925–933. https://doi.org/10.1016/j.jece.2013.08.005

Elbana TA, Sparks DL, Selim HM (2014) Transport of tin and lead in soils: miscible displacement experiments and second-order modeling. Soil Sci Soc Am J 78:701–712. https://doi.org/10.2136/sssaj2013.07.0265

Elbana TA, Bakr N, George B, Elbana M (2017a) Assessment of marginal quality water for sustainable irrigation management: case study of Bahr El-Baqar area, Egypt. Water Air Soil Pollut 228:214. https://doi.org/10.1007/s11270-017-3397-2

Elbana TA, Bakr N, Elbana M (2017b) Reuse of treated wastewater in Egypt: challenges and opportunities. In: Negm AM (ed) Unconventional water resources and agriculture in Egypt. Hdb Env Chem. Springer International Publishing. https://doi.org/10.1007/698_2017_46

El-Bana TA, Gaber HM, Bahnassy MH, Suliman AS (2006) Quality assessment of water resources in Northern Nile Delta: a case study in Kafr El-Sheikh Governorate. Egypt J Soil Sci 46(4):409–424

El-Geundi MS, Nassar MM, Farrag TE, Ahmed MH (2012) Removal of an insecticide (methomyl) from aqueous solutions using natural clay. Alexandria Eng J 51(1):11–18. https://doi.org/10.1016/j.aej.2012.07.002

El-Kabbany S, Rashed MM, Zayed MA (2000) Monitoring of the pesticide levels in some water supplies and agricultural land, in El-Haram, Giza (ARE). J Hazard Mater A72:11–21. https://doi.org/10.1016/S0304-3894(99)00174-0

Elkhatib EA, Mahdy AM, ElManeah MM (2013) Effects of drinking water treatment residuals on nickel retention in soils: a macroscopic and thermodynamic study. J Soils Sediments 13:94–105. https://doi.org/10.1007/s11368-012-0577-y

Elnazer AA, Salman SA, Seleem EM, Abu El Ella AM (2015). Int J Ecol. Article ID 689420. http://dx.doi.org/10.1155/2015/689420

Elnemr M (2017) Applicability of sustainable agriculture in Egypt. In: The handbook of environmental chemistry. Springer, Berlin. https://doi.org/10.1007/698_2017_156

ElSayed EM, Prasher SO (2014) Sorption/desorption behavior of oxytetracycline and sulfachloropyridazine in the soil water surfactant system. Environ Sci and Pollut Res 21(5):3339–3350. https://doi.org/10.1007/s11356-013-2273-x

ElShafei GS, Nasr IN, Hassan ASM, Mohammad SGM (2009) Kinetics and thermodynamics of adsorption of cadusafos on soils. J Hazard Mater 172:1608–1616. https://doi.org/10.1016/j.jhazmat.2009.08.034

Elsokkary IH (1982) Contamination of soils and plants by mercury as influenced by the proximity to industries in Alexandria, Egypt. 23:55–60. https://doi.org/10.1016/0048-9697(82)90121-8

Fakhry A, Osman O, Ezzat H, Ibrahim M (2016) Spectroscopic analyses of soil samples outside Nile Delta of Egypt. Spectrochim Acta Part A Mol Biomol Spectrosc 168:244–252. https://doi.org/10.1016/j.saa.2016.05.026

FAO, Food and Agriculture Organization (2005) Fertilizer use by crop in Egypt. Rome, 50 pp

Farag RS, Abdel Latif MS, Abd El-Gawad AE, Dogheim SM (2011) Monitoring of pesticide residues in some Egyptian herbs, fruits and vegetables. Int Food Res J 18:659–665

Feleafel MN, Mirdad ZM (2013) Hazard and effects of pollution by lead on vegetable crops. Agric Environ Ethics 26:547–567. https://doi.org/10.1007/s10806-012-9403-1

Gad Alla SA, Almaz MM, Thabet WM, Nabil MM (2015) Evaluation of pesticide residues in some egyptian fruits. Int J Environ 4(1):87–97

Gemail KMS (2012) Monitoring of wastewater percolation in unsaturated sandy soil using geoelectrical measurements at Gabal el Asfar farm, northeast Cairo, Egypt. Environ Earth Sci 66:749–761. https://doi.org/10.1007/s12665-011-1283-6

Han FX (2007) Binding and distribution of trace elements among solid-phase components in arid zone soils. In: Alloway BJ, Trevors JK (eds) Biogeochemistry of trace elements in arid environments. Springer, The Netherlands, pp 131–167. https://doi.org/10.1007/978-1-4020-6024-3_5

Hassan SK, Khoder MI (2017) Chemical characteristics of atmospheric $PM_{2.5}$ loads during air pollution episodes in Giza, Egypt. Atmos Environ 150:346–355. https://doi.org/10.1016/j.atmosenv.2016.11.026

Heathwaite AL, Quinn PF, Hewett CJM (2005) Modelling and managing critical source areas of diffuse pollution from agricultural land using flow connectivity simulation. J Hydrol 304:446–461. https://doi.org/10.1016/j.jhydrol.2004.07.043

Hou H, Takamatsu T, Koshikawa MK, Hosomi MM (2005) Migration of silver, indium, tin, antimony, and bismuth and variations in their chemical fractions on addition to uncontaminated soils. Soil Sci 170:624–639. https://doi.org/10.1097/01.ss.0000178205.35923.66

Hussein IAEG (2011) Desertification process in Egypt. In: Brauch HG, Oswald Spring Ú, Mesjasz C, Grin J, Kameri-Mbote P, Chourou B, Dunay P, Birkmann J (eds), Coping with global environmental change, disasters and security—threats, challenges, vulnerabilities and risks. Hexagon Series on Human and environmental security and peace, vol 5. Springer, Berlin, pp 863–874. https://doi.org/10.1007/978-3-642-17776-7_50

Ibrahim MIM, Mohamed NAEM (2016) Towards sustainable management of solid waste in Egypt. Procedia Environ Sci 34:336–347. https://doi.org/10.1016/j.proenv.2016.04.030

Johnsen AR, Wick LY, Harms H (2005) Principles of microbial PAH-degradation in soil. Environ Pollut 133:71–84. https://doi.org/10.1016/j.envpol.2004.04.015

Kabata-Pendias A (2011) Trace elements in soils and plants, 4th edn. CRC Press, Boca Raton

Kabata-Pendias A, Mukherjee AB (2007) Trace elements from soil to human. Springer, Berlin, 556 pp

Kandil MM, El-Aswad AF, Koskinen WC (2015) Sorption-desorption of imidacloprid onto a lacustrine Egyptian soil and its clay and humic acid fractions. J Environ Sci Health B 50(7):473–483. https://doi.org/10.1080/03601234.2015.1018758

Kodešová R, Kočárek M, Kodeš V, Drábek O, Kozák J, Hejtmánková K (2011) Pesticide adsorption in relation to soil properties and soil type distribution in regional scale. J Hazard Mater 186(1):540–550. https://doi.org/10.1016/j.jhazmat.2010.11.040

Lal R (2008) Soils and sustainable agriculture. A review. Agron Sustain Dev 28:57–64. https://doi.org/10.1051/agro:2007025

Landrigan PJ, Fuller R, Acosta NJR, Adeyi O, Arnold R, Basu N et al (2017) The lancet commission on pollution and health. Lancet 2017, published online Oct 19. http://dx.doi.org/10.1016/S0140-6736(17)32345-0

Loch JPG, Greve PA, van der Berg S (1990) Accumulation and leaching of the fungicide fentin acetate and intermediates in sandy soils. Water Air Soil Pollut 53:119–129. https://doi-org.libezp.lib.lsu.edu/10.1007/BF00154996

Mahmoud EK, Ghoneim AM (2016) Effect of polluted water on soil and plant contamination by heavy metals in El-Mahla El-Kobra, Egypt. Solid Earth 7:703–711. https://doi.org/10.5194/se-7-703-2016

Mansour SA (2004) Pesticide exposure-Egyptian scene. Toxicology 198:91–115. https://doi.org/10.1016/j.tox.2004.01.036

Mansour SA (2008) Environmental Impact of pesticides in Egypt. In: Whitacre DM (ed) Reviews of environmental contamination and toxicology, vol 196. Springer, New York, pp 1–51. https://doi.org/10.1007/978-0-387-78444-1_1

Masoud AA (2014) Groundwater quality assessment of the shallow aquifers west of the Nile Delta (Egypt) using multivariate statistical and geostatistical techniques. J Afr Earth Sci 95:123–137. https://doi.org/10.1016/j.jafrearsci.2014.03.006

Melegy AA, Cvečková V, Krčmová K, Rapant S (2010) Environmental risk assessment of some potentially toxic elements in El-Tabbin region (Cairo, Egypt). Environ Earth Sci 61:429–439. https://doi.org/10.1007/s12665-009-0355-3

Morillo E, Villaverde J (2017) Advanced technologies for the remediation of pesticide-contaminated soils. Sci Total Environ 586:576–597. https://doi.org/10.1016/j.scitotenv.2017.02.020

Mortvedt JJ (1995) Heavy metal contaminants in inorganic and organic fertilizers. Fertilizer Res 43:55–61. https://doi.org/10.1007/BF00747683

Muhammetoğlu H, Muhammetoğlu A, Soyupak S (2002) Vulnerability of groundwater to pollution from agricultural diffuse sources: a case study. Water Sci Technol 45(9):1–7

Nziguheba G, Smolders E (2008) Inputs of trace elements in agricultural soils via phosphate fertilizers in European countries. Sci Total Environ 390:53–57. https://doi.org/10.1016/j.scitotenv.2007.09.031

Omran EE (2017) Land and groundwater resources in the Egypt's Nile Valley, Delta, and its fringes. In: The handbook of environmental chemistry. Springer, Berlin. https://doi.org/10.1007/698_2017_64

Osman KT (2014) Soil degradation, conservation and remediation. Springer, Dordrecht, pp 149–226. https://doi.org/10.1007/978-94-007-7590-9

Osman MA, Kishk FM, Moussa AA, Gaber HM (2017) Sorption and desorption of atrazine on natural bentonite and organically modified bentonite. Alexandria Sci Exch J 38(1):271–283. https://doi.org/10.21608/asejaiqjsae.2017.3475

Özcan Z, Kentel E, Alp E (2017) Evaluation of the best management practices in a semi-arid region with high agricultural activity. Agric Water Manag 194:160–171. https://doi.org/10.1016/j.agwat.2017.09.007

Peer WA, Baxter IR, Richards EL, Freeman JL, Murphy AS (2005) Phytoremediation and hyperaccumulator plants. In: Tamas MJ, Martinoia E (eds) Molecular biology of metal homeostasis and detoxification. Topics in current genetics, vol 14. Springer, Berlin, pp 299–340. https://doi.org/10.1007/4735_100

Pignatello JJ (2006) Fundamental issues in sorption related to physical and biological remediation of soils. In: Twardowska I, Allen HE, Häggblom MM, Stefaniak S (eds) Soil and water pollution monitoring, protection and remediation. NATO science series, vol 69. Springer, Dordrecht, pp 41–68. https://doi.org/10.1007/978-1-4020-4728-2_3

Rashad M, Shalaby EA (2007) Dispersion and deposition of heavy metals around two municipal solid waste (MSW) dumpsites, Alexandria, Egypt. Am Eurasian J Agric Environ Sci 2(3):204–212

Rashed MN (2010) Monitoring of contaminated toxic and heavy metals, from mine tailings through age accumulation, in soil and some wild plants at Southeast Egypt. J Hazard Mater 178:739–746. https://doi.org/10.1016/j.jhazmat.2010.01.147

Rinklebe J, Shaheen SM (2017) Geochemical distribution of Co, Cu, Ni, and Zn in soil profiles of fluvisols, luvisols, gleysols, and calcisols originating from Germany and Egypt. Geoderma 307:122–138. https://doi.org/10.1016/j.geoderma.2017.08.005

Rüdel H (2003) Case study: bioavailability of tin and tin compounds. Ecotoxicol Environ Saf 56:180–189. https://doi.org/10.1016/S0147-6513(03)00061-7

Sabiha-Javied, Mehmood T, Chaudhry MM, Tufail M, Irfan N (2009) Heavy metal pollution from phosphate rock used for the production of fertilizer in Pakistan. Microchem J 91:94–99. https://doi.org/10.1016/j.microc.2008.08.009

Sadegh-Zadeh F, Wahid SA, Jalili B (2017) Sorption, degradation and leaching of pesticides in soils amended with organic matter: a review. Adv Environ Technol 2:119–132. https://doi.org/10.22104/AET.2017.1740.1100

Salman SA, Elnazer AA, El Nazer HA (2017) Integrated mass balance of some heavy metals fluxes in Yaakob village, South Sohag, Egypt. Int J Environ Sci Technol 14:1011–1018. https://doi.org/10.1007/s13762-016-1200-3

Shaheen SM (2009) Sorption and lability of cadmium and lead in different soils from Egypt and Greece. Geoderma 153(1–2):61–68. https://doi.org/10.1016/j.geoderma.2009.07.017

Shaheen SM, Tsadilas CD, Mitsibonas T, Tzouvalekas M (2009) Distribution coefficient of copper in different soils from Egypt and Greece. Commun Soil Sci Plant Anal 40(1):214–226. https://doi.org/10.1080/00103620802625625

Shaheen SM, Tsadilas CD, Rupp H, Rinkleb J, Meissner R (2015) Distribution coefficients of cadmium and zinc in different soils in mono-metal and competitive sorption systems. J Plant Nutr Soil Sci 178:671–681. https://doi.org/10.1002/jpln.201400475

Shaheen SM, Kwon EE, Biswas JK, Tack FMG, Ok YS, Rinkleb J (2017) Arsenic, chromium, molybdenum, and selenium: Geochemical fractions and potential mobilization in riverine soil profiles originating from Germany and Egypt. Chemosphere 180:553–563. https://doi.org/10.1016/S0147-6513(03)00061-7

Shahid M, Dumat C, Khalid S, Niazi NK, Antunes PMC (2016) Cadmium bioavailability, uptake, toxicity and detoxification in soil-plant system. In: de Voogt P (eds) Reviews of environmental contamination and toxicology. Reviews of environmental contamination and toxicology (Continuation of residue reviews), vol 241. Springer, Cham, pp 73–137

Shamrukh M, Corapcioglu MY, Hassona FAA (2001) Modeling the effect of chemical fertilizers on ground water quality in the Nile valley aquifer, Egypt. Ground Water 39(1):59–67. https://doi.org/10.1111/j.1745-6584.2001.tb00351.x

Shetaya WH, Marzouk ER, Mohamed EF, Elkassas M, Bailey EH, Young SD (2018) Lead in Egyptian soils: origin, reactivity and bioavailability measured by stable isotope dilution. Sci Total Environ 618:460–468. https://doi.org/10.1016/j.scitotenv.2017.11.040

Shokr MS, El Baroudy AA, Fullen MA, El-beshbeshy TR, Ramadan AR, Abd El Halim A et al (2016) Spatial distribution of heavy metals in the middle Nile delta of Egypt. Int Soil Water Conserv Res 4:293–303. https://doi.org/10.1016/j.iswcr.2016.10.003

Taha SM, Amer ME, Elmarsafy AE, Elkady MY (2014) Adsorption of 15 different pesticides on untreated and phosphoric acid treated biochar and charcoal from water. J Environ Chem Eng 2:2013–2025. https://doi.org/10.1016/j.jece.2014.09.001

Usman ARA (2008) The relative adsorption selectivities of Pb, Cu, Zn, Cd and Ni by soils developed on shale in New Valley, Egypt. Geoderma 144:334–343. https://doi.org/10.1016/j.geoderma.2007.12.004

Vidali M (2001) Bioremediation. An overview. Pure Appl Chem 73 (7):1163–1172. https://doi.org/10.1351/pac200173071163

Wahba MAS (2017) Assessment of options for the sustainable use of agricultural drainage water for irrigation in Egypt by simulation modelling. Irrig Drain 66:118–128. https://doi.org/10.1002/ird.2029

Wong MH (2003) Ecological restoration of mine degraded soils, with emphasis on metal contaminated soils. Chemosphere 50:775–780. https://doi.org/10.1016/S0045-6535(02)00232-1

Xu J, Bravo AG, Lagerkvist A, Bertilsson S, Sjöblom R, Kumpiene J (2015) Sources and remediation techniques for mercury contaminated soil. Environ Int 74:42–53. https://doi.org/10.1016/j.envint.2014.09.007

Zidan MS, Dawoud MA (2013) Agriculture use of marginal water in egypt: opportunities and challenges. In: Shahid SA et al (eds) Developments in soil salinity assessment and reclamation: innovative thinking and use of marginal soil and water resources in irrigated agriculture. Springer Science + Business Media, Dordrecht, pp 661–679. https://doi.org/10.1007/978-94-007-5684-7_43

12 Soils and Humans

Hassan El-Ramady, Tarek Alshaal, Tamer Elsakhawy, Alaa El-Dein Omara, Neama Abdalla, and Eric C. Brevik

Abstract

Soil has a great and holy position worldwide. This position has been acquired from the importance of soil in saving food, feed, fuel, and fibre for animals and humans. Egypt was and still one of the most important countries, which soils played a crucial role in the Egyptian civilization. Therefore, very strong link between soils and humans has been reported based on the great roles of soils in plant and human nutrition. On the other hand, there are several anthropogenic activities, which cause many problems for soils such as pollution, degradation, and erosion. There are direct and/or indirect effects of soils on human health as well as plants. Therefore, this chapter is an attempt to emphasize the great roles of soils in plant and human health as well as the security of soils under pollution conditions.

Keywords

Soils • Human health • Human creation
Plant nutrition • Soil security

H. El-Ramady (✉) · T. Alshaal
Soil and Water Department, Faculty of Agriculture, Kafrelsheikh University, Kafr El Sheikh, Egypt
e-mail: ramady2000@gmail.com

T. Elsakhawy · A. E.-D. Omara
Soils, Water and Environment Research Institute (SWERI), ARC, Sakha, Egypt

N. Abdalla
Plant Biotechnology Department, Genetic Engineering Division, National Research Center, Giza, Egypt

E. C. Brevik
Department of Natural Sciences, Dickinson State University, Dickinson, ND, USA

12.1 Introduction

Soils and humans were and still are one of the most important relationships, and scientists are seeking for new approaches and facts in this relationship (Brevik and Burgess 2012, 2014; Winiwarter 2014; Brevik and Sauer 2015; Brevik et al. 2017; Carré et al. 2017). This interaction between soils and humans includes many fields or issues such as (1) the role of soil in human emergence or creation, (2) its role in plant and human nutrition or biofortification, (3) its role in human health, and (4) the role of this interaction in soil security. Definitely, the plant and human nutrition mainly depends on soils and their status. In other words, right plant nutrition is important for proper and safe human health. This could be proved through the fact that "elements or nutrients, which could be found in soil, plant, and human including carbon, nitrogen, phosphorus, potassium; are the same and beneficial for all" as reported by Osman (2013). Only 11 elements, nearly constitute 99.9% of the human body (as atoms) including the major elements (C, H, O, and N) that represent 99% and minor elements (P, K, Ca, Mg, S, Na, and Cl) that represent 0.9% of the body (Combs 2005), and with the exception of C, H, and O, these elements come from the soil. Therefore, a strong interdependence between humans and soil exists. As mentioned before, the history of human civilization is based mainly on the strong relation between soil and humans as well as the influence of humans on soil quality, and humans' perception of how to manage soil resources (Lal 2005).

Therefore, this chapter will focus on the relationship between soils and humans including the role of soils in supporting proper human health, causing diseases for humans, supplying nutrition for both plants and humans through biofortification, and finally, soil pollution and security as it relates to human health. These previous issues, nowadays, are very important in drawing the plan for human health and maintaining the security of all our lives.

H. El-Ramady et al. (eds.), *The Soils of Egypt*, World Soils Book Series,
https://doi.org/10.1007/978-3-319-95516-2_12

12.2 Soils and Human Health

Human health depends on and still supports by soil and its characterization (Balks and Zabowski 2016; Prasad and Shivay 2016). So, it is very common to hear "*feed the soil to feed the human*" as shown in Fig. (12.1). It is also reported that approximately 78% of the average per capita calorie consumption worldwide comes from crops grown directly in soils, and another nearly 20% comes from terrestrial food sources that rely indirectly on soils (Brevik 2013a). The relationship between soils and human health and its potential has been recognized for thousands of years and is recorded from the time of the Pharaohs. There are many direct and indirect impacts of soils on human health including (1) the role of soils in food security (food availability and quality), (2) effect of various soil chemicals and pathogens on humans, and (3) acting as natural filters to remove different contaminants from water. Thus, there are several complicated soil variables which makes the investigations of soils and human health difficult and also means that the complete study of soils and human health should involve many different specialties such as soil scientists, medical professionals, toxicologists, anthropologists, etc. These specialty groups traditionally do not work together on research projects and do not always effectively communicate with one another (Pepper et al. 2009; Brevik and Hartemink 2010; Brevik 2013a, b; Pepper 2013; Shishkov and Kolev 2014; Brevik and Burgess 2014; Brevik and Sauer 2015; El-Ramady et al. 2015; Brevik et al. 2017).

Recently, a large number of studies regarding the effects of soils on human health have been published including issues such as exposures to radioactive elements in soils (Balonov et al. 1999; Dushenkov et al. 1999), the effects of pollution with heavy metals in soils (Alloway 1995;

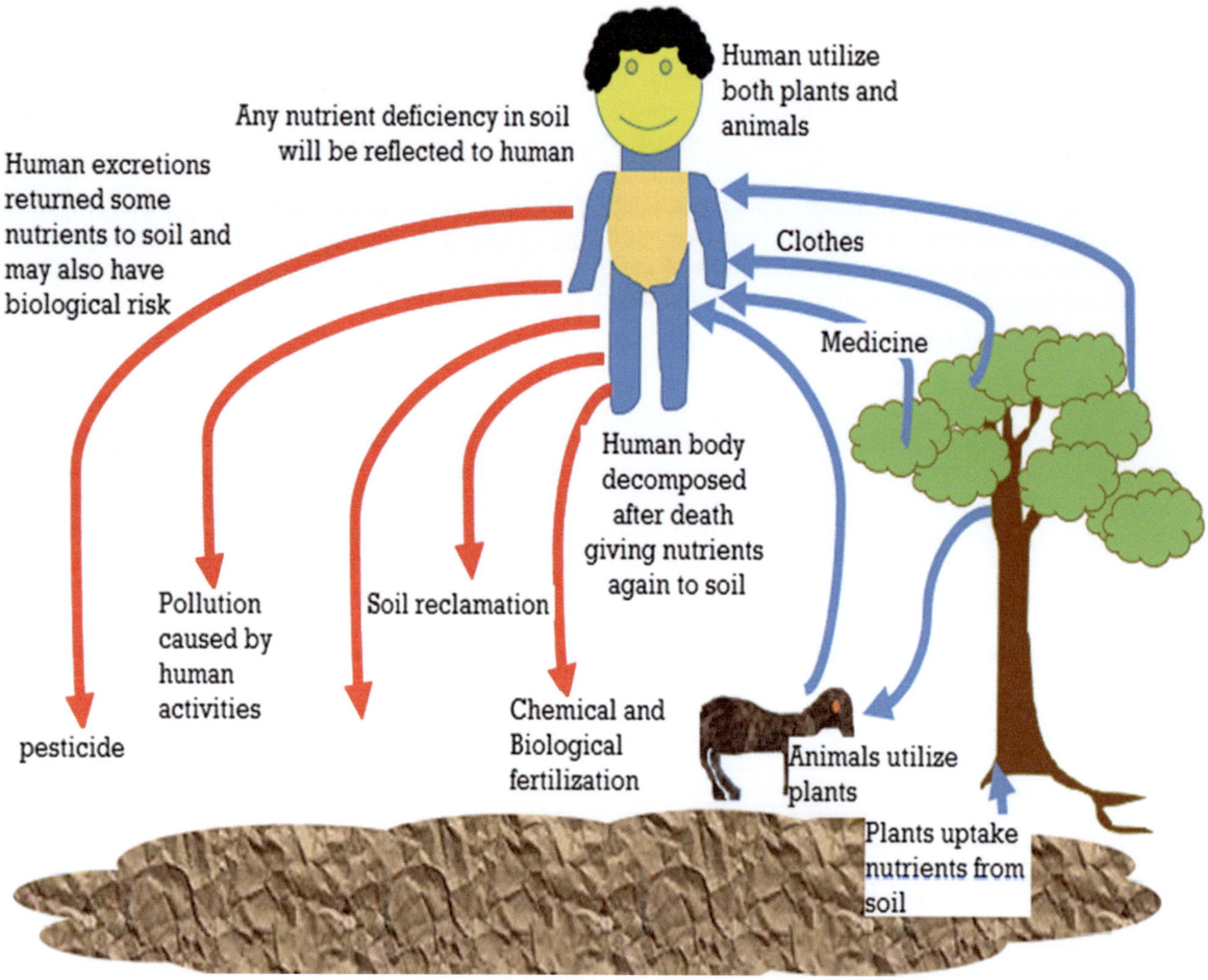

Fig. 12.1 The strong relationship between soils and humans in general include two sides: the first represents the harmful or negative action (destroying the soil through human activities like pollution) and the second includes the positive activities, which leads to increase soil health or quality. Source: authors

Albering et al. 1999; Gu et al. 2016; Li et al. 2017; Tepanosyan et al. 2017; Zhang et al. 2017) and organic contaminants in soils (Pohl et al. 1995; Simcox et al. 1995; Burgess 2013; Burras et al. 2013; Liu et al. 2016; Cai et al. 2017; Wei et al. 2017), geophagy or medical geology (Selinus et al. 2005; Hooda and Henry 2007; Henry and Cring 2013), biofortification (Cakmak 2002, 2008; El-Ramady et al. 2015a; El-Ramady et al. 2016a, b; dos Reis et al. 2017), and many other subjects (Lichtfouse 2012; Bini and Bech 2014; Alshaal et al. 2017; Bourliva et al. 2017; dos Reis et al. 2017).

On the other hand, soils are the main source for production of antibacterial agents and cancer drugs, with 78 and 60% of these drugs having a soil origin, respectively (Pepper et al. 2009). Furthermore, several other medicines have been derived from soils and soil organisms. Geophagy (the practice of eating soil) was common behavior in the rural parts of Egypt, especially for pregnant women. These women suffer from a deficiency of certain elements, so they eat clay particles that may supply nutrients such as calcium and iron. This phenomenon (geophagy) has attracted the interest of several geographers and anthropologists for many years (Hooda and Henry 2007; Henry and Cring 2013; Brevik and Sauer 2015). So, new approaches concerning the effects of soils on human health include investigating the transfer of nutrients from soils to plants to humans (e.g., Kabata-Pendias and Mukherjee 2007; Cakmak 2008; Spiegel et al. 2009; Roivainen et al. 2012; Simon 2014; El-Ramady et al. 2015b, 2016a; Ávila et al. 2017), from soils to plants to animals to humans (e.g., Jones 2005; Klasing et al. 2005; Abreu et al. 2014; Krajcarová et al. 2016; Saha et al. 2017), or even directly from soils to humans (e.g., Abrahams et al. 2006; Hooda and Henry 2007; Young 2007; Angelone and Udovic 2014; Brevik and Sauer 2015; Diami et al. 2016; Baran et al. 2017; Saha et al. 2017).

The effects of soils on human health can be either positive or negative. Positive effects of soils include (1) soils are the main source for the production of food, feed, fiber, and fuel for humans (Khalifa and Moussa 2017), (2) soils are the natural filter in removing and/or degrading a number of pollutants (Saha et al. 2017), (3) production of drugs and vaccines extracted from soil microbes (Bush and Page 2017), and (4) soil plays the main role in food security through production of sufficient amounts of nutritious cultivated crops (Bashour 2017; Blanco et al. 2017; Zdruli et al. 2017). On the other hand, negative impacts of soils on human health include its containing and leading to exposure to toxic levels of heavy metals, airborne dust, chemicals, and/or pathogens that have the potential to harm human health. It is worth mentioning that direct and/or indirect soil processes can lead to reduced crop production that will impact on food security and through that human health (Brevik and Burgess 2014).

In Egypt, like other nations, soils are considered the backbone of food production and its security, where soils were and still are the main source for agricultural production. Geohelminths distribution is one of the serious environmental problems in Egypt, and many researchers have investigated this problem in different Egyptian Governorates like Sharkyia (e.g., Etewa et al. 2016; Farghly et al. 2016) and Sohag (Melegy et al. 2014). The natural radioactivity levels in Egyptian soils are considered one of the most pressing problems facing people in locations like Assiut (El-Gamal et al. 2013), Port Said (Attia et al. 2015), Red Sea (Issa et al. 2015), Aswan (El-Gamal 2017), and Minia (El-Gamal 2017). Soil pollution is one of the most serious environmental problems in Egypt, and many researchers have investigated this problem in different Egyptian soils (e.g., Salman et al. 2017; Negm and Eltarabily 2017) and the Nile River (Abdel-Satar et al. 2017; Elhaddad and Al-Zyoud 2017) as well as aquifer system (e.g., El Baz 2015; Galal and Farahat 2015; Arnous and Hassan 2015; El-Kassas and Gharib 2016; Snousy et al. 2017).

12.3 The Role of Soil in Plant and Human Nutrition

The mineral elements that are needed for the growth of both plants and humans (or animals) are called the essential elements. Exactly which elements are essential is still under debate, although there is a common agreement for some nutrients by the plant nutrition scientists (e.g., Jones 2003; Kirkby 2012; El-Ramady et al. 2014a, b). This essentiality may also include new dimensions such as understanding the acquisition, accumulation, and transport of these nutrients in addition to their functions in plants. In other words, the latest available knowledge about the role of these elements play in plant physiology and molecular biology should be recognized (Fig. 12.2). This should also include analyzing the effects of the deficiency of different elements on both soils and the human diet or on both crop production and human health (Cakmak et al. 2017). Therefore, new dimensions of the essential mineral elements should be considered as regards to biofortification. Also, recent promising investigations of the role of elements should include the study of these elements in soil, plant, animal, and human nutrition.

The human body requires at least 29 mineral elements for survival (Welch and Graham 2004; White and Broadley 2005; Graham et al. 2007; Meena et al. 2016; Brevik et al. 2017), which are supplied by an appropriate diet. It is estimated that more than 60% of the world's people are iron (Fe) deficient, over 30% are zinc (Zn) and iodine (I) deficient and 15% selenium (Se) deficient (Thavarajah et al. 2014). This deficiency list also includes calcium (Ca), magnesium (Mg) and copper (Cu) as common in many developed and

Fig. 12.2 Proper plant nutrition with irrigation to modify surface irrigation in salt-affected rice paddy soils is presented in these photos from the Kafr El-Sheikh area. Photos by El-Ramady (2015)

developing countries (Meena et al. 2016). Therefore, there is a need to supply these nutrients. The process of increasing the density of minerals and vitamins in a crop through agronomic practices, plant breeding, or transgenic techniques is called biofortification. Regular consumption of biofortified staple crops can lead to improvements in human health and nutrition (Zhao et al. 2012; Bouis and Saltzman 2017). Common biofortified staple crops include wheat, rice, maize, cassava, sweet potato, etc. with either mineral elements (Fe, Zn, I, Se, Cu, Ca, Mg, etc.) or nutritional enhancers such as amino acids (methionine), folate, fatty acids, vitamins A and E (tocopherol), lycopene, flavonoids being enhanced (El-Ramady et al. 2015b, 2016a, b; Meena et al. 2016; Bouis and Saltzman 2017; dos Reis et al. 2017; Giuliano 2017; Goyer 2017; Gonzali et al. 2017; Mène-Saffrané and Pellaud 2017; Pearce 2017; Singh et al. 2017; Strobbe and Van Der Straeten 2017; Vasconcelos et al. 2017).

In Egypt, the common mineral element deficiencies include iron, zinc, selenium, and iodine. Iodine is often fortified in salt, whereas Fe, Zn, and Se can be biofortified in staple crops. The reader is referred to dos Reis et al. (2017) for more information on investigations into selenium deficiency and toxicity in Egypt and its effects on human health. Egypt has been listed on the HarvestPlus program biofortified global crop map as using staple foods to overcome micronutrient deficiencies (Bouis and Saltzman 2017). Further studies are needed concerning biofortification in Egypt and these investigations should include nutritional enhancers such as amino acids (methionine), folate and vitamins (like A and E) in addition to the previously mentioned mineral elements (Fe, Se, Zn, and I).

12.4 Soil Pollution and Its Security Towards Human Health

As mentioned before, soil pollution is considered as one of the most important environmental problems. So, human health also should be considered as a major challenge of global soil security taking into account that soil pollution has a global scale and soil also has full resilience in dealing with pollutants (Carré et al. 2017; Hirsch et al. 2017). Soils have the ability to reduce contaminants by biodegradation, filter and then store them or to transform them resulting in different metabolites, all of which contributes to human health (Figs. 12.3, 12.4 and 12.5). Therefore, soil security includes dimensions that consider the societal and scientific challenges that are related to soil-human health studies and soil security (Brevik et al. 2017; Carré et al. 2017). This also emphasizes that soil-human health studies are very complicated endeavors and should be included in more than one dimension as mentioned before (Brevik and Sauer 2015; Carré et al. 2017).

This relationship between soils and human health should be linked to soil security and the five dimensions (soil capability, condition, capital, connectivity, and codification) of soil security should be tied to soils and their impact on the health of humans (Brevik et al. 2017; Carré et al. 2017; Kim et al. 2017). The maintenance and improvement of different soil resources is called soil security, which is very close to other security concepts including food, water, and energy security. The different dimensions of soil security and their relationships to human health can be summarized as follows:

(1) **Soil capability and human health**: the ability of soils to produce enough high-quality food (food security) and to produce clean and safe water (water security) through soil biogeochemical degradation and filtration of wastes. The ability of soil to provide different essential nutrients to humans could be through direct and/or indirect pathways including providing nutrients from soils to plants and/or animals to humans. The maintenance of important soil properties (e.g., soil structure, thickness, fertility, aeration, biology, etc.) helps in maintaining the capacity of the soil to promote human health (Hirsch et al. 2017; Brevik et al. 2017; Bouma et al. 2017).

(2) **Soil condition and human health**: soil conditions including physical, chemical, and biological soil properties impact the nutritional quality of agricultural products. The condition dimension is closely related to soil health and healthy soils often produce abundant nutritious crops that can support the world's dietary needs and provide a positive contribution to human health (Brevik et al. 2017; McBratney et al. 2017a).

(3) **Soil capital and human health**: soil capital has a value that is evaluated in dollars and cents. This value includes several soil services, which could provide to human health (e.g., nutrients that are supplied), ecosystem services (e.g., neutralization of pollution), filtration of water, carbon sequestration and the supply of medications. All previous services might promote human health (Brevik et al. 2017; McBratney et al. 2017b).

(4) **Soil connectivity and human health**: this refers to the connection that societies form with soils, which influences how those soils are valued and managed. This, in turn, influences interactions between soils and surrounding agroecosystems. The connectivity of soils can also influence the loss of land through degradation processes, which may negatively impact on human health. Degraded soils represent a great threat to the stability of society due to deterioration of soil fertility and then human activity. Furthermore, both human

Fig. 12.3 Plants are an important part of the relationship between soil security and human health, as plants derive most of their nutrients from soils and grazing may support soil productivity, as resulted from manure of sheep and other farm animals in the middle photo, thus introducing sufficient nutrients to the food web. Photos in the Kafrelsheikh area by El-Ramady (2016)

Fig. 12.4 Non-healthy soils often suffer from a deficiency of essential elements to support healthy plants. The saline and alkaline soils in the Kafrelsheikh area have many problems that impede productivity and thus human health. Photos by El-Ramady (2016)

Fig. 12.5 A healthy soil, which has good biological, chemical, and physical properties, is able to produce safe and nutritious food that supports human health like in the Botanic Garden in Aswan, where a low level of pollution supports the security of soils. Photo by El-Ramady (2010)

health and soil security will be destroyed under highly polluted and degraded conditions (Brevik et al. 2017; Carré et al. 2017; Kim et al. 2017).

(5) **Soil codification and human health**: this dimension has typically focused on water and soil conservation more than directly on human health. Appropriate conservation practices can improve water and soil quality, leading to high-quality crop production in those soils and improving human health. This dimension refers to governmental programs, policies, and goals, which impacts the production conditions of soil and finally, all harvested crops is consumed by humans (Amundson 2017; Brevik et al. 2017; Koch 2017). It can be concluded that soils greatly contribute to the quality of the necessities of our life (water, air, and food), which are directly linked to the health of humans. This human health definitely depends on the soil security. This soil security also, in turn, will be affected by all factors and processes that lead to or contribute to the deterioration of soils, like pollution. Therefore, further investigations are crucial to focus on the direct and indirect link between soil security and human health under different conditions.

In Egypt, soil security should be considered for more than the current situation. For example, all advanced nations criminalize infringement on soils and punish the infringer. This advanced behavior protects both the government and people. In the USA, on the other hand, agricultural policies are based on encouraging proper maintenance of soils rather than punishment for improper soil management. There are punishments for pollution of soils in certain circumstances, such as with industrial contamination. In Egypt, the main problem in this context is represented in the urban sprawl that is devouring vast areas of agricultural land in the Nile Delta (Fig. 12.6). The radioactivity of soils also has been become a serious problem in Egypt. The radioactive materials found in Egyptian soils may result from some industries

Fig. 12.6 The urban sprawl that is devouring vast areas of agricultural land in the Nile Delta represents one of the biggest challenges in Egypt. This urban crawl and resulting pollution could threaten soil security and then human health. Photo for Tanta City by El-Ramady (2015)

Table 12.1 List of some phosphate rocks, phosphogypsum, and phosphate fertilizers used in Egypt including trade names, plants, and the radioactive activity concentration (Bq kg^{-1}) as adapted from El-Bahi et al. (2017)

Plant (or Region)	Activity concentration (Bq kg^{-1})				
	^{40}K	^{226}Ra	^{232}Th	^{235}U	^{238}U
Phosphate rocks					
Abu-Tartor	891	1496	200	86.2	2032
	893	974	83.2	53.0	1151
Wadi Qena	610	1226	118.4	65.8	1694
Abu-Zabal Factory (El-Sebaeya)	687	422	51.8	33.4	687
	572	463	32.4	26.6	639
El-Nasr Company (El-Sebaeya)	796	1141	107.4	60.1	1378
Single super phosphate (SSP)					
El-Nasr Company (El-Sebaeya)	901	466	N.D	22.6	491
Kafr El-Zayat Factory	891	351	53.6	28.1	594
Tri granular phosphate fertilizer					
Abu-Zabal Factory (El-Sebaeya)	737	302	193.9	10.9	1849
Phosphogypsum					
Abu-Zabal Factory (El-Sebaeya)	785	397	11.5	51.7	283
	954	702	147.3	42.6	1041

N.D Not Detected

(El-Zakla 2013; Attia et al. 2015), mineral fertilizers like phosphate fertilizers (El-Bahi et al. 2017) (Table 12.1), natural materials like phosphate rock (Issa et al. 2015; El-Bahi et al. 2017), or phosphogypsum (El-Didamony et al. 2013; El-Bahi et al. 2017), and other sources (Badawy et al. 2013; El-Gamal et al. 2013; Mohamed et al. 2016).

According to many studies, it is very important to build and develop a database for a radio-ecological atlas as soon as possible for Egypt.

12.5 Conclusion

The relationship between soils and human health has been represented in public regulations including the traditions, morals, ethics, and customs that have acquired crucial importance in civil societies (Shishkov and Kolev 2014). The relationship between soils and humans has been recognized as having critical importance over time as an intransitive global issue. It is also worth mentioning that the relationship between soils and humans is primarily a relationship between man and earth. This fact penetrates the development of mankind over all ages, civilizations and historical social formations (Brevik et al. 2018). Thus, the nature of this relationship is very complex and soil is essential for the reproduction of biota including human beings. Soils are also a natural resource sui generis for humans and are essential for the creation, existence, development, and prosperity of the human race. Soil security is an approach that shows the close link between soil services and human ability to solve different key issues for sustainable development including the security of food, water, energy, climate regulation, and biodiversity. An important cycle for the relationship between soil security and human health starts with plants, which derive most of their nutrients from soils, continues to animals that can eat those plants, and both plants and animals supplying those nutrients to the food web (including humans). Therefore, healthy soils are essential for the health of plants and animals, including humans, which eat the plant products grown in those soils or other products from further up the food web.

Acknowledgements The authors thank the outstanding contributions of STDF research teams (Science and Technology Development Fund, Egypt) and MBMF/DLR (the Federal Ministry of Education and Research of the Federal Republic of Germany), (Project ID 5310) for their help. Support from the German-Egyptian Research Fund (GERF) is gratefully acknowledged.

References

Abdel-Satar AM, Ali MH, Goher ME (2017) Indices of water quality and metal pollution of Nile River, Egypt. Egypt J Aquat Res 43:21–29. https://doi.org/10.1016/j.ejar.2016.12.006

Abrahams PW, Follansbee MH, Hunt A, Smith B, Wragg J (2006) Iron nutrition and possible lead toxicity: an appraisal of geophagy undertaken by pregnant women of U.K. Asian communities. Appl Geochem 21:98–108

Abreu MM, Bech J, Carvalho LC, Santos E (2014) Potential hazardous elements fluxes from soil to plants and the food chain. In: Bini C, Bech J (eds) PHEs, environment and human health. Springer Science + Business Media, Dordrecht, pp 309–337. https://doi.org/10.1007/978-94-017-8965-3_8

Albering HJ, van Leusen SM, Moonen EJC, Hoogewerff JA, Kleinjans JCS (1999) Human health risk assessment: a case study involving heavy metal soil contamination after the flooding of the River Meuse during the winter of 1993–1994. Environ Health Persp 107:37–43

Alloway BJ (1995) Soil processes and the behavior of heavy metals. In: Alloway BJ (ed) Heavy metals in soils, 2nd edn. Blackie Academic and Professional, London, pp 11–37

Alshaal T, Al-Saeedi AH, Shalaby T, Elsakhawy T, Omara AED, Gad A, Hamad E, El-Ghamry A, Mosa A, Amer M, Abdalla N, El-Ramady H (2017) The rhizosphere and plant nutrition under climate change. In: Naeem M et al (eds) Essential plant nutrients. Springer International Publishing AG, Switzerland. https://doi.org/10.1007/978-3-319-58841-4_11,

Amundson R (2017) Whose security is important? Communicating environmental risk about soil to a diverse audience. In: Field DJ et al (eds) Global soil security, progress in soil science. Springer International Publishing, Switzerland, pp 383–387. https://doi.org/10.1007/978-3-319-43394-3_34

Angelone M, Udovic M (2014) Potentially harmful elements in urban soils. In: Bini C, Bech J (eds) PHEs, environment and human health. Springer Science + Business Media, Dordrecht, pp 221–251. https://doi.org/10.1007/978-94-017-8965-3_6

Arnous MO, Hassan MAA (2015) Heavy metals risk assessment in water and bottom sediments of the eastern part of Lake Manzala, Egypt, based on remote sensing and GIS. Arab J Geosci 8:7899–7918. https://doi.org/10.1007/s12517-014-1763-6

Attia TE, Shendi EH, Shehata MA (2015) Assessment of natural and artificial radioactivity levels and radiation hazards and their relation to heavy metals in the industrial area of Port Said city, Egypt. Environ Sci Pollut Res 22:3082–3097. https://doi.org/10.1007/s11356-014-3453-z

Ávila PF, da Silva EF, Candeias C (2017) Health risk assessment through consumption of vegetables rich in heavy metals: the case study of the surrounding villages from Panasqueira mine, Central Portugal. Environ Geochem Health 39:565–589. https://doi.org/10.1007/s10653-016-9834-0

Badawy WM, Eissa HS, Belal A, Mohamed ES (2013) Estimation of some radiological parameters due to gamma radioactivity in soil samples from some Southern Governorates–Egypt. J Nucl Res Dev 5:37–44

Balks MR, Zabowski D (2016) Celebrating soil: discovering soils and landscapes. Springer International Publishing, Switzerland. https://doi.org/10.1007/978-3-319-32684-9

Balonov MI, Krisyuk EM, Ramel C (1999) Environmental radioactivity, population exposure and related health risks in the east Baltic region, Scand. J Work Environ Health 25:17–32

Baran A, Wieczorek J, Mazurek R, Urbanski K, Klimkowicz-Pawlas A (2017) Potential ecological risk assessment and predicting zinc accumulation in soils. Environ Geochem Health. https://doi.org/10.1007/s10653-017-9924-7

Bashour I (2017) Soils and food security in the Arab world. In: Murad S et al (eds) Water, energy & food sustainability in the middle east. Springer International Publishing AG, Switzerland, pp 283–298. https://doi.org/10.1007/978-3-319-48920-9_13

Bini C, Bech J (2014) PHEs, environment and human health: potentially harmful elements in the environment and the impact on human health. Springer Science + Business Media, Dordrecht. https://doi.org/10.1007/978-94-017-8965-3

Blanco A, Salazar MJ, Cid CV, Pignata ML, Rodriguez JH (2017) Accumulation of lead and associated metals (Cu and Zn) at different growth stages of soybean crops in lead-contaminated soils: food security and crop quality implications. Environ Earth Sci 76:182. https://doi.org/10.1007/s12665-017-6508-x

Bouis HE, Saltzman A (2017) Improving nutrition through biofortification: a review of evidence from HarvestPlus, 2003 through 2016. Glob Food Secur 12:49–58

Bouma J, van Ittersum MK, Stoorvogel JJ, Batjes NH, Droogers P, Pulleman MM (2017) Soil capability: exploring the functional potentials of soils. In: Field DJ et al (eds) Global soil security, progress in soil science. Springer International Publishing, Switzerland, pp 27–44. https://doi.org/10.1007/978-3-319-43394-3_3

Bourliva A, Papadopoulou L, Aidona E, Giouri K, Simeonidis K, Vourlias G (2017) Characterization and geochemistry of technogenic magnetic particles (TMPs) in contaminated industrial soils: Assessing health risk *via* ingestion. Geoderma 295:86–97

Brevik EC (2013a) Soils and human health—an overview. In: Brevik EC, Burgess LC (eds) Soils and human health. CRC Press, Boca Raton, pp 29–56

Brevik EC (2013b) Climate change, soils, and human health. In: Brevik EC, Burgess LC (eds) Soils and human health. CRC Press, Boca Raton, pp 345–383

Brevik EC, Burgess LC (2012) Soils and human health. CRC Press, Boca Raton

Brevik EC, Burgess LC (2014) The influence of soils on human health. Nat Educ Knowl 5(12):1. http://www.nature.com/scitable/knowledge/library/the-influenceof-soils-on-human-health-127878980

Brevik EC, Sauer TJ (2015) The past, present, and future of soils and human health studies. Soil 1:35–46. https://doi.org/10.5194/soil-1-35-2015

Brevik EC, Steffan JJ, Burgess LC, Cerdà A (2017) Links between soil security and the influence of soil on human health. In: Field DJ et al (eds) Global soil security, progress in soil science series. Springer International Publishing, Switzerland, pp 261–274. https://doi.org/10.1007/978-3-319-43394-3_24

Brevik EC, Homburg JA, Sandor JA (2018) Soils, climate, and ancient civilizations. In: Horwath W, Kuzyakov Y (eds) Changing soil processes and ecosystem properties in the anthropocene. Developments in soil science series. Elsevier, Amsterdam

Burgess LC (2013b) Organic pollutants in soil. In: Brevik EC, Burgess LC (eds) Soils and human health. CRC Press, Boca Raton, pp 83–106

Burras CL, Nyasimi M, Butler L (2013) Soils, human health, and wealth: A complicated relationship. In: Brevik EC, Burgess LC (eds) Soils and human health. CRC Press, Boca Raton, pp 215–226

Bush K, Page MGP (2017) What we may expect from novel antibacterial agents in the pipeline with respect to resistance and pharmacodynamic principles. J Pharmacokinet Pharmacodyn 44:113–132. https://doi.org/10.1007/s10928-017-9506-4

Cai CY, Li JY, Wu D, Wang XL, Tsang DCW, Li XD, Sun JT, Zhu LZ, Shen HZ, Tao S, Liu WX (2017) Spatial distribution, emission source and health risk of parent PAHs and derivatives in surface soils from the Yangtze River Delta, eastern China. Chemosphere 178:301–308

Cakmak I (2002) Plant nutrition research: priorities to meet human needs for food in sustainable ways. Plant Soil 247:3–24

Cakmak I (2008) Enrichment of cereal grains with zinc: agronomic or genetic biofortification? Plant Soil 302:1–17

Cakmak I, McLaughlin MJ, White P (2017) Zinc for better crop production and human health. Plant Soil 411:1–4. https://doi.org/10.1007/s11104-016-3166-9

Carré F, Caudeville J, Bonnard R, Bert V, Boucard P, Ramel M (2017) Soil contamination and human health: a major challenge for global soil security. In: Field DJ et al (eds) Global soil security, progress in soil science. Springer International Publishing Switzerland, pp 275–295. https://doi.org/10.1007/978-3-319-43394-3_25

Combs GF Jr (2005) Geological impacts on nutrition. In: Selinus O (eds) Essentials of medical geology. Elsevier, Amsterdam, pp 161–177

Diami SM, Kusin FM, Madzin Z (2016) Potential ecological and human health risks of heavy metals in surface soils associated with iron ore mining in Pahang, Malaysia. Environ Sci Pollut Res 23:21086–21097. https://doi.org/10.1007/s11356-016-7314-9

dos Reis AR, El-Ramady H, Santos EF, Gratão PL, Schomburg L (2017) Overview of selenium deficiency and toxicity worldwide: affected areas, selenium-related health issues, and case studies. In: Pilon-Smits EAH et al (eds) Selenium in plants, Plant ecophysiology, vol 11. Springer International Publishing AG, Switzerland, pp 209–230. https://doi.org/10.1007/978-3-319-56249-0_13

Dushenkov S, Mikheev A, Prokhnevsky A, Ruchko M, Sorochinsky B (1999) Phytoremediation of radiocesium-contaminated soil in the vicinity of Chernobyl, Ukraine. Environ Sci Technol 33:469–475

El Baz SM (2015) Benthic foraminifera as bioindicators of heavy metal pollution in Lake Burullus. Egypt Arab J Geosci 8:5491–5509. https://doi.org/10.1007/s12517-014-1611-8

El-Bahi SM, Sroor A, Mohamed GY, El-Gendy NS (2017) Radiological impact of natural radioactivity in Egyptian phosphate rocks, phosphogypsum and phosphate fertilizers. Appl Radiat Isot. https://doi.org/10.1016/j.apradiso.2017.02.031

El-Didamony H, Ali MM, Awwad NS, Attallah MF, Fawzy MM (2013) Radiological characterization and treatment of contaminated phosphogypsum waste. Radiochemistry 55(4):454–459. https://doi.org/10.1134/S106636221304019X

El-Gamal AA (2017) Distribution of natural radioactivity in the Egyptian part of the Nile River from Aswan to El-Minia. In: Negm AM (ed) The Nile River, Handbook of Environmental Chemistry, vol 56. Springer International Publishing AG, Switzerland, pp 187–205. https://doi.org/10.1007/698_2016_94

El-Gamal H, Farid ME, Abdel Mageed AI, Hasabelnaby M, Hassanien HM (2013) Assessment of natural radioactivity levels in soil samples from some areas in Assiut, Egypt. Environ Sci Pollut Res 20:8700–8708. https://doi.org/10.1007/s11356-013-1844-1

Elhaddad E, Al-Zyoud S (2017) The quality assessment of pollution of Rosetta branch, Nile River, Egypt. Arab J Geosci 10:97. https://doi.org/10.1007/s12517-017-2870-y

El-Kassas HY, Gharib SM (2016) Phytoplankton abundance and structure as indicator of water quality in the drainage system of the Burullus Lagoon, southern Mediterranean coast, Egypt. Environ Monit Assess 188:530. https://doi.org/10.1007/s10661-016-5525-7

El-Ramady HR, Alshaal TA, Amer M, Domokos-Szabolcsy É, Elhawat N, Prokisch J, Fári M (2014a) Soil quality and plant nutrition. In: Ozier-Lafontaine H, Lesueur-Jannoyer M (eds) Agroecology and global change, sustainable agriculture reviews, vol 14. Springer International Publishing, Switzerland, pp 345–447. https://doi.org/10.1007/978-3-319-06016-3_11

El-Ramady HR, Alshaal TA, Shehata SA, Domokos-Szabolcsy É, Elhawat N, Prokisch J, Fári M, Marton L (2014b) Plant nutrition: from liquid medium to micro-farm. In: Ozier-Lafontaine H, Lesueur-Jannoyer M (eds) Agroecology and global change, sustainable agriculture reviews, vol 14. Springer International Publishing, Switzerland, pp 449–508. https://doi.org/10.1007/978-3-319-06016-3_12

El-Ramady H, Abdalla N, Alshaal T, Domokos-Szabolcsy É, Elhawat N, Prokisch J, Sztrik A, Fári M, El-Marsafawy S, Shams MS (2015a) Selenium in soils under climate change, implication for human health. Environ Chem Lett 13(1):1–19. https://doi.org/10.1007/s10311-014-0480-4

El-Ramady HR, Domokos-Szabolcsy É, Shalaby TA, Prokisch J, Fári M (2015b) Selenium in agriculture: water, air, soil, plants, food, animals and nanoselenium. In: Lichtfouse E et al (eds) CO_2 sequestration, biofuels and depollution, environmental chemistry for a sustainable world, vol 5. Springer International Publishing, Switzerland, pp 153–232. https://doi.org/10.1007/978-3-319-11906-9_5

El-Ramady H, Abdalla N, Taha HS, Alshaal T, El-Henawy A, Faizy SE-DA, Shams MS, Youssef SM, Shalaby T, Bayoumi Y, Elhawat N, Shehata S, Sztrik A, Prokisch J, Fári M, Domokos-Szabolcsy É, Pilon-Smits EA, Selmar D, Haneklaus S, Schnug E (2016a) Selenium and nano-selenium in plant nutrition. Environ Chem Lett 14(1):123–147. https://doi.org/10.1007/s10311-015-0535-1

El-Ramady H, Alshaal T, Abdalla N, Prokisch J, Sztrik A, Fári M, Domokos-Szabolcsy É (2016b) Selenium and nano-selenium biofortified sprouts using micro-farm systems. In: Bañuelos GS, Lin ZQ, Guilherme LRG, dos Reis AR (eds) Proceedings of the 4th international conference on selenium in the environment and human health, Sao Paulo, 18–21 Oct 2015. CRC, Taylor & Francis Group, London, pp 189–190. https://doi.org/10.13140/rg.2.1.1065.9925

El-Zakla T (2013) Evaluation of natural radioactivity levels and radiation hazards of some industrial Egyptian ores. Radiochemistry 55(2):236–242. https://doi.org/10.1134/S1066362213020161

Etewa SE, Abdel-Rahman SA, Abd El-Aal NF, Fathy GM, El-Shafey MA, Ewis AMG (2016) Geohelminths distribution as affected by soil properties, physicochemical factors and climate in Sharkyia governorate Egypt

Farghly AM, Mohamed SMA, Abdel-Rahman SA, Mohammed FE, El-Bahaie ES, El-Shafey MA (2016) The relation between the prevalence of soil transmitted parasites in the soil and among school children in Zagazig district, Sharkyia Governorate, Egypt. J Parasit Dis 40(3):1021–1029. https://doi.org/10.1007/s12639-014-0627-z

Galal TM, Farahat EA (2015) The invasive macrophyte *Pistia stratiotes* L. as a bioindicator for water pollution in Lake Mariut, Egypt. Environ Monit Assess 187:701. https://doi.org/10.1007/s10661-015-4941-4

Giuliano G (2017) Provitamin a biofortification of crop plants: a gold rush with many miners. Curr Opin Biotechnol 44:169–180

Gonzali S, Kiferle C, Perata P (2017) Iodine biofortification of crops: agronomic biofortification, metabolic engineering and iodine bioavailability. Curr Opin Biotechnol 44:16–26

Goyer A (2017) Thiamin biofortification of crops. Curr Opin Biotechnol 44:1–7

Graham RD, Welch RM, Saunders DA, OrtizMonasterio I, Bouis HE, Bonierbale M, de Haan S, Burgos G, Thiele G, Liria R, Craig A, Meisner CA, Beebe SE, Potts MJ, Kadian M, Hobbs PR, Gupta RK, Twomlow S (2007) Nutritious subsistence food systems. Adv Agron 92:1–74

Gu YG, Gao YP, Lin Q (2016) Contamination, bioaccessibility and human health risk of heavy metals in exposed-lawn soils from 28 Urban parks in southern China's largest city, Guangzhou. Appl Geochem 67:52–58

Henry JM, Cring FD (2013) Geophagy: an anthropological perspective. In: Brevik EC, Burgess LC (eds) Soils and human health. CRC Press, Boca Raton, pp 179–198

Hirsch PR, Jhurreea D, Williams JK, Murray PJ, Scott T, Misselbrook TH, Goulding KWT, Clark IM (2017) Soil resilience and recovery: rapid community responses to management changes. Plant Soil 412:283–297. https://doi.org/10.1007/s11104-016-3068-x

Hooda P, Henry J (2007) Geophagia and human nutrition. In: MacClancy J, Henry J, MacBeth H (eds) Consuming the inedible. Neglected dimensions of food choice. Berghahn Books, New York, pp 89–98

Issa SAM, Mostafa AMA, Lotfy AM (2015) Radiological impacts of natural radioactivity in phosphate rocks from El-Sibaiya and Red Sea coast mines, Egypt. J Radioanal Nucl Chem 303:53–61. https://doi.org/10.1007/s10967-014-3312-x

Jones JB Jr (2003) Agronomic handbook—management of crops, soils, and their fertility. CRC Press, Boca Raton

Jones B (2005) Animals and medical geology. In: Selinus O, Alloway B, Centeno JA, Finkelman RB, Fuge R, Lindh U, Smedley P (eds) Essentials of medical geology. Elsevier, Amsterdam, pp 513–526

Kabata-Pendias A, Mukherjee AB (2007) Trace elements from soil to human. Springer, Berlin

Khalifa HE, Moussa HA (2017) Soil and agriculture after the Aswan High Dam. In: Satoh M, Aboulroos S (eds) Irrigated agriculture in Egypt. Springer International Publishing, Switzerland, pp 81–124. https://doi.org/10.1007/978-3-319-30216-4_5

Kim SC, Lim KJ, Yang JE (2017) The measurement of soil security in terms of human health: examples and ideas. In: Field DJ et al (eds) Global soil security, progress in soil science. Springer International Publishing, Switzerland, pp 297–303. https://doi.org/10.1007/978-3-319-43394-3_26

Kirkby E (2012) Introduction, definition and classification of nutrients. In: Marschner P (ed) Marschner's mineral nutrition of higher plants. Elsevier Ltd., San Diego, pp 1–5. https://doi.org/10.1016/b978-0-12-384905-2.00001-7

Klasing KC, Goff JP, Greger JL, King JC, Lall SP, Lei XG, Linn JG, Nielsen FH, Spears JW (2005) Mineral tolerance of animals. The National Academies Press, Washington, DC

Koch A (2017) Soil security for agricultural productivity: the policy disconnect and a promising future. In: Field DJ et al (eds) Global soil security, progress in soil science, Springer International Publishing, Switzerland, pp 425–435. https://doi.org/10.1007/978-3-319-43394-3_39

Krajcarová L, Novotný K, Chattová B, Elster J (2016) Elemental analysis of soils and Salix polaris in the town of Pyramiden and its surroundings (Svalbard). Environ Sci Pollut Res 23:10124–10137. https://doi.org/10.1007/s11356-016-6213-4

Lal R (2005) Human society and soil. In: Lal (ed) Encyclopedia of soil science. Taylor & Francis, UK, pp 838–841. https://doi.org/10.1081/e-ess-120042700

Li H, Ji H, Shi C, Gao Y, Zhang Y, Xu X, Ding H, Tang L, Xing Y (2017) Distribution of heavy metals and metalloids in bulk and particle size fractions of soils from coal-mine brown field and implications on human health. Chemosphere 172:505–515

Lichtfouse E (2012) Organic fertilisation, soil quality and human health. Sustainable agriculture reviews series, vol 9. Springer Science + Business Media B.V, The Netherlands. https://doi.org/10.1007/978-94-007-4113-3

Liu G, Niu J, Guo W, An X, Zhao L (2016) Ecological and health risk-based characterization of agricultural soils contaminated with polycyclic aromatic hydrocarbons in the vicinity of a chemical plant in China. Chemosphere 163:461–470

McBratney AB, Field DJ, Jarrett LE (2017a) General concepts of valuing and caring for soil. In: Field DJ et al (eds) Global soil security, progress in soil science. Springer International Publishing, Switzerland, pp 101–108. https://doi.org/10.1007/978-3-319-43394-3_9

McBratney AB, Morgan CLS, Jarrett LE (2017b) The value of soil's contributions to ecosystem services. In: Field DJ et al (eds) Global soil security, progress in soil science. Springer International Publishing, Switzerland, pp 227–235. https://doi.org/10.1007/978-3-319-43394-3_20

Meena NL, Gupta OP, Sharma SK (2016) Nutritional enhancers/promoters in biofortification. In: Singh U et al (eds) Biofortification of food crops. Springer India, New Delhi, pp 349–357. https://doi.org/10.1007/978-81-322-2716-8_25

Melegy AA, Shaban AM, Hassaan MM, Salman SA (2014) Geochemical mobilization of some heavy metals in water resources and their impact on human health in Sohag Governorate, Egypt. Arab J Geosci 7:4541–4552. https://doi.org/10.1007/s12517-013-1095-y

Mène-Saffrané L, Pellaud S (2017) Current strategies for vitamin E biofortification of crops. Curr Opin Biotechnol 44:189–197

Mohamed ES, Alia AM, El Shirbenya MA, Abd El Razekb AA, Savinc IY (2016) Near infrared spectroscopy techniques for soil contamination assessment in the Nile Delta. Eurasian Soil Sci 49 (6):632–639. https://doi.org/10.1134/S1064229316060065

Negm AM, Eltarabily MGA (2017) Modeling of fertilizer transport through soil, case study: Nile Delta. In: Negm AM (ed) The Nile Delta, Handbook of the Environmental Chemistry, vol 55. Springer International Publishing, Switzerland, pp 121–157. https://doi.org/10.1007/698_2016_88

Osman KT (2013) Plant nutrients and soil fertility management. In: Osman KT Soils: principles, properties and management. Springer Science + Business Media, Dordrecht, pp 129–159. https://doi.org/10.1007/978-94-007-5663-2_10

Pearce EN (2017) Iodine deficiency disorders and their elimination. Springer International Publishing AG, Switzerland. https://doi.org/10.1007/978-3-319-49505-7

Pepper IL (2013) The soil health: human health nexus. Crit Rev Env Sci Tec 43:2617–2652

Pepper IL, Gerba CP, Newby DT, Rice CW (2009) Soil: a public health threat or savior? Crit Rev Env Sci Tech 39:416–432

Pohl H, DeRosa C, Holler J (1995) Public health assessment for dioxin exposure from soil. Chemosphere 31:2437–2454

Prasad R, Shivay YS (2016) Sulphur in soil, plant and human nutrition. Proc Natl Acad Sci India Sect B Biol Sci. https://doi.org/10.1007/s40011-016-0769-0

Roivainen P, Makkonen S, Holopainen T, Juutilainen J (2012) Element interactions and soil properties affecting the soil-to-plant transfer of six elements relevant to radioactive waste in boreal forest. Radiat Environ Biophys 51:69–78. https://doi.org/10.1007/s00411-011-0393-6

Saha JK, Selladurai R, Coumar MV, Kundu MLDS, Patra AK (2017) Soil pollution—an emerging threat to agriculture. Environmental chemistry for a sustainable world series, vol 10. Springer Nature Singapore Pte Ltd, Singapore. https://doi.org/10.1007/978-981-10-4274-4

Salman SA, Elnazer AA, El Nazer HA (2017) Integrated mass balance of some heavy metals fluxes in Yaakob village, south Sohag, Egypt. Int J Environ Sci Technol 14:1011–1018. https://doi.org/10.1007/s13762-016-1200-3

Selinus O, Alloway B, Centeno JA, Finkelman RB, Fuge R, Lindh U, Smedley P (2005) Essentials of medical geology. Elsevier, Amsterdam

Shishkov T, Kolev N (2014) Soils and humans. In: Shishkov T, Kolev N (eds) The soils of Bulgaria, World soils book series. Springer Science + Business Media, Dordrecht, pp 199–203. https://doi.org/10.1007/978-94-007-7784-2_16

Simcox NJ, Fenske RA, Wolz SA, Lee IC, Kalman DA (1995) Pesticides in household dust and soil: exposure pathways for children of agricultural families. Environ Health Persp 103:1126–1134

Simon L (2014) Potentially harmful elements in agricultural soils. In: Bini C, Bech J (eds) PHEs, environment and human health. Springer Science + Business Media, Dordrecht, pp 85–150. https://doi.org/10.1007/978-94-017-8965-3_3

Singh D, Rajawat MVS, Kaushik R, Prasanna R, Saxena AK (2017) Beneficial role of endophytes in biofortification of Zn in wheat genotypes varying in nutrient use efficiency grown in soils sufficient and deficient in Zn. Plant Soil. https://doi.org/10.1007/s11104-017-3189-x

Snousy MG, Zawrah MF, Abdel-Moghny T, Ebiad MA, Rashad AM, Khalil MM, Abu El Ella EM, El-Sayed E, Tantawy MA (2017) Mobility and fate of pollutants in the aquifer system of the Northwestern Suez Gulf, Egypt. In: Pim de Voogt (ed) Reviews of environmental contamination and toxicology, vol 240. Springer International Publishing, Switzerland. https://doi.org/10.1007/398_2016_5

Spiegel H, Sager M, Oberforster M, Mechtler K, Stüger HP, Baumgarten A (2009) Nutritionally relevant elements in staple foods: influence of arable site versus choice of variety. Environ Geochem Health 31:549–560

Strobbe S, Van Der Straeten D (2017) Folate biofortification in food crops. Curr Opin Biotechnol 44:202–211

Tepanosyan G, Maghakyan N, Sahakyan L, Saghatelyan A (2017) Heavy metals pollution levels and children health risk assessment of Yerevan kindergartens soils. Ecotoxicol Environ Saf 142:257–265

Thavarajah D, Thavarajah P, Gupta DS (2014) Pulses biofortification in genomic era: multidisciplinary opportunities and challenges. In: Gupta S et al (eds) Legumes in the Omic Era. Springer Science + Business Media, New York, pp 207–220. https://doi.org/10.1007/978-1-4614-8370-0_10

Vasconcelos MW, Gruissem W, Bhullar NK (2017) Iron biofortification in the 21st century: setting realistic targets, overcoming obstacles, and new strategies for healthy nutrition. Curr Opin Biotechnol 44:8–15

Wei J, Zhang X, Liu X, Liang X, Chen X (2017) Influence of root components of celery on pyrene bioaccessibility, soil enzymes and microbial communities in pyrene and pyrene-diesel spiked soils. Sci Total Environ 599–600:50–57

Welch RM, Graham RD (2004) Breeding for micronutrients in staple food crops from a human nutrition perspective. J Exp Bot 55:353–364

White PJ, Broadley MR (2005) Biofortifying crops with essential mineral elements. Trends Plant Sci 10:586–593

Winiwarter V (2014) Environmental history of soils. In: Agnoletti M, Neri Serneri S (eds) The basic environmental history, environmental history, vol 4. Springer International Publishing, Switzerland. https://doi.org/10.1007/978-3-319-09180-8_3

Young S (2007) Evidence for the consumption of the inedible. Who, what, when, where and why? In: MacClancy J, Henry J, MacBeth H (eds) Consuming the inedible. Neglected dimensions of food choice. Berghahn Books, New York, pp 17–30

Zdruli P, Lal R, Cherlet M, Kapur S (2017) New world atlas of desertification and issues of carbon sequestration, organic carbon stocks, nutrient depletion and implications for food security. In: Erşahin S et al (eds) Carbon management, technologies, and trends in mediterranean ecosystems, The Anthropocene: Politik–Economics–Society–Science, vol 15. https://doi.org/10.1007/978-3-319-45035-3_2

Zhang H, Huang B, Dong L, Hu W, Akhtar MS, Qu M (2017) Accumulation, sources and health risks of trace metals in elevated geochemical background soils used for greenhouse vegetable production in southwestern China. Ecotoxicol Environ Saf 137:233–239

Zhao L, Yuan L, Wang Z, Lei T, Yin X (2012) Phytoremediation of zinc-contaminated soil and zinc-biofortification for human nutrition. In: Yin X, Yuan L (eds) Phytoremediation and biofortification, Springer briefs in green chemistry for sustainability, pp 33–57. https://doi.org/10.1007/978-94-007-1439-7_3

13 Future Soil Issues

Hassan El-Ramady, Tarek Alshaal, Hamada Abdelrahman, and Omar El-Hady

Abstract

Soils are among the key resources of sustainable development in Egypt. There would be no development, in any nation, without policies and implementation for soil protection, conservation, and sustainability. This is due to the role soils play in almost all fields, including agriculture and its subsectors (farming of animals and plants to produce food, feed, fiber, fuel, etc.), as well as the industrial sector. Egypt faces, currently, great and serious challenges related to the changes in land use, new challenges for soil sciences scientists. Other important future soil issues include the role of soils in global climate changes mitigation/adaptation, establishment of soil protection law, and enforcing it. To solve emerging soil-related problems in Egypt, potential contributions from soil scientists, policymakers, and society are expected. Therefore, this chapter is an attempt to focus on emerging concern on soil and to suggest suitable solutions under the Egyptian conditions.

Keywords

Egypt • Soil issues • Soil scientists • Emerging problems Global soils

H. El-Ramady (✉) · T. Alshaal
Soil and Water Department, Faculty of Agriculture, Kafrelsheikh University, Kafr El Sheikh, Egypt
e-mail: ramady2000@gmail.com

H. Abdelrahman
Soils Department, Faculty of Agriculture, Cairo University, Giza, Egypt

O. El-Hady
Biological and Agricultural Division, Soils and Water Use Department, National Research Centre, Dokki, Giza, Egypt

13.1 Introduction

It is well known that soils are the main source for supporting human health and his habitation, providing ecosystem services, storing carbon, and regulating emissions of greenhouse gases. Unprecedented pressures on soils from pollution, degradation, urbanization, and mismanagement are the main threats for both agro-ecological balances and food security. These will enforce us to manage and preserve soils sustainably for future generations (Viscarra Rossel et al. 2016). Soils, globally, share common future concerns including the role of soils in climate changes mitigation and/or adaptation, linking soils to ecosystem services, the global soil biodiversity, different priorities for research in soil ecology, etc. (e.g., Adhikari and Hartemink 2016; Láng et al. 2016; Trolard et al. 2016; Bach and Wall 2017; Coyle et al. 2017; Eisenhauer et al. 2017; Ginzky et al. 2017; Guimarães et al. 2017; Streck and Gay 2017).

Several works addressed the future of global soils such as the current and future soil researches (Bockheim and Hartemink 2017), different soil issues according to Bangladesh, Italy, the Philippines, Spain, Greece, and the USA, respectively (Huq and Shoaib 2013; Terribile et al. 2013; Carating et al. 2014; Merino et al. 2016; Yassoglou et al. 2017; Drohan 2017). Generally, several research issues could be addressed in relation to soil micromorphology, soil mineralogy, spatial variability, humic substances, monitoring of soil pollution and related biochemistry, soil biodiversity, and soil modeling. It is sufficiently clear that addressing soil-related problems for a certain nation must be implemented and adapted to the current and future needs of this nation (Terribile et al. 2013). Globally, soil science is facing many challenges regarding the future of this science. These challenges are serious threats to global food production such as soil erosion, driving the global agriculture to protect soil and its ecosystem services, sustaining both education and training of soil sciences, promoting public soil education and protecting human behavior toward soils, and funding

H. El-Ramady et al. (eds.), *The Soils of Egypt*, World Soils Book Series,
https://doi.org/10.1007/978-3-319-95516-2_13

problems for soil research and its management (Drohan 2017). For Egypt, more soil-related challenges can be, further, listed including soil pollution, urban sprawl, soil degradation, soil salinization, and soil desertification. Therefore, the main objective of this chapter is to highlight the future of soil issues or problems under global and Egyptian conditions.

13.2 Changes of Soil and Land Uses in Egypt

Soil and land resources are very limited and should be protected and sustained in Egypt. The main problem in Egypt is represented in overexploitation of the Egyptian lands and soils. Furthermore, the management and sustainability of these soil resources should be the main guarantee for the optimal and safe production. Hence, new modern agricultural practices should be used including new crop varieties, using nonconventional fertilizers, developed irrigation methods, and advanced agrochemicals such as nanomaterials and nanobiopesticides. Egypt is considered one of the countries with the highest intensification of the cropping system; however, the sustainability in using soil and land resources is in continuous, which is mainly due to mismanagement and degradation of soil and land resources (Figs. 13.1 and 13.2). These previous figures represent some common uses of land (e.g., parks, agriculture, museums, resorts, river streams, urbanization, etc.). Accordingly, Egypt was forced to follow proper and sustainable approaches to manage different agronomic practices as well as preventing/prohibiting the conversion of agricultural land to other uses.

Soil and land use could be considered a local and an environmental issue. Therefore, several studies have been focused on planned and unplanned changes in land use as a major problem and central issues in Egyptian environmental change studies (e.g., Omran 2009, 2012; Hereher 2015; Arnous et al. 2017; Eid et al. 2017; Hassan and Omran 2017; Frihy 2017; Hereher 2017a, b; Mahmoud and Divigalpitiya 2017). In order to establish local land use policy, monitoring land use and/or land cover changes is a very necessary process for the development plan. Thus, planning, management, and decision-making of urban changes are mainly dependent on useful data availability in Egypt (Hassan and Omran 2017). Moreover, soil and land use have been changed in Egypt, including all activities such as agriculture, industry, and urbanization. Therefore, information on the rate of changes and use of land resources are very critical for a proper planning management (Omran 2012; Hassan and Omran 2017).

Concerning land and soil uses in Egypt, several studies have been recently published, which cover different areas such as the Nile Delta (e.g., Moghanm 2015; AbdelRahman et al. 2017; Eid et al. 2017; Hereher 2017b), Eastern Desert (e.g., Belal et al. 2014; El Bastawesy and Abu El Ella 2017), Western Desert (e.g., Halmy et al. 2015; Hereher 2015; Rashed 2016), Sinai (e.g., Mohamed et al. 2014; Yossif and Ebied 2016; Arnous et al. 2017; Omran 2017), as well as some Egyptian governorates like Kafr El-Sheikh (Abdel Kawy 2013; Shaheen et al. 2013; Hossen and Negm 2016; El-Amier et al. 2017; El-Zeiny and El-Kafrawy 2017), Port Said (Hassan and Omran 2017), Assuit (Kamel and Abu El Ella 2016; Mahmoud and Divigalpitiya 2017), Gharbia (Masoud 2015; Masoud et al. 2017), Daqahlia (Hegazy and Kaloop (2015), Damietta (El-Gammal et al. 2014), Fayoum (Abd-Elgawad et al. 2013), Mersa Matruh (Abdulaziz and Faid 2015), and Sohag (Kamel and Abu El Ella 2016; Youssef et al. 2017).

As clearly evident that soil and land use have been changed in Egypt, the applications of Global Information System (GIS), remote sensing tools, and digitized color photography in multipurpose fields have been used to define, estimate, and monitor soil and land use changes. These use of GIS and other tools mainly focused on the following issues:

- Soil and land use changes in space and time,
- Soil classification,
- Soil informatics and its geostatistics,
- Agricultural management plans at the farm level,
- Short- and long-term studies of soil and land use,
- Land deterioration at different levels including the agricultural, ecological (biodiversity) and hydrological areas,
- Risk assessment for soil/land subject to natural hazards and/or human-induced hazard/degradation,
- Monitoring the decline in soil fertility, and
- Monitoring soil contamination and pollution.

Egypt faces a great challenge concerning soil and land use and related changes. Without proper and sustainable management, serious and aggravated problems will overwhelm Egypt. Thus, advanced and new approaches, such as the use of GIS and remote sensing and environmental nanotechnology, should be employed to help overcoming the expected challenges. Egypt should formulate and enact a "*Soil Protection Policy*" at the national level relying on the following considerations, as suggested by the Indian policy for the protection of soils:

1. Establishment of framework for monitoring soil quality.
2. Creation of reliable database on status of soil quality in different agro-ecological regions of the country.
3. Popularization of soil testing program for polluted land areas.

Fig. 13.1 More several uses (urbanization) and misuses for lands have been recorded as sprawl urbanization in Tanta as misuse of soils in Egypt. Photos El-Ramady from 2013–2015

Fig. 13.2 Distinguished features for the wonderful coastal beach of Ageeba in Mersa Matrouh district (about 300 km west of Alexandria) as a very famous resort in Egypt. Photos by El-Ramady, July 2017

4. Appropriation of prudent steps to shield fertile soil from physical, chemical, and biological pressures.
5. Generation of awareness among all sections of society about the vital functions of soil in the ecosystem and its vulnerability.
6. Sustainable, careful, and cautious use of soil should only be permitted.
7. Conservation of fertile soils during any developmental activities.
8. Remediation of contaminated soils must be accomplished to ascertain whether all risks have been eliminated.
9. Responsibility soil protection should also lie on the owners of land.
10. Enforcement of legal standards to ensure protection of soil resources, in terms of quantity and quality.
11. Institutional framework for soil protection should be enhanced and endowed with more budgetary and human resources.
12. Cooperation among the departments related to land use planning, agriculture, and water must be established for protection of soil resources.
13. Environmental impact assessment of any industrial, infrastructural, mining projects should compulsorily include soil health evaluation.
14. Creation and strengthening of infrastructural facility for soil pollution assessment.

13.3 Future Issues in Soil Sciences in Egypt

No doubt that all basic issues of soils, which are indeed important and fundamental for soil sciences and for the progress of these sciences, should be developed within the framework of society's needs. Therefore, soil scientists must have a very active role in ameliorating their country potentialities and mitigating their country limitations. This role could lead to feasible contributions of soil scientists toward overcoming the constraints of their society (Terribile et al. 2013). Concerning the future issues in soil sciences, the following scientific issues can be included, as suggested by Merino et al. (2016) for soils in Spain:

- Time course of soil properties after land abandonment,
- Soil responses to afforestation,
- Biogeochemical cycles in transformed and abandoned soils,
- Nutrient imbalances in agricultural and forest soils,
- Carbon sequestration in abandoned soils,
- Pollution and balances of gases with greenhouse effect from soils,
- Soil–water efficiency in intensive agriculture,
- Soil–water efficiency in rangelands
- Soil salinization is semi-arid and arid areas,
- Soil acidification and impoverishment in soils submitted to percolation,
- Effectiveness of forests for soil protection,
- Soil contamination in areas supporting intensive agriculture (greenhouses), and
- Soil reclamation and decontamination of contaminated areas.

The future issues of soil in Egypt could include the following subsections: (i) emerging soil issues, (ii) the role of soils in global climate changes, (iii) the global soil protection and its law, (iv) the new needs and potential contribution from soil scientists, and (v) the future soil issues to solve emerging problems.

13.3.1 Emerging Soil Issues

In Egypt, emerging soil issues include soil pollution and degradation, land use change and its impacts, soil fertility, loss of soil biodiversity, desertification and soil salinization, and impact of climate changes and water crisis. Regarding soil degradation in Egypt, several issues could be listed including (1) using of marginal lands, (2) degradation of irrigated farmlands due to the use of low quality irrigation water, (3) degradation of rain-fed farmlands in northern coastal belt and northern Sinai, due to inefficient harvesting of rainfall (100–250 mm), (4) pollution of waters and lands, and (5) the deterioration of diversified and valued natural fauna and flora resources (Darwish et al. 2013).

13.3.2 The Role of Soils in Global Climate Changes

Due to their specific features, soils are affected by changing in climate and could also contribute to climate change. Soils could be affected by climate changes through change in rainfall patterns, which would increase soil erosion and its runoff, decrease the available water content, organic matter supply, and chemical depletion as well as soil structure degradation and decreased its porosity. As mentioned before, soil plays a very important role in provisioning different vital ecosystem services and sustainable development goals such as production of food, feed, fiber, fuel (or biomass), biodiversity, storage of carbon, storage, and supply of water, added cultural and social services. As consequence, soils may become depleted from their essential components and consequently lose their vital functions. Soils are the second largest reservoir of C after oceans; however, soils could release large amounts of greenhouse gases, e.g., carbon dioxide, methane, or nitrous oxide, into the atmosphere. Regarding the impact of climate change on soils, changes in precipitation pattern or temperature affect crops production and the increase in catastrophic weather events can wipe out harvests on large areas of land (Streck and Gay 2017).

The interaction between soil and climate change is very strong. So, soil conservation, management, and its degradation are nowadays depending on the global climate changes. No doubt that the three main global issues of climate changes, water, and energy crisis threaten most countries worldwide. These problems, and climate changes, in particular, became a research hot spot worldwide and consequently several publications have been produced worldwide on local and global levels and for all fields (e.g., Berga 2016; Keyvanshokouhi et al. 2016; Li and Fang 2016; Sohngen and Tian 2016; Yigini and Panagos 2016; Adams et al. 2017; Bojko and Kabala 2017; Chilkoti et al. 2017; Coyle et al. 2017; Ćirić et al. 2017; Durán et al. 2017; Epule et al. 2017; Gao et al. 2017; Islam and Nursey-Bray 2017; Olaya-Abril et al. 2017; Li and Fang 2017; Liu et al. 2017; Lozano-García et al. 2017; Mäkinen et al. 2017; Muñoz--Rojas et al. 2017; Moreno-Cruz and Singh et al. 2017; Moreno-Cruz and Smulders 2017; Soleimani et al. 2017; Souza et al. 2017; Zhang et al. 2017).

Soils have a crucial role in mitigation and adaptation of climate changes. The most common gas emissions from soil, which are associated with agricultural practices, include (1) carbon dioxide and methane emissions from drained peatlands and manure on fields and rice fields and (2) nitrous oxide emissions from the application of N-fertilizer. Reducing the gas emission from the soil is one of the foreseen mitigation strategies. Mitigation strategies or reduction of soil emission include (i) the enhancement of rice cultivation methods, (ii) restoration of cultivated organic or peaty soils and degraded lands, (iii) better livestock and manure management practices, and (vi) the development of N-fertilizer application techniques that reduce nitrous oxide emissions (Berga 2016; Streck and Gay 2017).

Concerning the role of soils in the adaptation of climate changes, the main different adaptation strategies to climate changes in the agricultural sector include (1) crop relocation, (2) the adjustment of planting dates and crop variety, (3) soil carbon sequestration, and (4) improvement of land management. Soil carbon sequestration could increase the soil ability to withstand erosion, hold moisture, and enrich the biodiversity of the ecosystem. Thus, soil carbon sequestration gives soils the ability to extract carbon from the atmosphere and store it in soil restoring combines adaptation with mitigation benefits (Yigini and Panagos 2016; Lozano-García et al. 2017; Zhang et al. 2017). On the other hand, the loss of carbon from soils often leads to the degradation of soils, through acceleration of erosion and mineralization processes as well as land use change and desertification (Bojko and Kabala 2017; Islam and Nursey-Bray 2017; Streck and Gay 2017).

In Egypt, the most features of climate changes include the raising of sea level and thus the risk of sinking of the Nile Delta, higher temperature, and water crisis and crop requirements. As on a global level, Egypt will suffer from the changes in climate in particular the agricultural sector. It could be concluded the global impact of warming on the agricultural sector in Egypt as follows:

- Increased evapotranspiration and crop water requirements,
- Reduction of crops yield,
- Plant disease and pests,
- Drought and decreased precipitation,
- Sea level rising and changing land use, and
- Seawater intrusion and secondary salinization (Mahmoud 2017).

13.3.3 Global Soil Protection and Its Law

It is well known that the United Nations (UN) General Assembly has declared the year of 2015 as the first "*International Year of Soils*" as well as the December 5 of every year is known as the "*World Soil Day*"' Globally, a significant progress has been achieved regarding the protection of soils as documented by the UN. During the past four decades, the global soil protection law has been developed in several phases. These phases were based on the international conferences started with the UN Conference on the Human Environment in Stockholm in 1972 as the first phase of international soil protection law. In 1981, the action for protecting soils was adopted by the "*World Soil Charter*" from the FAO (Food and Agriculture Organization) and the "*World Soils Policy*" from the United Nations Environment Program (UNEP) in 1982. Both action of FAO and UNEP were reflected and put into concrete terms by the UNEP "*Environmental Guidelines for the formulation of National Soil Policies*". By October 28, 1982, the "*World Charter for Nature*" was adopted by the UN General Assembly and established soil protection objectives through the inclusion of land in its general principles of conservation (Boer et al. 2017).

Concerning the second phase of international soil protection law, the UN Conference on "*Environment and Development*" was held in Rio de Janeiro in 1992. This conference has declared three legally binding agreements (*known as the Rio Conventions*) about soil protection and its sustainable use. These agreements included the UN Framework Convention on Climate Change, the Convention on Biological Diversity (signed both in Rio in 1992), and the

UN Convention to Combat Desertification (adopted in Paris in 1994). The World Summit has been held in the Johannesburg, South Africa on "*Sustainable Development*". This declaration stated that "*desertification claims more and more fertile land*". The third phase of international soil protection law included the 2012 UN Conference on Sustainable Development (Rio+20 Conference) stating about "*The Future We Want*". The fourth phase presented by the UN General Assembly about "*the 2030 Agenda for Sustainable Development*" by September 25, 2015. This agenda stated that "*land degradation neutrality is a manifestation of the increasing political awareness that soil needs to be protected more effectively*" as mentioned before (Flasbarth 2017; Boer et al. 2017; Ginzky et al. 2017).

Definitely, Egypt has several laws for soil protection since the Pharaohs era till now but the main problem in this context is the activation of these laws. Therefore, Egypt faced and still facing very serious challenge concerning the protection of her soils and her heritage.

13.3.4 New Needs and Potential Contribution from Soil Scientists

The future of soil science research commonly is expressed on the ability of soil scientists to address several different challenges related to agriculture, forestry, and environment. These issues have been drawn the attention to the dynamic and multifunctional role of soils. On the other hand, the complexity of these previous challenges definitely requires advanced technical tools (Terribile et al. 2013). As mentioned before, the soil needs and its potential contribution for any nation mainly depend on the needs of the society of this nation. That means all threats for soils, sustainable soil management, land uses, and all other soil issues should be planned. These plans also should focus on the soil protection because of the importance of soil as a natural resource and for the potential productivity as well as for all other ecosystem functions. Emerging soil issues should include new and old needs of the society and the potential contribution by soil scientists with respect to these needs. Accordingly, Egyptian soil scientists have a great mission in developing and advancing soil sciences and their issues, which could be achieved through the following issues:

1. Improvement of digital soil assessment,
2. Support and improving sustainable production,
3. Increase the C-stock in the agro-ecosystems,
4. Protect groundwater from pollution,
5. Use of marginal lands for biomass production,
6. Implement food quality and traceability system,
7. Improve procedures for soil remediation, and
8. Mitigation and adaption to the projected climate changes.

13.3.5 Future Soil Issues to Solve Emerging Problems

There are many future trends in soil sciences. These trends require soil professionals to diversify their expertise, maintaining and deep understanding of soil science disciplines, e.g., chemistry, physics, and biology, to tackle new needs, technologies and new mathematical and conceptual tools. These trends also should implement more studies in basic and applied soil sciences among soil scientists to create a large synergistic interconnection between soil science branches. The future of soil sciences should be much productive in knowledge relevant to societal needs and should empower soil sciences in society. Future soil scientists must have not only an integrated and comprehensive view but also an improved interaction between the soil sciences and policy-makers, as well as stakeholders (Terribile et al. 2013). Four main key points were identified to help the movement of soil sciences in this direction as follows:

(1) *Creating awareness for the policy cycle as a basis for the planning for soil research,*
(2) *Providing a range of options to choose from when dealing with certain sustainability problems rather than offering straightforward solutions,*
(3) *Following the knowledge chain when mobilizing knowledge, and*
(4) *Improving communication* (Bouma 2010; Terribile et al. 2013).

13.4 Conclusion

It could be concluded that soils are nonrenewable natural resources and the main component of the world's land. Due to its vital roles in the Earth's ecosystem, soils are the main source for agricultural biodiversity, the major supplier of the world's food, and the main regulator of the global climate. As a result of the accelerated development activities, very serious disrupt in ecological soil functions has been occurred. Furthermore, it is estimated that about one-third of the arable land worldwide is affected by degradation processes. In Egypt like other countries, there is nowadays an increasing concern and political awareness toward soil degradation processes, loss of soil biodiversity, the adverse effects of climate changes, and the conservation of soils through sustainable management. Many international agreements have been signed and coordinated by the UN establishing soil protection and its conservation measures to control soil erosion and improve soil fertility.

Acknowledgements Authors thank members of Egyptian Soil Science Society (ESSS).

References

Abdel Rahman MAE, Shalaby A, Belal A (2017) A GIS based model for land evaluation mapping: a case study North Delta Egypt. Egypt J Soil Sci 55(4). https://doi.org/10.21608/ejss.2017.255.1043

Abd-Elgawad M, Shendi MM, Sofi DM, Abdurrahman HA, Ahmed AM (2013) Geographical Distribution of soil salinity, alkalinity, and calcicity within Fayoum and Tamia Districts, Fayoum Governorate, Egypt. In: Shahid SA et al (eds) Developments in soil salinity assessment and reclamation: innovative thinking and use of marginal soil and water resources in irrigated agriculture. Springer Science+Business Media Dordrecht, pp 219–236. https://doi.org/10.1007/978-94-007-5684-7_14

Abdulaziz AM, Faid AM (2015) Evaluation of the groundwater resources potential of Siwa Oasis using three-dimensional multilayer groundwater flow model, Mersa Matruh Governorate, Egypt. Arab J Geosci 8:659–675. https://doi.org/10.1007/s12517-013-1199-4

Adams VM, Álvarez-Romero JG, Capon SJ, Crowley GM, Pressey RL (2017) Making time for space: the critical role of spatial planning in adapting natural resource management to climate change. Environ Sci Policy 74:57–67

Adhikari K, Hartemink AE (2016) Linking soils to ecosystem services—A global review. Geoderma 262:101–111

Ansari AA, Gill SS, Gill R, Lanza GR, Newman L (2017) Phytoremediation Management of Environmental Contaminants, vol 5. Springer International Publishing AG. https://doi.org/10.1007/978-3-319-52381-1

Arnous MO, El-Rayes AE, Helmy AM (2017) Land-use/land-cover change: a key to understanding land degradation and relating environmental impacts in Northwestern Sinai, Egypt. Environ Earth Sci 76:263. https://doi.org/10.1007/s12665-017-6571-3

Bach EM, Wall DH (2017) Trends in global biodiversity: soil biota and processes. In: Reference module in earth systems and environmental sciences 2017. Elsevier Inc. https://doi.org/10.1016/B978-0-12-409548-9.09822-5

Banwart SA, Bernasconi SM, Blum WEH, de Souza DM, Chabaux F, Duffy C, Kercheva M, Krám P, Lair GJ, Lundin L, Menon M, Nikolaidis NP, Novak M, Panagos P, Ragnarsdottir KV, Robinson DA, Rousseva S, de Ruiter P, van Gaans P, Weng L, White T, Zhang B (2017) Soil Functions in Earth's critical zone: key results and conclusions. Adv Agron 142:1–27. https://doi.org/10.1016/bs.agron.2016.11.001

Belal AA, Moghanm FS, Mohamed EE (2014) Evaluation of some agricultural expansion areas in the eastern desert of Egypt using GIS. Egypt J Soil Sci 54(3):209–227

Berga L (2016) The role of hydropower in climate change mitigation and adaptation: a review. Engineering 2(3):313–318

Bhattacharyya P, Roy KS, Neogi S (2017) Changes in soil–plant–microbes interactions in anticipated climatic change conditions. In: Rakshit A et al (eds) Adaptive soil management: from theory to practices. Springer Nature Singapore Pte Ltd. pp 261–275. https://doi.org/10.1007/978-981-10-3638-5_13

Blume HP, Brümmer GW, Fleige H, Horn R, Kandeler E, Kögel--Knabner I, Kretzschmar R, Stahr K, Wilke BM (2016) Threats to the soil functions. In: Blume H-P et al (eds) Scheffer/Schachtschabel soil science. Springer-Verlag Berlin Heidelberg, pp 485–559. https://doi.org/10.1007/978-3-642-30942-7_10

Bockheim JG, Hartemink AE (2017) Current and future soil research. In: Bockheim JG, Hartemink AE (eds) The soils of Wisconsin, world soils book series. Springer International Publishing AG, pp 223–228. https://doi.org/10.1007/978-3-319-52144-2_18

Boer BW, Ginzky H, Heuser IL (2017) International soil protection law: history, concepts and latest developments. In: Ginzky H et al (eds) International Yearbook of Soil Law and Policy 2016, International Yearbook of Soil Law and Policy. Springer International Publishing AG, pp 49–72. https://doi.org/10.1007/978-3-319-42508-5_7

Bojko O, Kabala C (2017) Organic carbon pools in mountain soils—sources of variability and predicted changes in relation to climate and land use changes. CATENA 149(Part 1):209–220

Bouma J (2010) Implications of the knowledge paradox for soil science. Adv Agron 106. ISSN: 0065-2113. https://doi.org/10.1016/s0065-2113(10)06004-9

Bouma J, Kwakernaak C, Bonfante A, Stoorvogel JJ, Dekker LW (2015) Soil science input in transdisciplinary projects in the Netherlands and Italy. Geoderma Reg 5:96–105. https://doi.org/10.1016/j.geodrs.2015.04.002

Carating RB, Galanta RG, Bacatio CD (2014) Soil issues and challenges. In: Carating RB et al (eds) The soils of the Philippines, World Soils Book Series. Springer Science+Business Media Dordrecht, pp 273–304. https://doi.org/10.1007/978-94-017-8682-9_6

Chilkoti V, Bolisetti T, Balachandar R (2017) Climate change impact assessment on hydropower generation using multi-model climate ensemble. Renewable Energy 109:510–517

Ćirić VI, Drešković N, Mihailović DT, Mimić G, Đurđević V (2017) Which is the response of soils in the Vojvodina Region (Serbia) to climate change using regional climate simulations under the SRES-A1B? CATENA 158:171–183

Coyle DR, Nagendra UJ, Taylor MK, Campbell JH, Callaham MA (2017) Soil fauna responses to natural disturbances, invasive species, and global climate change: Current state of the science and a call to action. Soil Biol Biochem 110:116–133

Darwish Kh, Safaa M, Momou A, Saleh SA (2013) Egypt: land degradation issues with special reference to the impact of climate change. In: Heshmati GA, Squires VR (eds) Combating Desertification in Asia, Africa and the Middle East: Proven practices. Springer Science+Business Media Dordrecht, p 6. https://doi.org/10.1007/978-94-007-6652-5

Drohan PJ (2017) Future challenges for soil science research, education, and soil survey in the USA. In: West LT et al (eds) The Soils of the USA, World Soils Book Series. Springer International Publishing AG. pp 373–384. https://doi.org/10.1007/978-3-319-41870-4_19

Durán J, Morse JL, Rodríguez A, Campbell JL, Groffman PM (2017) Differential sensitivity to climate change of C and N cycling processes across soil horizons in a northern hardwood forest. Soil Biol Biochem 107:77–84

Eid EM, Moghanm FS, Shaltout KH (2017) Effect of the different types of land-use on the distribution of soil organic carbon in north Nile Delta, Egypt. Rend. Fis. Acc. Lincei. https://doi.org/10.1007/s12210-017-0621-3

Eisenhauer N, Antunes PM, Bennett AE, Birkhofer K, Bissett A, Bowker MA, Caruso T, Chen B, Coleman DC, de Boer W, de Ruiter P, DeLuca TH, Frati F, Griffiths BS, Hart MM, Hättenschwiler S, Haimi J, Heethoff M, Kaneko N, Kelly LC, Leinaas HP, Lindo Z, Macdonald C, Rillig MC, Ruess L, Scheu S, Schmidt O, Seastedt TR, van Straalen NM, Tiunov AV, Zimmer M, Powell JR (2017) Priorities for research in soil ecology. Pedobiologia J SoilEcol 63:1–7. https://doi.org/10.1016/j.pedobi.2017.05.003

El Bastawesy M, Abu El Ella EM (2017) Quantitative estimates of flash flood discharge into waste water disposal sites in Wadi Al Saaf, the Eastern Desert of Egypt. J Afr Earth Sci. https://doi.org/10.1016/j.jafrearsci.2017.03.008

El-Amier YA, Elnaggar AA, El-Alfy MA (2017) Evaluation and mapping spatial distribution of bottom sediment heavy metal contamination in Burullus Lake, Egypt. Egypt J Basic Appl Sci 4 (1):55–66

El-Gammal MI, Ali RR, Eissa R (2014) Land use assessment of barren areas in Damietta Governorate, Egypt using remote sensing. Egypt J Basic Appl Sci 1(3–4):151–160

El-Zeiny A, El-Kafrawy S (2017) Assessment of water pollution induced by human activities in Burullus Lake using Landsat 8 operational land imager and GIS. Egypt J Remote Sens Space Sci 20(Supplement 1):s49–s56

Epule TE, Ford JD, Lwasa S, Lepage L (2017) Climate change adaptation in the Sahel. Environ Sci Policy 75:121–137

Flasbarth J (2017) Soils need international governance: a european perspective for the first volume of the international yearbook of soil law and policy. In: Ginzky H et al (eds) International Yearbook of Soil Law and Policy 2016, International Yearbook of Soil Law and Policy. Springer International Publishing AG, pp 15–19. https://doi.org/10.1007/978-3-319-42508-5_3

Frihy OE (2017) Evaluation of future land-use planning initiatives to shoreline stability of Egypt's northern Nile delta. Arab J Geosci 10:109. https://doi.org/10.1007/s12517-017-2893-4

Gao J, Wang E, Ren W, Liu X, Yang Y (2017) Effects of simulated climate change on soil microbial biomass and enzyme activities in young Chinese fir (*Cunninghamia lanceolata*) in subtropical China. Acta Ecol Sin 37(4):272–278

Ginzky H, Heuser IL, Qin T, Ruppel OC, Wegerdt P (2017) International yearbook of soil law and policy 2016. In: International yearbook of soil law and policy series, vol 2016. Springer International Publishing AG. https://doi.org/10.1007/978-3-319-42508-5

Guimarães RML, Lamandé M, Munkholm LJ, Ball BC, Keller T (2017) Opportunities and future directions for visual soil evaluation methods in soil structure research. Soil Tillage Res 173:104–113. https://doi.org/10.1016/j.still.2017.01.016

Halmy MWA, Gessler PE, Hicke JA, Salem BB (2015) Land use/land cover change detection and prediction in the north-western coastal desert of Egypt using Markov-CA. Appl Geogr 63:101–112

Hassan MAER, Omran EE (2017) Modelling of land-use changes and their effects by climate change at the southern region of Port Said governorate, Egypt. Model Earth Syst Environ 3:13. https://doi.org/10.1007/s40808-017-0276-1

Hegazy IR, Kaloop MR (2015) Monitoring urban growth and land use change detection with GIS and remote sensing techniques in Daqahlia governorate Egypt. Int J Sustain Built Environ 4(1): 117–124

Hereher M (2015) Environmental monitoring and change assessment of Toshka lakes in southern Egypt using remote sensing. Environ Earth Sci 73(7):3623–3632

Hereher M (2017a) Effect of land use/cover change on land surface temperatures—The Nile Delta, Egypt. J Afr Earth Sci 126:75–83

Hereher ME (2017b) Effects of land use/cover change on regional land surface temperatures: severe warming from drying Toshka lakes, the Western Desert of Egypt. Nat Hazards. https://doi.org/10.1007/s11069-017-2946-8

Hossen H, Negm A (2016) Change detection in the water bodies of Burullus Lake, Northern Nile Delta, Egypt, Using RS/GIS. Procedia Eng 154:951–958

Huq SMI, Shoaib JUM (2013) Future soil issues. In: Huq SMI, Shoaib JUM (eds) The soils of Bangladesh, world soils book series 1. Springer Science+Business Media Dordrecht, p 131. https://doi.org/10.1007/978-94-007-1128-0_14

Islam MT, Nursey-Bray M (2017) Adaptation to climate change in agriculture in Bangladesh: the role of formal institutions. J Environ Manage 200:347–358

Kamel M, Abu El Ella EM (2016) Integration of remote sensing & GIS to manage the sustainable development in the Nile Valley Desert fringes of assiut-sohag governorates, Upper Egypt. J Indian Soc Remote Sens 44(5):759–774. https://doi.org/10.1007/s12524-015-0529-2

Kawy WAMA (2013) Using GIS modeling to assess the agricultural sustainability in Kafr El-Sheikh governorate, Nile Delta, Egypt. Arab J Geosci 6:733–747. https://doi.org/10.1007/s12517-011-0377-5

Keyvanshokouhi S, Cornu S, Samouëlian A, Finke P (2016) Evaluating SoilGen2 as a tool for projecting soil evolution induced by global change. Sci Total Environ 571:110–123

Láng V, Fuchs M, Szegi T, Csorba Á, Michéli E (2016) Deriving world reference base reference soil groups from the prospective Global Soil Map product: a case study on major soil types of Africa. Geoderma 263:226–233

Li Z, Fang H (2016) Impacts of climate change on water erosion: a review. Earth Sci Rev 163:94–117

Li Z, Fang H (2017) Modeling the impact of climate change on watershed discharge and sediment yield in the black soil region, northeastern China. Geomorphology, 293(Part A): 255–271

Liu Y-R, Delgado-Baquerizo M, Trivedi P, He J-Z, Singh BK (2017) Identity of biocrust species and microbial communities drive the response of soil multi-functionality to simulated global change. Soil Biol Biochem 107:208–217

Lozano-García B, Muñoz-Rojas M, Parras-Alcántara L (2017) Climate and land use changes effects on soil organic carbon stocks in a Mediterranean semi-natural area. Sci Total Environ 579:1249–1259

Mahmoud MA (2017) Impact of climate change on the agricultural sector in Egypt. In: Negm AM (ed) Conventional water resources and agriculture in Egypt, Handbook of Environmental Chemistry. Springer International Publishing AG. https://doi.org/10.1007/698_2017_48

Mahmoud H, Divigalpitiya P (2017) Modeling future land use and land-cover change in the asyut region using markov chains and cellular automata. In: Isello A et al (eds) Smart and sustainable planning for cities and regions, green energy and technology. Springer International Publishing Switzerland, pp 99–112. https://doi.org/10.1007/978-3-319-44899-2_7

Mäkinen H, Kaseva J, Virkajärvi P, Kahiluoto H (2017) Shifts in soil–climate combination deserve attention. Agric For Meteorol 234–235:236–246

Masoud AA (2015) Geotechnical evaluation of the alluvial soils for urban land management zonation in Gharbiya governorate, Egypt. J Afr Earth Sci 101:360–374

Masoud AA, El Bouraie MM, El-Nashar W, Mashaly H (2017) Spatiotemporal evaluation of the groundwater quality in Gharbiya Governorate, Egypt. Environ Sci Pollut Res 24:8256–8278. https://doi.org/10.1007/s11356-017-8445-3

Merino A, Moreno G, Navarro FB, Gallardo JF (2016) Future issues. In: Gallardo JF (ed) The soils of Spain, world soils book series. Springer International Publishing Switzerland. https://doi.org/10.1007/978-3-319-20541-0_6

Moghanm FS (2015) Assessment of sustainable agricultural land management by using GIS techniques in North Delta, Egypt. Egypt J Soil Sci 55(4):409–424

Mohamed ES, Saleh AM, Belal AA (2014) Sustainability indicators for agricultural land use based on GIS spatial modeling in North of Sinai-Egypt. Egypt J Remote Sens Space Sci 17(1):1–15

Moreno-Cruz JB, Smulders S (2017) Revisiting the economics of climate change: the role of geoengineering. Res Econ 71(2):212–224

Muñoz-Rojas M, Abd-Elmabod SK, Zavala LM, De la Rosa D, Jordán A (2017) Climate change impacts on soil organic carbon stocks of Mediterranean agricultural areas: a case study in Northern Egypt. Agr Ecosyst Environ 238:142–152

Olaya-Abril A, Parras-Alcántara L, Lozano-García B, Obregón-Romero R (2017) Soil organic carbon distribution in Mediterranean areas under a climate change scenario via multiple linear regression analysis. Sci Total Environ 592:134–143

Omran ESE (2009) A proposed simplified method to improve land-use mapping accuracy. Agric Res J 9(3):123–132

Omran ESE (2012) Detection of land-use and surface temperature change at different resolutions. J Geogr Inf Syst 4(3):189–203. https://doi.org/10.4236/jgis.2012.43024

Omran EE (2017) Land and groundwater assessment for agricultural development in the Sinai Peninsula, Egypt. In: Negm AM (ed) Conventional water resources and agriculture in Egypt, Handbook of Environmental Chemistry. Springer International Publishing AG. https://doi.org/10.1007/698_2017_43

Rashed HSA (2016) Evaluation of sustainable land management on some selected soils of Siwa Oasis. Egypt J Soil Sci 56(3):453–470

Shaheen SM, Abo-Waly ME, Ali RA (2013) Classification, characterization, and management of some agricultural soils in the North of Egypt. In: Shahid SA et al (eds) Developments in soil classification, land use planning and policy implications: innovative thinking of soil inventory for land use planning and management of land resources. Springer Science+Business Media Dordrecht, pp 417–448. https://doi.org/10.1007/978-94-007-5332-7_23

Singh AS, Zwickle A, Bruskotter JT, Wilson R (2017) The perceived psychological distance of climate change impacts and its influence on support for adaptation policy. Environ Sci Policy 73:93–99

Sohngen B, Tian X (2016) Global climate change impacts on forests and markets. For Policy Econ 72:18–26

Soleimani A, Hosseini SM, Bavani ARM, Jafari M, Francaviglia R (2017) Simulating soil organic carbon stock as affected by land cover change and climate change, Hyrcanian forests (northern Iran). Sci Total Environ 599–600:1646–1657

Souza GM, Ballester MVR, de Brito Cruz CH, Chum H, Van der Wielen L (2017) The role of bioenergy in a climate-changing world. In: Environmental development (in press)

Streck C, Gay A (2017) The role of soils in international climate change policy. In: Ginzky H et al (eds) International yearbook of soil law and policy 2016. Springer International Publishing AG, pp 105–128. https://doi.org/10.1007/978-3-319-42508-5_10

Terribile F, Basile A, Bonfante A, Carbone A, Colombo C, Langella G, Iamarino M, Manna P, Minieri L, Vingiani S (2013) Future soil issues. In: Costantini EAC, Dazzi C (eds) The soils of Italy, world soils book series. Springer Science+Business Media Dordrecht. https://doi.org/10.1007/978-94-007-5642-7_11

Trolard F, Bourrie G, Baillieux A, Buis S, Chanzy A, Clastre P, Closet J-F, Courault D, Dangeard M-L, Di Virgilio N, Dussouilliez P, Fleury J, Gasc J, Geniaux G, Jouan R, Keller C, Lecharpentier P, Lecroart J, Napoleone C, Mohammed G, Olioso A, Reynders S, Rossi F, Tennant M, Lopez JV (2016) The PRECOS framework: measuring the impacts of the global changes on soils, water, agriculture on territories to better anticipate the future. J Environ Manage 181:590–601

Viscarra Rossel RA, Behrens T, Ben-Dor E, Brown DJ, Dematte JAM, Shepherd KD, Shi Z, Stenberg B, Stevens A, Adamchuk V, Aıchi H, Barthes BG, Bartholomeus HM, Bayer AD, Bernoux M, Bottcher K, Brodsky L, Du CW, Chappell A, Fouad Y, Genot V, Gomez C, Grunwald S, Gubler A, Guerrero C, Hedley CB, Knadel M, Morras HJM, Nocita M, Ramirez-Lopez L, Roudier P, Campos EM, Rufasto Sanborn P, Sellitto VM, Sudduth KA, Rawlins BG, Walter C, Winowiecki LA, Hong SY, Ji W (2016) A global spectral library to characterize the world's soil. Earth Sci Rev. https://doi.org/10.1016/j.earscirev.2016.01.012

Yassoglou N, Tsadilas C, Kosmas C (2017) Soil research. In: Yassoglou N et al (eds) The soils of Greece, world soils book series. Springer International Publishing AG, pp 97–100. https://doi.org/10.1007/978-3-319-53334-6_11

Yigini Y, Panagos P (2016) Assessment of soil organic carbon stocks under future climate and land cover changes in Europe. Sci Total Environ 557–558:838–850

Yossif TMH, Ebied GM (2016) Monitor of changes of land uses and soil characteristics at North West Area of Sinai using satellite imagery, Egypt. Egypt J Soil Sci 56(3):471–487

Youssef AM, El-Shater A, El-Khashab MH, El-Haddad BA (2017) Coupling of field investigations and remote sensing data for karst hazards in Egypt: case study around the Sohag City. Arab J Geosci 10:235. https://doi.org/10.1007/s12517-017-3029-6

Zhang L, Wang G, Zheng Q, Liu Y, Fan X (2017) Quantifying the impacts of agricultural management and climate change on soil organic carbon changes in the uplands of Eastern China. Soil Tillage Res 174:81–91

Index

H. El-Ramady et al. (eds.), *The Soils of Egypt*, World Soils Book Series,
https://doi.org/10.1007/978-3-319-95516-2

MIX
Papier aus verantwortungsvollen Quellen
Paper from responsible sources
FSC® C105338

If you have any concerns about our products,
you can contact us on
ProductSafety@springernature.com

In case Publisher is established outside the EU,
the EU authorized representative is:
Springer Nature Customer Service Center GmbH
Europaplatz 3, 69115 Heidelberg, Germany

Printed by Libri Plureos GmbH
in Hamburg, Germany